A TEXT-BOOK ON CHEMISTRY.

FOR THE USE OF

SCHOOLS AND COLLEGES.

BY

HENRY DRAPER, M.D.,

PROFESSOR ADJUNCT OF CHEMISTRY AND NATURAL HISTORY
IN THE UNIVERSITY OF NEW YORK.

With more than Three Hundred Illustrations.

NEW YORK:
HARPER & BROTHERS, PUBLISHERS,
FRANKLIN SQUARE.
1866.

PREFACE.

THIS text-book on Chemistry is intended for the use of Schools and Colleges. It embodies the valuable parts of the work on the same subject published by my father in 1846, which satisfied so completely the public wants that it passed through more than forty editions.

It has been found necessary to make this book larger than that by a hundred pages, and to incorporate a very considerable number of new illustrations, in order to bring the subject fully up to the present time. A free use has been made of all the most recent authorities, both in English and other languages, and it is believed that nothing essential for the student has been omitted.

The form the book has taken has been determined by long experience in teaching, not only at the University, but also in the many Colleges and Schools where the former text-book has been and is used. The subject is presented in the most practical way, all needless technicalities are avoided, and the facts are stated in plain language suited to the class-room.

At the bottom of each page a series of questions will be found; they are intended to assist a teacher in pointing out the more essential facts. The copiousness of illustration will also prove to be of material advantage in schools where there is a deficiency of apparatus for experiment.

HENRY DRAPER, M.D.

University of New York, 1866.

CONTENTS.

Lecture		Page
I.	History of Chemistry	1
II.	History of Chemistry	7
III.	Heat	11
IV.	Expansion of Gases and Liquids	15
V.	Expansion of Liquids and Solids	22
VI.	Expansion of Solids	27
VII.	Specific Heat of Bodies	31
VIII.	Specific Heat and Latent Heat	36
IX.	Latent Heat	42
X.	Vaporization	47
XI.	Ebullition	53
XII.	Vaporization	58
XIII.	Evaporation and Interstitial Radiation	65
XIV.	Conduction	72
XV.	Radiation	76
XVI.	Theory of the Exchanges of Heat	83
XVII.	Of Light	87
XVIII.	The Constitution of Solar Light	93
XIX.	Wave Theory of Light	100
XX.	Wave Theory of Light	103
XXI.	Wave Theory of Light	107
XXII.	Production of Light	111
XXIII.	Chemical Action of Light	116
XXIV.	Photography	119
XXV.	Electricity	124
XXVI.	Theory of Electrical Induction	129
XXVII.	Laws of the Distribution of Electricity and Theories of Electricity	133
XXVIII.	Electrical Induction and Faraday's Theory	138
XXIX.	Voltaic Electricity	147
XXX.	Forms of the Voltaic Battery	150
XXXI.	The Electro-Chemical Theory	157
XXXII.	Ohm's Theory of the Voltaic Pile	165
XXXIII.	Electro-Dynamics	178
XXXIV.	The Nomenclature	191
XXXV.	The Symbols	195
XXXVI.	The Laws of Combination	199
XXXVII.	Constitution of Bodies	204
XXXVIII.	Chemical Affinity	212
XXXIX.	Pneumatic Chemistry—Oxygen	218
XL.	Oxygen—*continued*	224
XLI.	Hydrogen	229
XLII.	Water	234
XLIII.	Nitrogen—Atmospheric Air	241
XLIV.	Atmospheric Air	247

Lecture		Page
XLV.	Atmospheric Air	253
XLVI.	Compounds of Nitrogen and Oxygen	258
XLVII.	Compounds of Nitrogen and Oxygen	261
XLVIII.	Sulphur	265
XLIX.	Compounds of Sulphur and Oxygen	269
L.	Sulphur—Phosphorus	271
LI.	Compounds of Phosphorus and Oxygen—Chlorine	275
LII.	Chlorine	280
LIII.	Chlorine—Iodine	284
LIV.	Bromine—Fluorine—Carbon	289
LV.	Carbonic Acid	295
LVI.	Cyanogen—Boron—Silicon—Ammonium	300
LVII.	Chemical Properties of the Metals	309
LVIII.	Potassium	313
LIX.	Sodium—Lithium—Cæsium—Rubidium—Barium	319
LX.	Strontium—Calcium—Magnesium, etc.	326
LXI.	Manganese—Iron	338
LXII.	Iron—Nickel—Cobalt—Zinc	343
LXIII.	Cadmium—Tin—Chromium, etc	349
LXIV.	Arsenic	355
LXV.	Arsenic—Antimony—Tellurium—Copper	359
LXVI.	Lead—Bismuth—Silver	365
LXVII.	Mercury—Gold—Palladium—Platinum, etc	371
LXVIII.	Organic Chemistry	377
LXIX.	Analysis of Organic Substances	384
LXX.	The Non-Nitrogenized Bodies	391
LXXI.	Action of Agents on the Starch Group	397
LXXII.	Metamorphoses of the Starch Group by Nitrogenized Ferments	402
LXXIII.	The Derivatives of the Fermentative Processes	407
LXXIV.	Derivative Bodies of Alcohol	413
LXXV.	Oxidation of Alcohol	417
LXXVI.	Derivatives of Acetyle—The Kakodyle Group	422
LXXVII.	The Wood-Spirit Group	426
LXXVIII.	The Potato-Oil Group—The Benzoyle Group	430
LXXIX.	The Salicyle and Cinnamyle Groups	435
LXXX.	The Nitrogenized Principles	439
LXXXI.	Bodies allied to Cyanogen	445
LXXXII.	Mellone—Urea	449
LXXXIII.	The Vegetable Acids	453
LXXXIV.	The Vegetable Alkaloids	458
LXXXV.	The Coloring Bodies	465
LXXXVI.	The Fatty Bodies	469
LXXXVII.	The Resins, Balsams, and Bodies arising in Destructive Distillation	474
LXXXVIII.	Animal Chemistry	480
LXXXIX.	Origin and Destiny of the Fats and Neutral Nitrogenized Bodies	483
XC.	Introduction and Transmission of Food through the System	487
XCI.	Nature of Respiration and Secretion	490

INTRODUCTION.

LECTURE I.

HISTORY OF CHEMISTRY. — *Chemistry the Science of Analysis and Synthesis.—Its early History.—Alchemy.—The Philosopher's Stone.—The Elixir of Life.—Discovery of the strong Acids.—The second Stage of Chemistry.— Cause of the Union of Bodies. — Doctrine of Acids and Alkalies. — Affinity. — The Electro-Chemical Theory.—Dalton's Atomic Theory.—Atoms and Interstices.—Atoms attracted together by Cohesion and repelled by Heat. — Experimental Illustrations.*

CHEMISTRY may be defined as the Science of Analysis and Synthesis. It deals with the atoms of which bodies are composed, and investigates the changes of form and properties they suffer by decompositions and combinations. It leaves to NATURAL PHILOSOPHY the description of the action of masses of matter upon one another.

Chemistry has existed from the earliest times, having been brought into Europe from the East. At first consisting of a few isolated facts, which, as their number increased, were grouped together by the aid of various hypotheses, it has in recent times acquired precision, and come to be regarded as one of the exact sciences. The great progress which it made in the Middle Ages was due to the impression then existing, that by its aid men might discover the means of transmuting the baser into the nobler metals, might learn how to turn lead into gold. The desire of acquiring wealth suddenly was the moving spring that tempted many to waste their entire possessions in long-continued experiments. Alchemy,

What are the objects of chemistry? What does Natural Philosophy treat of? What is the early history of chemistry? What is meant by Alchemy?

the Chemistry, the art of making the precious metals, as the science was termed by the Arabians, pursued another equally unattainable end, the search for the Elixir of Life, by which existence might be indefinitely prolonged. The alchemists supposed, as gold was the noblest of the metals, that if it could be obtained in a dissolved condition, potable gold, the problem would be solved.

In the prosecution of such investigations, conducted by many persons, and with the utmost ardor, great facts could not fail to be continually found out, facts which form the solid basis of modern chemistry. Of these, the most important was the discovery of the strong acids. Nitric acid was procured by Djafar, or Geber, toward the end of the eighth century, by distilling Cyprus vitriol, alum, and saltpetre together; and aqua regia was formed by the addition of sal ammoniac to that acid. In the ninth century, Rhazes described sulphuric acid, having prepared it by the distillation of green vitriol; he also obtained absolute alcohol by the action of quick-lime on spirit of wine.

When we recollect that previous to this time vinegar was the strongest acid known, it will be at once perceived how greatly the powers of the chemist were increased. He could decompose substances which had hitherto proved entirely intractable, and could form a large number of new compounds.

The discovery of aqua regia, the solvent for gold, by Geber, and the soon-proved inefficacy of potable gold to prolong life, gave a death-blow to that phase of alchemy. The transmutation of metals is still sought for, because no demonstration of its impossibility has been offered. Indeed, the mass of evidence looks the other way, and many eminent chemists are willing to admit that it may at any time be accomplished. The pursuit of the elixir of life ended in the introduction of chemistry into medicine, Basil Valentine publishing a treatise on the virtues of antimony in the fifteenth century, and Paracelsus introducing mercury as a specific in certain

What is the Elixir of Life? What great facts were discovered by the alchemists? Why was the discovery of the acids important? What is potable gold, and its use? Is transmutation of the metals possible?

affections in the sixteenth. Van Helmont, who first used the term gases, was the last alchemist of any note.

The second stage in the progress of chemistry may be regarded as originating in attempts to investigate the reason of the union of bodies. Alchemy left a mass of facts without order or arrangement, having no connection with one another. Sylvius, a Dutch physician, born in 1614, was the author of the doctrine of acids and alkalies, and their neutralization, his leading idea being that for substances to enter into union they must possess *opposite*, and not similar qualities. He showed that a compound may differ altogether in properties from the bodies from which it has been formed: thus, sulphuric acid, a sour, corrosive liquid, uniting with lime, a caustic solid, produces sulphate of lime, an inert body.

This doctrine of the affinity of substances of opposite natures was developed by Geoffroy, who constructed tables showing the graduated or elective affinity of bodies for one another, some exhibiting a weaker and others a stronger disposition to unite with a given substance. These tables, which are even now of considerable use, do not, however, represent the order of affinity, but only of decomposition, on account of the interference of certain external agents, such as temperature, solubility, cohesion.

Eventually, it was stated by Davy, in his electro-chemical theory, that the condition determining the union of bodies is that they must be in opposite electrical conditions. Thus, for example, potash is composed of potassium, an electro-positive metal, and oxygen, an electro-negative gas. On this theory, all bodies belong to one of two groups, electro-positive or electro-negative, depending on which pole they pass to when compounds containing them are decomposed by a voltaic current. Chemical affinity is therefore an electrical phenomenon. Although apparently simple, difficulties of a formidable kind present themselves to this hypothesis. The doctrine has received its utmost development by the researches of Faraday, to be described hereafter.

What was the second stage of chemistry? What is the doctrine of acids and alkalies? What are Geoffroy's tables? What do they indicate? What does the electro-chemical theory affirm? What two groups are bodies divided into?

Dalton's atomic theory accounts for the union of bodies in definite proportions by supposing that they consist of atoms, or indivisible portions, and that in the case of each substance the atom has a definite and special weight. Thus, the hydrogen atom being taken as the unit, the atom of oxygen is eight times as heavy, that of sulphur sixteen, that of gold a hundred and ninety-seven. The weight of an atom of water, consisting of one of hydrogen and one of oxygen, must be nine, and can not be either more or less. In the case of substances combining to form more than one compound, the proportions found in the various compounds bear a simple relation; thus, an atom of carbon weighing six may combine with one of oxygen weighing eight to form carbonic oxide, or with two of oxygen, weighing sixteen, to form carbonic acid, but no intermediate proportions are possible.

The atoms of which bodies consist are not in contact, but are separated by interstices, or intervals. The distance by which they are separated admits of change by external agents. If, for example, a metallic ball, *Fig.* 1, be so adjusted to a ring that, when cold, it will just pass through it, on subjecting the ball to heat it will be found to have dilated, and will no longer pass through, but will remain sustained. On allowing it to cool in that position, it contracts again to its former size, and will fall through the ring spontaneously. The heat has in this case partially overcome the mutual attraction of the atoms composing the metal; they have receded from one another, and an enlargement is the result. As the heat in the latter part of the experiment radiates away, cohesion again draws the atoms together to their original distance.

Fig. 1.

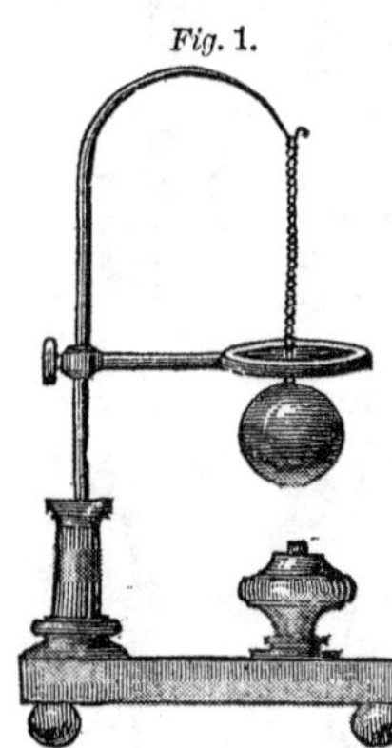

What does the atomic theory affirm? What is the weight of an atom of hydrogen? of oxygen? of water? How are atoms separated? Describe the apparatus *Fig.* 1. What was the effect of the heat?

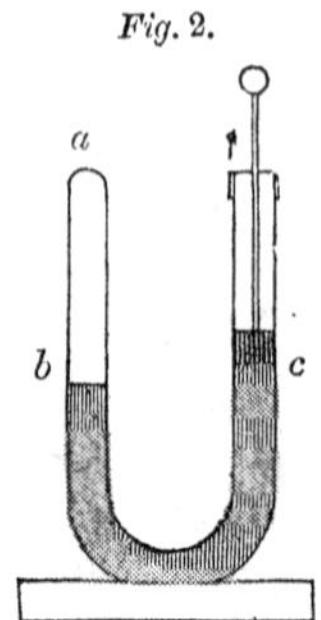

Fig. 2.

By the influence of pressure we may demonstrate the same fact, that atoms are separated by variable intervals. Let a glass tube, *a b c*, *Fig.* 2, bent as represented, be partly filled with water. On exerting pressure by the aid of the piston *c*, the column of air in *a b* may be made to diminish in size. The atoms of which the air consists have approached one another, and the size of the intervening interstices has been lessened.

Heat exercises a similar effect in the case of gases to that exhibited in solids. This is shown by the apparatus *Fig.* 3, in which *a* is a glass bulb terminating in a tube, *b*, dipping below the surface of some water in a vessel, *c*. On warming the bulb *a* by the hand, the liquid originally standing at the point *b* descends, indicating that the air in *a* has expanded. Its atoms have been forced apart by the heat.

Fig. 3.

By decreasing the pressure exerted by the atmosphere we may also cause an increase in the size of the interstices of a substance. In *Fig.* 4, *a* is a glass globe nearly filled with water, but containing a bubble of air at the upper part. Its neck, *b*, descends, air-tight, through the top of a bell-jar, *d*, and terminates in a vessel of water, *c*. On exhausting the air from the bell-jar by the air-pump, the bubble of air, *a*, will be seen to expand, and will eventually fill the entire bulb and tube. On restoring the pressure of the air, the bubble contracts to its former size.

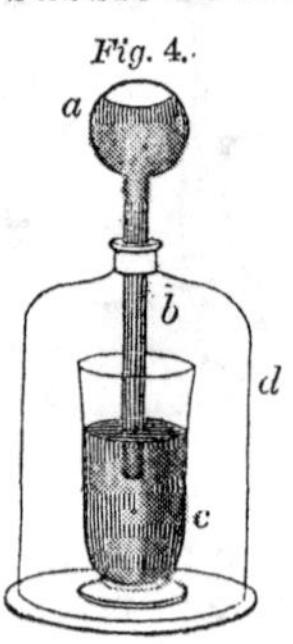

Fig. 4.

An illustration of cohesion, the force that opposes heat in the regulation of interstices, is seen in the experiment of the bullets. If two leaden bullets have each a bright spot pared upon it, and the two be then pressed

Describe *Fig.* 2. What fact does *Fig.* 3 illustrate? How may the interstices between atoms be enlarged? Describe *Fig.* 4. Describe the experiment of the leaden bullets.

closely together, they cohere, and can only be parted by the exertion of considerable force.

The distance through which these forces, attraction and repulsion, act, is very minute; for if, in the above experiment, the bullets be separated by the smallest interval, no attraction ensues; they must be forced together till less than a millionth of an inch intervenes, as was shown by Newton. The same fact is proved by the apparatus *Fig.* 5, which consists of a plate of glass, *a*, supported over some mercury in a cup, *e*. To the other arm of the balance, *b c*, is hung a pan, *d*, containing weights. As long as the glass and mercury are not in contact no attraction is exhibited, but immediately they touch a considerable weight must be put into *d* to draw them apart.

Fig. 5.

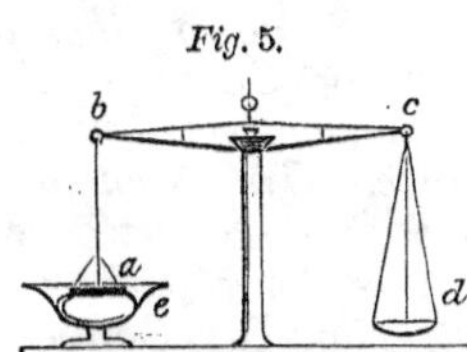

The atomic theory forms the foundation of modern chemistry, and is applicable not only to the combination of elements, but also to combinations of their compounds. As we shall see in organic chemistry, there are bodies which act like elements whose atoms or molecules are exceedingly complex, containing several elements and many atoms of each.

Through what distances do attraction and repulsion act? Describe *Fig.* 5. What is the present position of the atomic theory?

LECTURE II.

History of Chemistry.—*The third Stage of Chemistry.—Investigation of the Nature of Bodies.—Ancient Doctrines.—The Phlogistic Theory.—Lavoisier introduces the Use of the Balance.—The Nomenclature and Symbols.—The fourth Stage.—Assertion of the Indestructibility of Matter.—Correlation and Conservation of Force.—Illustrations of Convertibility of Force.—Atoms regarded as Centres of Force.*

At the same time that the investigations terminating in the atomic and electro-chemical theories were being carried on, the third stage of chemistry, the examination of the nature of bodies, was in progress.

In ancient times it was supposed that there were four elements, earth, air, fire, and water, and that from them arose all known substances. The alchemists introduced as a substitute the doctrine that the three components of all bodies were salt, sulphur, and mercury, and that their difference of properties arose from variation in the proportions. Hence transmutation was possible. To these elements Stahl added phlogiston, the principle of inflammability, and by its aid chemists were enabled to explain all phenomena until the beginning of this century. For example, when metallic earth, or calx, was brought into its reguline or metallic form, it was supposed to have imbibed a certain proportion of phlogiston. On subjecting the metal to the action of fire, if, as in the case of lead, it returned to the earthy condition, it was said that the phlogiston had been expelled. The reduction of what we now term oxides, by the aid of charcoal, was asserted to be due to the fact that charcoal contains a large proportion of phlogiston, and readily imparts it to the metal. The metals were compounds

What were the four elements of antiquity? What were the elements of the alchemists? What addition did Stahl make to them? How did he explain the action of fire on metals?

of calx and phlogiston, the calx or oxide being the simple body.

The phlogiston theory was destroyed by Lavoisier, who, by the introduction of the balance into such inquiries, showed that the metal gains in weight when it assumes the earthy appearance under the influence of fire and air, and that the air loses as much as the metal gains. The advocates of the old theory attempted to account for this fact by assuming that phlogiston was a principle of levity, and in combination rendered bodies lighter. It was regarded as identical with hydrogen, or phlogisticated air, as that gas was called.

With the introduction of the balance, chemistry became at once an exact science. The proportions in which elements combine to form compounds were ascertained with precision, and the experiments necessary for the completion of the atomic theory, by determining the atomic weights of bodies, were undertaken. An element was defined to be a body not decomposable by any known agency.

At the same epoch the defects in the system of naming substances were remedied by substituting for the old arbitrary titles, such as green vitriol, Epsom salts, a new nomenclature, and a language of symbols. Every substance was named according to its composition, and its constitution clearly expressed by a short formula. This improvement is due to a commission of French chemists, Lavoisier, De Morveau, Berthollet, and Fourcroy. Although the present wants of chemistry are hardly supplied by this nomenclature, yet it has worked its way into the arts, and been so generally adopted, that a better successor than any yet proposed will have to be found before it can be displaced.

The last stage through which chemistry has commenced to pass is that asserting the indestructibility of matter. In the old time, when a substance, as carbon, for example, was exposed to the action of fire and disappeared from view, it was supposed to have been destroyed or dissipated. At the same period it was

---

By whom was the balance introduced into chemistry? What consequences ensued? Why was phlogiston assumed to be a principle of levity? How was the nomenclature introduced? What is the last stage of chemistry?

imagined that new bodies could be created from nothing.

But the discovery of carbonic acid by Black and oxygen by Priestley led to another explanation of combustion, and proved that, although in that operation bodies may change their form, their atoms still exist, and may be recovered. In the burning of carbon the atoms are not destroyed, but only take on a gaseous form, and by the action of the leaves of plants, under the influence of sunlight, may be brought again into a visible condition. If, in a closed flask of air, counterpoised on the arm of a balance, a few grains of gunpowder be exploded, though the solid has disappeared from sight, the contents of the flask weigh as much as ever; that arm of the balance does not move upward.

To the doctrine of the indestructibility of matter has been added that of the correlation and conservation of force. By these terms it is implied that force is equally indestructible with matter, but that, in addition, it may present a variety of transformations. It may be stored up, or disappear in a latent state, as in the case of heat and light concealed in plants, and originally derived from the sun. Heat, light, electricity, motion, are all regarded as mutually convertible without loss. The conversion of motion into heat may be illustrated by an experiment. If a piece of metal tube, closed at the bottom, be so arranged, *Fig.* 6, as to be set in rapid rev-

Fig. 6.

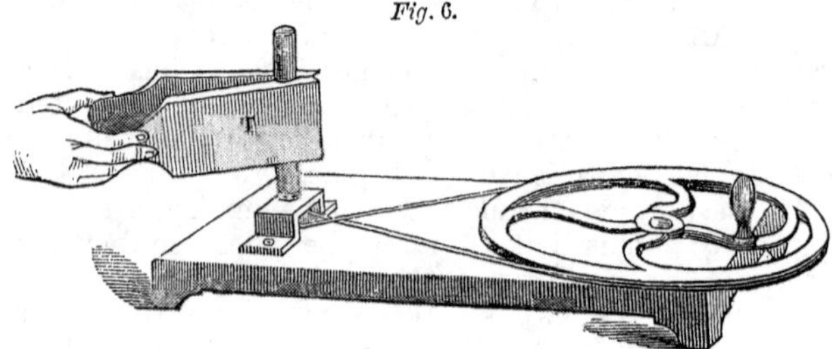

olution, on filling it with cold water, and causing its exterior to be compressed by a pair of hinged pieces of

How may it be shown that matter is not destructible? What is meant by correlation and conservation of force? Explain the experiment illustrated in *Fig.* 6.

wood, T, the friction will produce in a few minutes heat enough to boil the water. If a cork be inserted into the mouth of the tube, it will be violently expelled after a short interval.

Many experiments showing these conversions might be described. By a thermo-electric pile, heat may be converted into a current of electricity, magnetism produced in a coil, chemical affinity exhibited in the decomposition of solutions, and light given out in the spark.

The doctrine of the conservation of force is by no means new, having been asserted by the Arabians, who included the force exhibited in the animal functions, even intellection, in the same category.

In order to obtain a clear idea of the properties of the elements it will be first needful to enter somewhat in detail into a description of the forces acting upon them —heat, light, and electricity. Matter can not exist except under their influence, and they can not manifest themselves without it. It must, however, be borne in mind that eminent men have been found in ancient times, and also at the present epoch, who entirely deny the existence of matter; and, carrying the atomic theory one step farther than it is here set forth, assume that the so-called atoms are nothing but centres, or foci, of force.

What effects may be produced by a thermo-electric pile? Who were the inventors of the doctrine of the conservation of force? What are the forces of chemistry?

PART I.

THE FORCES OF CHEMISTRY.

LECTURE III.

HEAT.—*Two Hypotheses of the Nature of Heat, the Material and the Mechanical.—Influence of Heat on Inorganic and Organic Substances.—Transference and Equilibrium of Heat.—Heat affects the Magnitude of Bodies and their Form.—Affects Measures of Time and Space.—Determines the Distribution of Animals and Plants.*

THERE are two hypotheses of the nature of heat, the first of which regards it as a material substance, without weight, having a self-repulsive power, and an attraction for the particles of matter. The second supposes it to be a result of the vibration or movements of the atoms of which bodies are composed, and is termed the mechanical theory of heat. It also attempts to indicate with precision the exact mechanical equivalent of heat, expressing it in "foot-pounds," a term signifying the falling of one pound a foot. Thus it is said that the heat necessary to raise a pound of water one degree in temperature is equal to 772 foot-pounds.

So great is the control heat exercises over chemical changes that few experiments can be made in which transformations take place without disturbance of temperature, sometimes heat and sometimes cold being produced.

In the organic world, also, heat plays an equally important part. Life can only be sustained within a narrow range of temperature, the one hundred and eighty

---

How many hypotheses are there of the nature of heat? Describe each. What is the meaning of the term foot-pound? What quantity of heat is required to raise a pound of water a degree? Explain the control of heat over the organic world.

degrees which intervene between the boiling and freezing points of water; and, indeed, the limits are narrower than that, for in the higher tribes a very slight variation from a fixed degree causes the operations of the body to be seriously interfered with, or even stopped altogether.

When an ignited mass, as a red-hot ball, is placed in the middle of a room, common observation satisfies us that it rapidly loses its heat, its temperature descending until it becomes the same as that of the surrounding walls and other bodies. This loss is due to several causes. A part of the heat is carried away by contact with the stand which supports the ball, a part by certain motions established in the surrounding air, and a part by radiation. This removal passes under the name of *transference;* and as soon as the temperature has declined to that of the adjacent bodies, an *equilibrium* is said to have been attained.

There are two methods by which caloric can be transferred: 1st. By radiation; 2d. By convection. Of the former we have two varieties—general radiation and interstitial radiation.

Substances, no matter what their form, solid, liquid, or gaseous, expand by increase of heat. The experiment *Fig.* 1 shows this fact in the case of a metallic ball. That the same is true of liquids is proved by taking a glass tube, *a b*, *Fig.* 7, open at one end, but having a bulb blown on the other. The bulb and part of the tube to *b* being filled with water or any other liquid, a spirit-lamp is to be applied. The fluid soon commences to rise in the tube *b*, the dilatation increasing with the temperature.

Fig. 7.

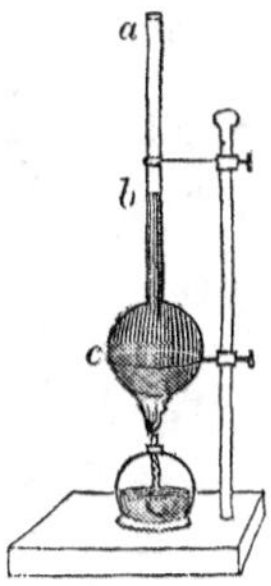

If the tube be emptied of the fluid, and inserted, as in *Fig.* 8, in a glass of water, the expansion of gaseous substances can be shown; for, on warming the bulb *a*, the air it contains expands, and escapes in bubbles through the water in *c*.

Describe the manner in which a red-hot ball cools. Define the term transference of heat. Define the term equilibrium of heat. By what methods may heat be transferred. Describe *Fig.* 7. Describe *Fig.* 8.

Fig. 8.

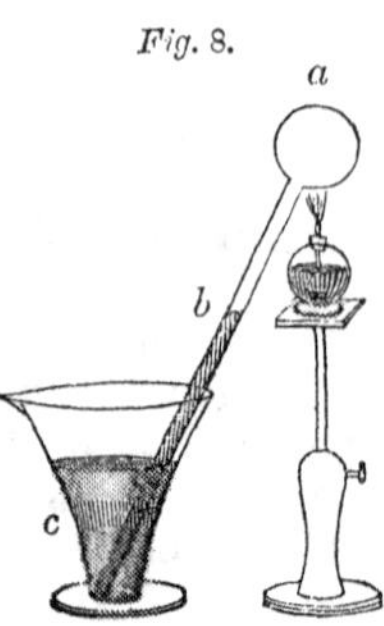

Solids, liquids, and gases expand, therefore, as their temperature rises, and contract as it falls. The size of all objects is determined by their temperature. A measure which is a yard long in summer is shorter in winter; a vessel which holds a gallon in winter holds more in summer. The appearance of stationary magnitude of such objects is altogether a delusion. They change with every hour of the day.

Heat also determines the *form* of bodies; that is, whether they shall be solid, liquid, or gaseous. A mass of ice raised to 32° melts into water, a liquid; and if the water be heated to 212° it turns into steam, a gas. The water is thus made to exhibit all the forms of matter by change of temperature.

In the same manner that it affects our measures of space, heat affects our measures of time. Clocks and watches measure time by the vibrations of pendulums or the oscillations of balance-wheels, the uniformity of the action of which depends on the uniformity of their size. When the temperature rises, the rod of a pendulum lengthens, and its vibrations are made more slowly; the clock to which it is attached loses time. When the temperature declines, the pendulum shortens; it beats too quickly, and the clock gains. Similar observations may be made in the case of watches. To obviate these difficulties many contrivances have been invented, such as the gridiron pendulum and the compensation balance-wheel. Advantage has also been taken of such substances as expand but little for a given elevation of temperature, and thus excellent clocks have been made, the pendulum-rods of which were formed of a slip of marble, or a rod of baked wood varnished.

Natural as well as artificial measures of time depend on heat. Our unit of time, the day, is the period occupied by one rotation of the earth on her axis. The time

What effect has heat on the size of bodies? What is meant by the term *form* of bodies? How may water be made to exhibit all the forms of matter? Describe the manner in which heat affects measures of time.

of rotation is influenced by the temperature of her mass; if it should fall she would become less in diameter, and would rotate more quickly. Thus, when we tie a weight to the end of a string, and, whirling it round in the air, permit the thread to twine round one of our fingers, as it shortens the revolution of the weight becomes quicker. From such reasoning we ascertain that the temperature of the globe has not fallen in 2000 years, because astronomical observations show that the length of the day has not shortened in that time $\frac{1}{100}$ of a second. There was, however, a period when the earth was a molten mass.

The distribution of heat on the earth's surface determines the distribution of animals and plants. In each climate animals whose constitution suits it have appeared. The monkey only lives in the torrid zone, and dies of consumption in our latitude; the white bear prefers the frigid regions. Man, from the control he acquires by diet, houses, clothing, and fire, can live in any region, although his physical appearance and mental powers are affected by a prolonged residence in an unpropitious and unusual climate. The types of plants vary entirely with change of climate, that is, change in temperature, vegetation being most luxuriant in the tropics, while on the snow of the polar regions the red snow alga only can live. The same changes in plants are seen in ascending a high mountain as in making a journey toward the pole, the warmth in both cases declining, and the vegetation becoming less and less vigorous.

How do we know that the earth's temperature has not fallen in 2000 years? Was it ever higher than at present? What effect has heat on the distribution of animals? What effect on plants? Why can man live in any climate?

LECTURE IV.

EXPANSION OF GASES AND LIQUIDS.—*Rudberg's Law.—Regularity of Gaseous Expansion.—Hot-air Balloon.—Changes in Temperature change the Volume of a given Weight of Air.—Sanctorio's Thermometer.—Is affected by the Pressure of the Air.—The Differential Thermometer.—Differing Expansion of Liquids.—Their irregular Expansion.—The Mercurial Thermometer.—The Fahrenheit, Centigrade, and Reaumur Scales.*

IF we compare together the three forms of bodies as respects their changes of volume under the influence of heat, we shall find that, for a given rise of temperature, gases expand the most, liquids intermediately, and solids least of all. To this rule but few exceptions are known; liquid carbonic acid, however, expands about four times as much as any gaseous body.

When heated from the freezing to the boiling point of water,

1000 cubic inches of iron become			. . . .	1004
1000 "	" water	"	. . . .	1045
1000 "	" air	"	. . . .	1365

Experiment has proved that gases differ among themselves in expansibility, though the differences are not to any great extent. Those which can be liquefied expand the most. For the permanently elastic gases, atmospheric air may be taken as the type. The experiments of Rudberg and later philosophers show that it expands $\frac{1}{491}$ of its volume at 32° for every degree of Fahrenheit's thermometer. As the same quantity of gas occupies very different volumes at different temperatures, it is necessary, in this and other such cases, to state some special temperature at which the estimate of its volume is made. The same gaseous mass occupies a much greater space

What is the relation of expansion of gases, liquids, and solids? Give an exception to the rule. Do gases expand alike? What is Rudberg's law? What substance is taken as the type of gases? Why is it necessary to state the temperature at which the estimate of volume is made?

at 75° than it does at 32°. In the instance before us we consider the original volume to be that which the gas would have at 32°, and, as has been said, every degree above that point will increase the volume by $\frac{1}{491}$ of the bulk it then possessed.

Gases expand with uniformity as their temperature increases. Ten degrees of heat produce the same relative effect, whether applied at a low or at a high temperature. This regularity probably arises from the want of cohesion which the gaseous particles exhibit. As we shall presently see, it is not observed in the case of liquids and solids.

The change in specific gravity of air, when it is warmed, is one of the causes of the rise of Montgolfier balloons. These, which were invented in France in the year 1782, consist of a bag or globe of light materials, such as paper or silk, with an aperture at the lower part, through which, by the aid of combustible material, as straw or shavings, the air in the interior may be rarefied. On a small scale, they may be made of thin tissue paper, pasted together so as to form a sphere of two or three feet in diameter, *Fig.* 9, an aperture being cut in the lower portion six inches or more in width, and beneath it a piece of sponge, soaked in spirits of wine, suspended. This being set on fire, the flame rarefies the air in the interior of the balloon, which, though it might be at first flaccid, soon dilates, and the whole apparatus will now rise in the air precisely on the same principle that a cork rises from the bottom of a vessel of water.

Fig. 9.

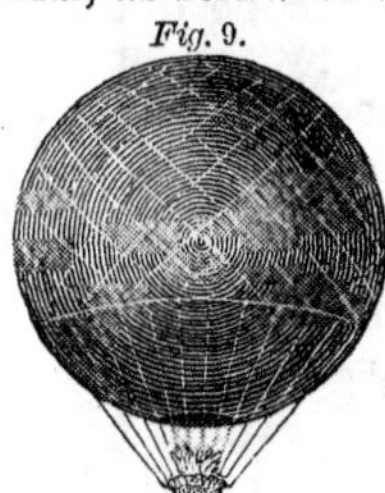

In addition to the heated air, the vapor of water that is produced from the burning body assists in the ascension, because it is lighter than air. By decline in temperature, the heated air collapses to its original size, the steam condenses, and the balloon descends to the surface of the earth again.

In the operation of cupping, the cupping-glass re-

Do gases expand with uniformity? Describe the construction of the Montgolfier balloon. Why does it rise in the air? What is the cause of its descent? Describe the operation of cupping.

ceives for a moment the flame of a spirit-lamp, and is then quickly applied to the surface of the skin. The vapor of water condensing, and the heated air contracting, a very good vacuum can be made.

As the volume of air changes so readily with the temperature, becoming less when cooled and greater when warmed, the amount of air that a given measure will hold is very different at various places on the earth's surface. A vessel that will hold an ounce' weight at the mean temperature of New-York will hold more in the cold polar regions, and less at the tropics. In the former the air is more dense, because it is in a contracted condition by reason of the low temperature, and therefore a greater weight is included under a given volume; in the latter, the reverse is the case. The influence of such changes in the bulk of the air is well seen in the Inca Indians, who inhabit the elevated plateaus near the city of La Paz, in South America. The size of their bodies is much greater than that of average men; and, on examination, this is found to be due to increase in the size of the chest, caused by the larger lungs that they require in order to secure an equal amount of oxygen.

As the expansion of atmospheric air takes place with regularity when the temperature rises, it is sometimes employed to measure temperature. The air-thermometer, called also Sanctorio's thermometer, but which was invented by Galileo about 1603, consists of a tube of glass, *Fig.* 10, terminated at its upper end by a bulb, *a*. The other end of the tube, being open, dips beneath the surface of some colored water in a cup or reservoir, *c*, which serves also as a foot or support to the instrument. The bulb and part of the tube are full of air; the remainder of the tube is occupied by the colored water, which, by its movements up and down, indicates changes of volume in the included air. To the side of the tube a scale of divisions, *b*, is affixed. The tube is not

Fig. 10.

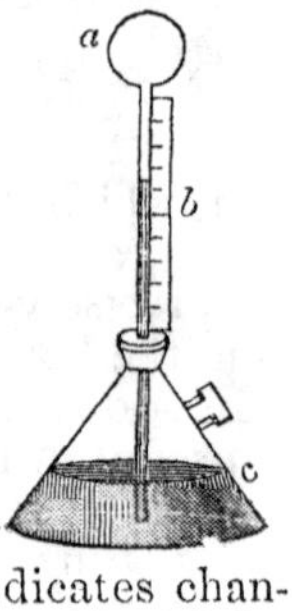

Why does the weight of air in a given measure vary in different places? What is the peculiarity of the Inca Indians? Describe Sanctorio's thermometer. Who was really its inventor?

arranged tightly in the reservoir, but there is a free passage for the air in and out of that part of the instrument. On touching the ball with the fingers, the air within it becomes warm, dilates, and depresses the liquid in the tube; or, on touching with any cold body, it contracts, and the liquid rises.

Fig. 11.

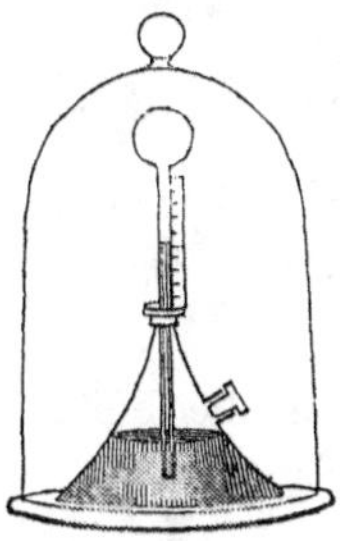

This form of thermometer is liable to a difficulty which renders it impossible to rely upon its indications, except under particular circumstances. It is affected by variations of atmospheric pressure, as well as by changes of heat. To prove that this is the case, place such a thermometer under the receiver of an air-pump, as shown in *Fig.* 11. On producing the slightest degree of rarefaction, the liquid in the tube is instantly depressed, and on restoring the pressure of the air it returns to its original position.

Fig. 12.

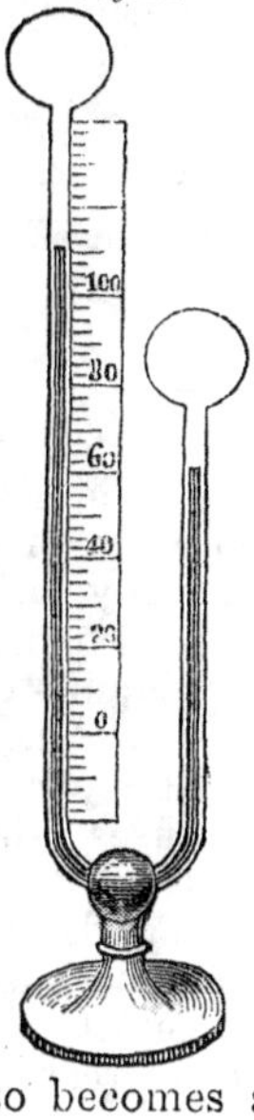

This difficulty is avoided by a change in the form of the instrument, by which atmospheric pressure is altogether avoided. A tube, bent as in *Fig.* 12, has a bulb blown on each extremity. The bulbs and tube are filled with air, except where there is a column of colored liquid that serves as an index, and whose movements are measured by the aid of a scale. This, the differential thermometer, is so called because it only indicates the difference in temperature between the two bulbs, and not their actual warmth. When both bulbs are heated, the liquid index is equally pressed on both sides, and does not move, and the same when both are cooled. By inserting a tight instead of a loose stopper at *c*, *Fig.* 10, that instrument also becomes a differential thermometer.

Describe the action of Sanctorio's thermometer. What is the difficulty with this thermometer? How may the effect of pressure on it be shown? How is this difficulty avoided? Describe the differential thermometer and its mode of action.

Different liquids expand differently for the same thermometric disturbance. This is easily shown by an apparatus, as in *Fig.* 13, in which we have two tubes, *a b*, with bulbs on their ends, dipping into a large vessel. The tubes and bulbs should be of the same size, and filled with the liquids to be tried to the same height. To each a scale is annexed. Let one, *a*, be filled with alcohol, and the other, *b*, with mercury; on pouring hot water into the vessel two phenomena are witnessed: 1st. Both liquids expand; 2d. They expand unequally when compared together, the mercury expanding least.

Fig. 13.

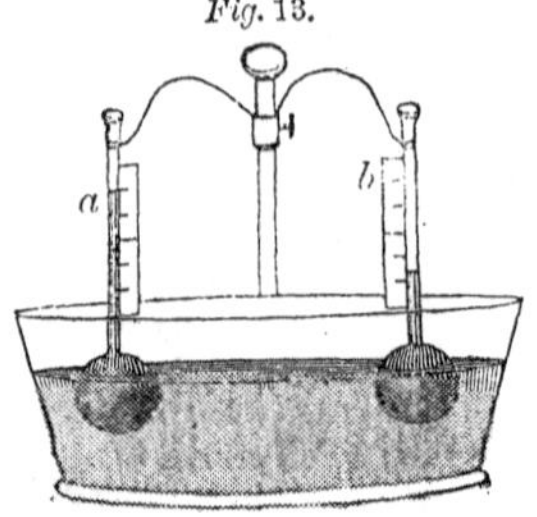

On being heated from the freezing to the boiling point of water, liquids expand as follows:

Expansion of Liquids.

At 32°.	At 212°.	Expansion.
1,000,000 parts of Mercury become	1,018,153	1 in 55
" " Water "	1,046,600	1 in 21.3
" " Oil "	1,080,000	1 in 12.5
" " Alcohol "	1,111,000	1 in 9

The most volatile liquids are the most expansible. This is shown by those which have arisen from the condensation of gases, as cyanogen, sulphurous acid, and especially carbonic acid. The latter, warmed from 32° to 86°, expands four times as much as atmospheric air.

In liquids of analogous chemical constitution the expansion is nearly the same, if the comparison is made, not at the same temperature, but at corresponding temperatures; that is, at equal distances from the boiling point. If the same precaution is adopted, many dissimilar liquids present close resemblances.

Unlike gases, all liquids expand irregularly as their temperature rises, a given amount of heat producing a much greater effect at a high than at a low tempera-

How may it be proved that different liquids expand differently? What is the rate of expansion of mercury? water? oil? and alcohol? What kind of liquids are the most expansible? What is the relation of expansion between liquids of analogous composition?

ture. Ten degrees of heat applied to a given liquid at 200° will produce a greater expansion than if applied at 100°. The reason appears to be that, as a liquid dilates, its cohesive force becomes less, because its particles are being removed farther from each other; and as the cohesive force weakens, its antagonistic power, heat, produces a greater effect.

Fig. 14.

Advantage is taken of the properties of liquids in the making of thermometers. For these purposes alcohol and mercury are the fluids selected. The mercurial thermometer, *Fig.* 14, consists of a fine capillary tube of a lenticular section, with a bulb blown at one end. The bulb and part of the tube are to be filled with quicksilver, and the air expelled from the rest of the tube by warming the bulb until the metal rises by expansion to the top of the tube, and at that moment hermetically sealing the glass by melting the end of it with a blowpipe. As the thermometer cools, the mercury retreats from the top of the tube, and leaves a vacuum above it.

Fig. 15.

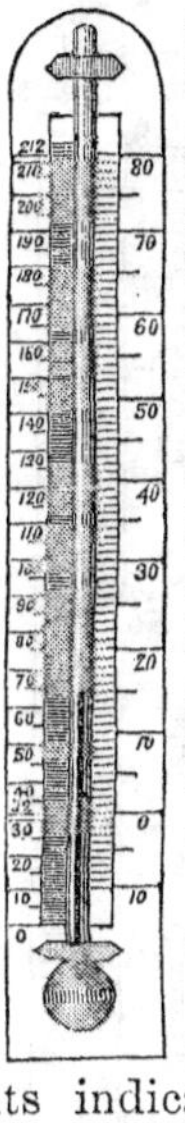

It remains now to annex such a scale to the instrument as may make its indications comparable with other instruments. To effect this, the thermometer is plunged into a vessel containing melting ice or snow, and opposite the point at which the quicksilver stands is marked 32°. It is then transferred to another vessel, *Fig.* 15, in which water is rapidly boiling, and in

---

Do liquids expand with regularity? What liquids are used for thermometers? Describe the mercurial thermometer and the method of making it. How is the scale adjusted? What point is marked 32°?

which it is surrounded on all sides by steam, and the point opposite which the mercury then stands is marked 212°. The intervening space is divided into 180 equal parts; these are degrees, and similar divisions are made in the scale for all points above 212° and below 32°. The zero point, or cipher, is therefore 32 degrees below the freezing point of water.

Fig. 16.

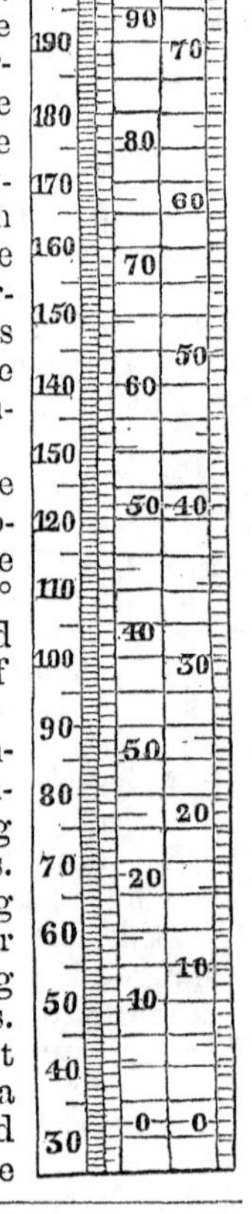

It has been observed that, in the course of time, the fixed points of some thermometers change. This is due to the pressure of the air acting on the bulb, the thin glass of which yields to a certain extent, and the liquid consequently rises in the tube. The same effect will often take place instantaneously by exposing a thermometer to a high temperature. It is therefore necessary to verify, from time to time, the graduations of these instruments.

The zero point of the thermometer scale is not to be regarded as indicating the total absence of heat. Observations have been made in cold climates of degrees 50° below zero, and by the aid of liquefied protoxide of nitrogen and bisulphide of carbon, −220° has been reached.

The Florentine academicians, who introduced the liquid thermometer, employed an arbitrary scale, merely dividing the tube into a number of equal parts. Celsius, a Swede, proposed the melting point of ice and the boiling point of water as standard fixed points, the interval being divided into a certain number of parts. Unfortunately this interval is by different nations differently divided: in America and England, Fahrenheit's scale, described above, is used; in France, and on the

---

What point on the thermometer is marked 212°? How is the space between 32° and 212° divided? What change may occur in thermometers? What is meant by the zero? What is the lowest temperature yet reached? Who proposed the standard fixed points?

Continent generally, the Centigrade scale is preferred. It marks the melting point of ice as 0°, and the boiling of water 100°. In Germany and Russia Reaumur's scale is employed; it has the same zero as the Centigrade, but the boiling of water is marked 80°. It is essential, therefore, in speaking of thermometric degrees, to state what scale is meant, and this is accomplished by putting the letter F, or C, or R after the specified degree. In this book the Fahrenheit division is always used, except in a few cases. In *Fig.* 16 the three scales are, for the sake of comparison, represented side by side.

LECTURE V.

EXPANSION OF LIQUIDS AND SOLIDS.—*Importance of the Thermometer.—Advantages of Quicksilver and Alcohol.—Maximum Density of Water and of other Liquids.—Connection with the Duration of the Seasons.—Ground Ice.—Expansion of Solids.—The Pyrometer.—Force of Metallic Expansion and Contraction.—Its Use in the Arts.*

As all measures of space and time are affected by variations of temperature, the thermometer, which determines those variations, must necessarily be one of the fundamental instruments of physical science. If we state that a given object is a foot long, we must specify the temperature at which the measure was taken, for at a lower temperature it will be less, and at a higher more than a foot. In constructing bridges, etc., where long masses of iron are used, provision must be made for expansion by diurnal and seasonal variations of heat. Each tube of the Britannia Bridge across the Menai Straits is liable in the course of the day to change three inches in length.

Quicksilver is, for several reasons, eminently fitted to be a thermometric fluid. 1st. It can be easily obtained

---

Describe the Fahrenheit, Centigrade, and Reaumur scales. How are they distinguished from one another in writing? What scale is used in this book? What effect does temperature have on measures of space and time? Why is quicksilver suitable for making thermometers?

of standard purity. 2d. It expands with greater regularity than most liquids, and, when in a glass bulb, the irregular expansion of the glass almost exactly compensates the irregularity of the mercury, and hence the true temperature is accurately indicated. The total expansion of mercury, between 32° and 212°, is 1 part in 55.08; between 212° and 392°, 1 in 54.61; between 392° and 572°, 1 in 54.01. 3d. The range of temperature between boiling and solidification is from 662° to —39°, about seven hundred degrees. 4th. It does not soil or wet the tube in which it is contained, for it does not adhere to glass, as water or other fluids would do. 5th. It is more quickly affected by a given amount of heat than water or alcohol, as we shall see when speaking of the capacity of bodies for heat.

When, however, temperatures which approach or are below the freezing point of quicksilver require to be measured, alcohol is appropriately used, because at the lowest temperature yet reached it does not solidify. It may be tinged of any color that is desirable to render it visible.

If some water at 100° is taken and placed in a vessel in which its changes of volume can be observed, on reducing its temperature it will be found, under the general law, to cool as it contracts. As it passes through the various degrees down to a point between 39° and 40°, it steadily diminishes, but below that point, though the cooling may progress at the same rate as before, it begins to expand, and continues to do so until it reaches 32°, when it freezes. The same fact is demonstrated on warming water at 32°: it contracts till 39° is reached, and then expands.

It is obvious, therefore, if we take water at 39°, that it makes no difference whether it be warmed or cooled, it will expand. The liquid occupies at that temperature the smallest bulk, and is at its greatest density, for neither by cooling nor warming can we reduce its dimensions. This point is designated "*The point of maximum density of water.*"

Is its expansion regular? When must alcohol be used in thermometers? What peculiarities does water exhibit in cooling to 32°? What is observed on warming it from 32°? What is the point of maximum density of water?

Many other liquids have points of maximum density, which are reached before solidifying. In the act of turning into a solid, water undergoes a very great dilatation, equal to $\frac{1}{9}$th of its volume, and hence ice will float on water. Several melted metals exhibit the same phenomenon, and advantage is taken of the fact in the arts. The alloy of which types are formed, or stereotype plates cast, in the act of solidifying expands, and forces itself into every part of the mould into which it may be poured, copying it perfectly; the same is true of cast iron. But it is impossible to make good castings with lead, which contracts as it solidifies, and either separates from the mould, or leaves vacant spaces in it.

The fact that water possesses a point of maximum density is connected to a great extent with several remarkable natural phenomena. The freezing of water on the surface is one of these results. If the water contracted as it cooled, the colder portions would descend, and still water would commence to freeze at the bottom first, the solidification extending gradually upward. Collections of water would, during the course of a winter, become solid masses of ice, and would greatly retard the approach of spring, from the length of time they would require for thawing through the non-conducting water above. But, as things are now arranged, the coldest water is the lightest; it floats on the warm water below; solidification takes place only on the surface, and the layer of ice that forms protects the water below from farther refrigeration. When the warm weather of spring comes on, the ice on the surface is in the most favorable position for melting, and thus the point of maximum density of water comes to be connected with the duration of the seasons.

Under certain circumstances, ice, which is then called ground or anchor ice, does form at the bottom of water instead of on the surface. In clear, rapid, rocky streams, the turbulence of the current causes the whole mass of water to become so uniformly mixed that there is no expanded cold layer floating on the top when the

---

Have other liquids points of maximum density? Give examples. What has the maximum density of water to do with the seasons? Explain what occurs in the freezing of water. What is ground or anchor ice?

freezing point is attained. As radiation proceeds from the plants and rocks at the bottom of the stream, ice commences to form on their irregular surfaces in spongy masses, occasionally damming up the stream. Not infrequently it floats to the surface, bearing with it masses of rock.

When salt is added to water, the point of maximum density descends, until it eventually sinks below the freezing point. In the ocean the mass of water is so great that, although the maximum density is below 32°, the winter does not last sufficiently long to reduce the whole to the frozen condition.

It has already been shown by the instrument represented in *Fig.* 1 that solids dilate on being heated. The same results may be rendered apparent by the apparatus *Fig.* 17, which consists of a board having two up-

Fig. 17.

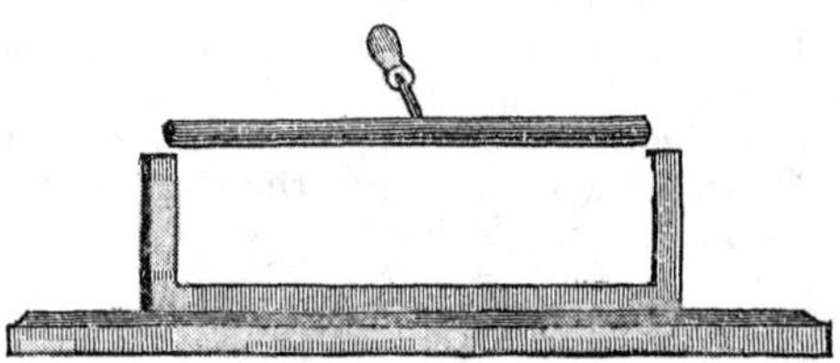

rights, between which a metallic bar, provided with a handle, fits so as to permit a rattling noise when it is moved in the direction of its length. If the bar be heated by pouring hot water upon it, it will dilate to such an extent as to be with difficulty forced into place, and will no longer rattle. Hearing often aids us to detect intervals invisible to the eye.

The pyrometer, of which we have several varieties, is represented in *Fig.* 18. It may serve to illustrate the fact that solid substances expand by heat. It consists essentially of a metallic bar, *a a*, resting at one end against an immovable prop, *e*; the other end bearing upon a lever, *b*. The extremity of this lever presses upon a second lever, *c*, which also serves as an index. Upon

---

Explain its formation. What effect has the addition of salt to water on the point of maximum density? Explain *Fig.* 17. What fact does it illustrate? What is the use of pyrometers? Explain the construction of *Fig.* 18. Describe its action.

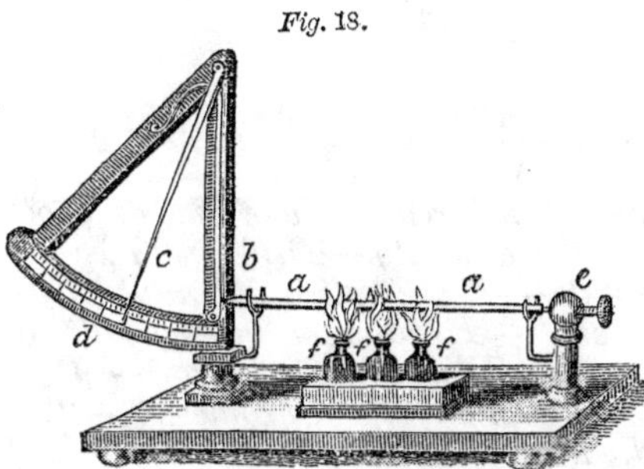

Fig. 18.

the index-lever a spring acts so as to oppose the lever *b*, and the point of the index ranges over a graduated scale, *d*.

If lamps be applied to the bar it expands, and the pressure, taking effect on the lever, puts it in motion, the index traversing over the scale. On removing the lamps the bar contracts, and the spring, pressing the lever in the opposite direction as soon as the bar is cold, brings the index back to the original point.

The force with which such expansions take place is enormous, being equal in amount to that which would be required to elongate or compress the material to the same extent by mechanical means. For a variation of 80° as between the cold of winter and heat of summer, a wrought-iron bar 10 inches long will vary in length $\frac{1}{200}$ of an inch, and will exert a strain equal to 50 tons on the square inch. This property is useful in the arts, where bands of iron are shrunk on wheels, cannon, etc. Being made somewhat too small to fit when cold, on heating they expand enough to be slipped on, and when cooled adhere tightly. For the same reason, boiler rivets are made red-hot.

How much force is required to expand a metal bar? How much does a bar 10 inches long expand for a temperature of 80°? Of what use is this fact in the arts?

LECTURE VI.

Expansion of Solids.—*They regain their Size on cooling.—They expand increasingly.—Different Expansion of various Solids.—The Gridiron Pendulum.—Table of Expansions.—Expansion of crystallized Carbonate of Lime.—Solid Thermometers.—Breguet's Thermometer.—Daniell's Pyrometer.—The Thermometer indicates Intensity of Heat.*

It is very commonly supposed that when solid bodies have been heated they do not return rigidly to their former dimensions on cooling. But a few facts dispose of this supposition. A bar of metal exposed to the weather is subjected to continual variations of temperature, warming and expanding when the sun shines on it, cooling and contracting at night. If it did not come back to its original size exactly, it would be seen to grow, and in the course of time it would be much larger. The iron railings around public parks and squares never increase and become too large for their situations. Solids, therefore, on cooling from a warmed state, regain the normal size. Lead, however, offers an exception; for, on account of its softness, a leaden pipe used for conveying steam will in a short time become perhaps several inches longer than at first. Leaden floorings of hot-water sinks, also, are thrown into ridges and puckers.

By linear dilatation is meant increase in one dimension, as in length; by cubic dilatation, increase in all three dimensions, length, breadth, and thickness. Knowing the amount of linear dilatation, the cubic dilatation is found nearly enough by multiplying by 3.

Solids expand increasingly as their temperature rises, in that respect resembling liquids, and, for the same reason, a diminution of the cohesive force, because of the increased separation of the atoms. Compared one

Do solids regain their size on cooling? What is the reason that lead is an exception to the rule? What is meant by linear dilatation? What by cubic dilatation? Do solids expand with uniformity?

with another, their rate of expansion is very different, a fact shown experimentally by riveting together two bars, one of iron, *a a*, and one of brass, *b b*, as in *Fig.* 19. At ordinary temperatures the compound bar is straight, but if hot water be poured on it, it curves as at *a c*, the brass being on the outside; while, if cooled below the original point, the curvature is in the other direction, as at *b d*, the brass being on the inside. These changes are due to the fact that brass expands and contracts more for a given disturbance of temperature than iron.

Fig. 19.

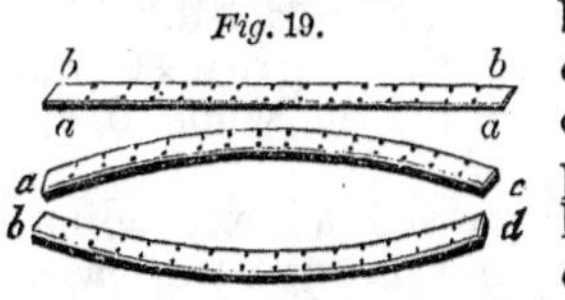

By the aid of proper metallic combinations, a bar can be constructed that will always remain of the same length, regardless of the temperature. The gridiron pendulum is made up of two bars of brass and three of iron. The expansions being in contrary directions, and the lengths being accurately proportioned to the ratio of expansion, the ball, or bob, remains at a constant distance from the point of support, and the time of vibration of the pendulum does not alter.

The following table exhibits the expansion in length of various substances, when heated from the freezing to the boiling point of water:

English Flint Glass...	1 in 1248	Gold......	1 in 682
Glass tube (French)..	1 in 1148	Copper...	1 in 582
Platinum................	1 in 1131	Brass.....	1 in 536
Palladium	1 in 1000	Silver....	1 in 524
Untempered Steel.....	1 in 926	Tin.......	1 in 516
Antimony	1 in 923	Lead......	1 in 351
Iron......................	1 in 846	Zinc......	1 in 340
Bismuth................	1 in 718		

Ice is much more expansible than the metals, surpassing even zinc. Glass and platinum may be fused together without parting as they cool, for their rates of expansion are nearly alike. The process of cutting glass with a hot iron depends on unequal expansion.

Do all solids expand alike? Describe the compound bar and its action. What is the construction of the gridiron pendulum? Describe its action. What is the table intended to illustrate? What is the amount of expansion of ice? Why can glass and platinum be fused together without separating on cooling?

Rupert's drops are tear-shaped portions of glass, which have been formed by dropping that substance while hot into water. The exterior being suddenly cooled, while the interior is yet fluid, is subjected to a great strain. If the tip end of the tail be broken off, the whole drop flies into powder.

Though a solid is usually regarded as expanding equally in all directions, this is not always the case. Those crystals which possess the property of double refraction, and in which all the sides and angles are not alike, change their shape when heated. In a crystal of calcareous spar the obtuse angles become more acute, and the inclination of the faces to each other is made 8½′ less by an elevation from 32° to 212°. The crystal elongates most in the direction of the optic axis, and contracts in directions at right angles. The general bulk, however, increases about 1 part in 510 for the above warming.

Some metallic bodies have points of maximum density in the solid state. Rose's fusible metal, a compound of lead, bismuth, and tin, when heated from 32° to 111°, expands, but after that contracts, and continues to do so till 156°, at which temperature it is less than at 32°. After that it again expands, and continues to do so until it melts at about 201°.

Liquid thermometers can not measure the highest degrees of temperature, because the vaporization of their contents would cause their destruction. For this reason, the expansion of one of the more infusible solids, as, for example, platinum, must be used, and some contrivance similar to *Fig.* 18 employed. The difficulty with such apparatus is that the expansion of a short bar of metal is so minute that some means of magnifying the effect is necessary, and at once irregularity is introduced by wheel-work and friction of levers. A compound strip of metal is free from these difficulties, and will, if long enough, indicate temperatures with great precision. Breguet's thermometer consists of a delicate slip of

---

What are Rupert's drops? Do solids expand equally in all directions? Describe the effect of heat on calcareous spar. State the peculiarities of fusible metal when heated. For what purposes are liquid thermometers inapplicable? What is the difficulty with solid thermometers?

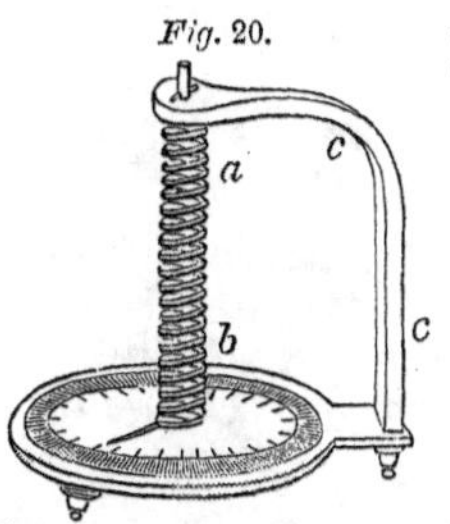

Fig. 20.

platinum soldered with gold to one of silver, and curved into the form of a spiral, *a b*, *Fig.* 20. It is fastened at its upper end to a metallic support, *c c*, and from its lower end an index projects, which plays over a graduated circle. As silver expands twice as much as platinum, when the temperature rises, curvature, with motion of the index, takes place; when it falls, motion in the opposite direction results. The principle is the same as that described in the case of the compound bar, *Fig.* 19. This thermometer is exceedingly rapid in its indications, because the mass of the spiral is so small, as compared with the mercury in the bulb of a thermometer. By breathing upon it, it will in an instant rise to above 90°.

For the highest temperatures Daniell's pyrometer is employed. It consists of a bar of platinum inclosed in a cylinder of black-lead earthenware. When it is heated, as the platinum expands more than the earthenware, it presses an index forward, the index remaining protruded when the pyrometer is taken out of the fire. The amount of protrusion is then measured by a scale, which converts it into degrees. By its aid it has been shown that brass melts at 1869°, silver at 1873°, copper at 1996°, gold at 2200°, and cast-iron at 2786°. The highest heat of a wind-furnace is 3280°.

The thermometer does not in reality measure the amount of heat in a body. If immersed in a glass and a bucketful of the same well-water, it stands at the same point, but of course there is much more heat in the latter. It measures the intensity, that is, the quantity contained in a space equal in volume to the mercury in the instrument itself. Besides this, though it may stand at the same height in the same amount of two liquids, it does not follow that they contain the same absolute quantity of heat, as we shall see in the next lecture.

Describe Breguet's thermometer, *Fig.* 20. Give an illustration of its sensitiveness. State the reason for it. What is the construction of Daniell's pyrometer? Give the melting points of brass, silver, copper, gold, iron. What does the thermometer indicate?

LECTURE VII.

SPECIFIC HEAT OF BODIES.—*Methods of Calorimetry.—Warming.—Melting.—Cooling.—Mixture.—Specific Heat of various Bodies.—Effect of Physical Condition on specific Heat.—Black's Theory of Capacity for Heat.—The Dynamical Theory.—Rumford's Experiment.*

MANY years ago it was discovered by Boyle that if two vessels of the same size and form were filled with different liquids, and placed before the fire so as to receive its heat equally, their temperature did not rise similarly. If one was filled with water and the other with quicksilver, the temperature of the latter would rise much more rapidly than that of the former. The same quantity of heat will raise the temperature of mercury twice as high as that of an equal volume of water.

From a series of similar experiments it has been proved *that different bodies require different amounts of heat to warm them equally.*

CALORIMETRY.

There are several different methods by which the specific heat of a body may be determined, such as, 1st, by warming; 2d, by melting; 3d, by cooling; 4th, by mixture.

The first is that just mentioned as the experiment of Boyle, and consists in exposing the same weight of the substances to be tried to a uniform source of heat, as, for example, a bath of hot water, and observing how high the temperature rises in a given time. It will be found that it takes thirty times as long to warm water as mercury, and hence the specific heat of water is thirty times greater.

What was Boyle's discovery? When two vessels are filled, one with mercury, the other with water, which warms most rapidly? What is the comparative effect of the same quantity of heat on water and mercury? What has been proved to be the law of warming? What methods are there of determining the specific heat of bodies? Describe the method of Boyle.

The second process is accomplished by the aid of the calorimeter, the principle of which is illustrated in *Fig.* 21. A solid block of ice, *a a*, is taken, and in it a cavity, *b*, is made, the mouth of which may be covered with a slab of ice, *c c*. To determine the relative specific heats of water and mercury, take a flask, *d*, and place in it an ounce weight of water, and, by immersing the flask in a bath of hot water, raise its temperature to a given point; for example, 200°. Then place it in the cavity, *b*, and put on the cover, *c c*. As soon as the temperature of the flask has fallen to 32°, a certain portion of the surrounding ice will be found to have melted; this is to be poured out and measured.

Fig. 21.

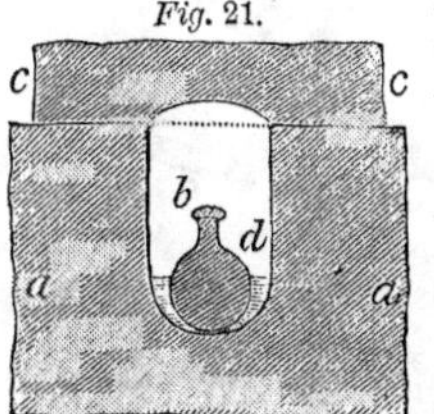

In the same way expose an ounce weight of mercury to the same process. The quantity of water melted will be only $\frac{1}{30}$ of the amount in the previous case. A given weight of water will therefore melt thirty times as much ice as the same weight of quicksilver in cooling through the same number of degrees.

Lavoisier's calorimeter, *Fig.* 22, acts on the same principle. It consists of a set of tin vessels within one another. In the central one, *a* is the substance to be examined. Between this and the next is a quantity of broken ice, the water from the melting of which flows out through a stopcock, *c*, into a graduated glass. The vessel *b* is surrounded by another, *d*, to avoid the melting of its ice by the warm external air. The water from this is carried off by the stopcock, *e*. Though excellent in principle, the difficulties in the practical working of this instrument are great.

Fig. 22.

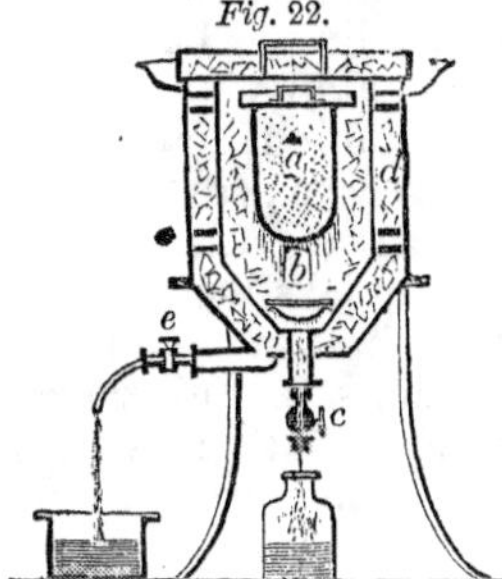

The mercurial calorimeter of Favre and Silberman,

Describe *Fig.* 21, and the method of using it. What is the relative specific heat of water and mercury? Describe Lavoisier's calorimeter.

Fig. 23, is, in reality, an enlarged mercurial thermome-

Fig. 23.

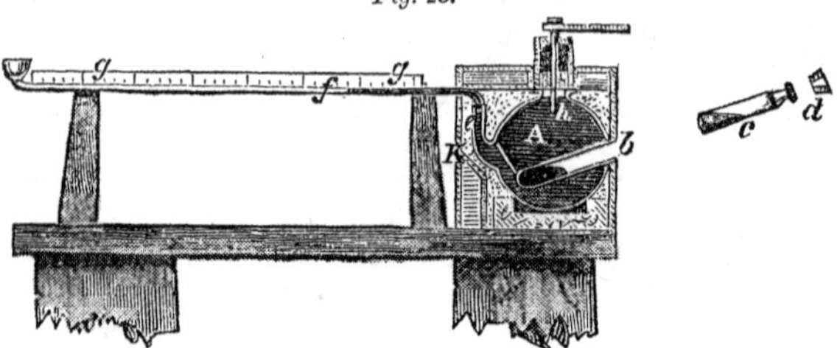

ter, the bulb of which may receive the substances to be experimented upon. It consists of a glass globe, A, with three apertures. Into one of them, *b*, is fixed a platinum tube, and into this tube a bottle, *c*, fits, and is inclosed by a cork, *d*. In the tube *b* is a small quantity of mercury, to secure the speedy conduction of heat. The aperture *e* terminates in a stem like a thermometer tube, and by the movements of the mercury, *f*, along the scale, *g g*, the dilatation of the mercury in A is measured. A piston moved by a screw, *h*, serves to bring the mercury, *f*, to the zero of the scale. The globe is incased in a box, K, lined with swans' down, to avoid disturbance from exterior cooling. The value of the degrees is determined experimentally.

The third process, the method by cooling, consists in ascertaining the length of time necessary to cool through a definite interval. Water, which has a great specific heat, takes a length of time to cool; quicksilver, on the contrary, a much less time. The method of cooling requires several precautions; among others, the substances must be placed in vacuo. With liquids it gives very good results, but with solids the differing rates of radiation and conduction render it objectionable.

The fourth, the method by mixture, is the best. It is easily understood. If a pint of water at 40° be mixed with a pint of water at 100°, the temperature will be the mean, that is, 70°. But if a pint of mercury at 100°

Describe the calorimeter of Favre and Silberman. What is it in reality? In what does the method of cooling consist? What precautions does it require? How is the method by mixture conducted?

be mixed with a pint of water at 40°, the temperature of the mixture will be 60°; so that the forty degrees lost by the mercury have only raised the water twenty degrees. If equal *weights* instead of equal *volumes* of the substances are used, the results are more striking, the proportion being then as 1 to 30 instead of 1 to 2.

The method of mixture is equally applicable to solids. If a pound of copper at 300° be plunged into a pound of water at 50°, the resulting temperature is 72°, from which it appears that the specific heat of water is about ten times that of copper.

By resorting to the above methods the specific heat of a number of substances has been ascertained, but as it is not the absolute quantity of heat that is determined, it is necessary to have some standard of comparison. For solids and liquids water has been chosen, its specific heat being the highest, while for gaseous bodies air is used.

Fig. 24.

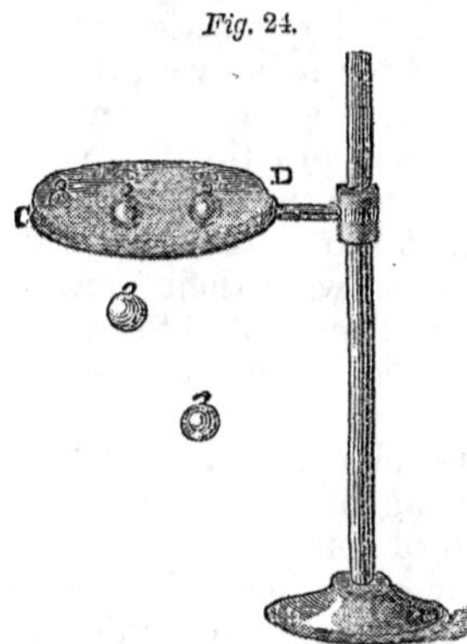

The varying specific heat of bodies may be illustrated by *Fig.* 24, which consists of a cake of wax, C, supported on a stand, D, and a number of equally warmed balls of various substances. Those which have a high specific heat can melt their way through the wax, and fall to the table, and those which have less only melt partly through.

Specific Heats of Equal Weights between 32° *and* 212°

Water	1.00000	Brass	0.09391
Oil of Turpentine	0.42593	Silver	0.05701
Charcoal	0.24150	Tin	0.05623
Glass	0.19768	Mercury	0.03332
Iron	0.11379	Platinum	0.03243
Zinc	0.09555	Gold	0.03244
Copper	0,09515	Lead	0.03140

How do we know that the specific heat of water is ten times that of copper? Why is a standard of comparison for specific heat necessary? What are the substances used as standards? Describe *Fig.* 24. What does it illustrate?

The specific heat of a substance varies with its physical condition. Water, when solidified into ice, has only one half the specific heat, as 505 to 1000. In the form of steam there is also a diminution, the ratio being as 480 to 1000 for equal weights. Mechanical compression also alters the specific heat, copper, which had stood at 0·950, by hammering being reduced to 0·936. On annealing, it, however, regained its primitive condition. In dimorphous bodies, the densest form has the lowest specific heat, diamond, for instance, being 0·1468; graphite is 0·2018, and charcoal 0·2415.

Dr. Black, who was one of the early investigators of heat phenomena, introduced a term, "Capacity of bodies for heat," implying the idea that this principle, entering their pores, was taken up by different bodies to different amounts, in the same way that two sponges of different densities will hold different quantities of water. On this supposition, the lighter a body is, the greater should be its capacity for heat, because its atoms are more widely separated. But in practice this is found not to be the case, oil, that floats on water, not having half the capacity for heat that the water has.

The specific heat of bodies increases with their temperature. On Black's doctrine, this is accounted for by saying that the atomic interstices become larger, and there is more room for the heat. In the process of compression by hammering, heat is evolved, and, according to Black, this is due to the forcing of the atoms together, which causes the exudation of some of the heat. But it can not be supposed that the great heat which is produced can all arise from such a source. According to the dynamical theory of heat, it results from a conversion of mechanical motion into molecular motion, heat being regarded not as a substance, but as a motion of the atoms of bodies.

An experiment by Rumford illustrates this point. He took a hollow cylinder of iron, into which a plunger was fitted, and caused to press against the bottom. A

How may the specific heat of a substance be varied? What is meant by capacity for heat? Give an illustration. Why is the specific heat of bodies decreased by hammering, on Black's theory? How is it accounted for on the dynamical theory? Describe Rumford's experiment.

box containing 18¾ pounds of water surrounded the cylinder; its temperature, which was, at the beginning of the experiment, 60°, was marked by a thermometer. The cylinder was turned by a horse, and in an hour after the friction had commenced the temperature was 107°. At the end of two hours and a half the water actually boiled. The quantity of metal abraded, and from which, according to the material theory, this great amount of heat must have been produced, was only a few hundred grains. In addition to this, he found that the chips had the same capacity for heat as before, and hence concluded that the mechanical motion had been converted into heat.

There is also a very significant experiment of Davy which indicates the same fact. Ice has only one half the capacity for heat that water possesses; and, in addition, an immense amount of heat must be consumed in changing it to the liquid condition. But Davy, by friction of two pieces of ice together, produced water, and yet the original quantity of heat in the ice was but a small fraction of that in the water. The extra amount must have been furnished by the mechanical motion used.

LECTURE VIII.

SPECIFIC HEAT AND LATENT HEAT.—*Variability of Specific Heat under Compression and Dilatation.—Theory of the Formation of Clouds.—The Fire Syringe.—Cold of the Upper Regions of the Air.—Connection between Specific Heats and Atomic Weights.—Latent Heat.—Melting Points of various Bodies.—Latent Heat of Water and other Bodies.*

THE capacity of gases for heat is determined by warming a known weight by means of a spiral tube immersed in hot oil, and then conducting the gas through a vessel surrounded by a known weight of water. The inquiry is attended by very great difficulties.

When the volume of a gas increases its capacity for

What did Rumford conclude? What was Davy's experiment? What conclusion did he come to? How is the capacity of gases for heat determined?

heat increases, and a diminution of volume is attended with a diminution of capacity. If a Breguet's thermometer be placed under the receiver of an air-pump, which is rapidly exhausted, *Fig.* 25, a sudden reduction of temperature is indicated. As the rarefaction proceeds the capacity for heat increases, an increase which is satisfied at the expense of a portion of the sensible heat.

Fig. 25.

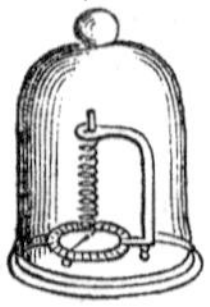

On the same principle, the appearance of the fog or cloud, which comes when moist air is rarefied, is explained. The quantity of vapor that can exist in a given space depends on the temperature. If the space is cooled, a portion of the vapor will condense. When, by suddenly rarefying air, we increase its capacity for heat, the temperature falls, and part of its moisture assumes the form of drops. If a bell-jar is taken, the inside having been rinsed out with water and placed on the air-pump stand, on exhausting, a mist makes its appearance, which immediately clears up on readmitting the air. A candle placed behind the jar will show the effect to a large audience.

Fig. 26.

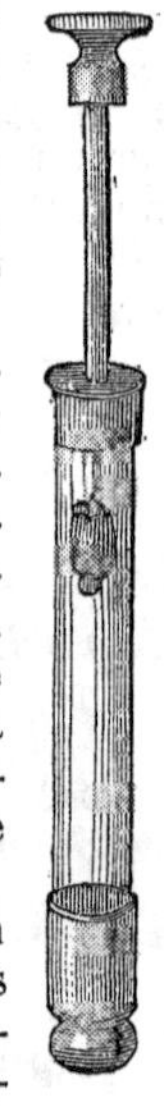

When air is suddenly compressed, its capacity for heat diminishes. This is experimentally shown by the fire-syringe, *Fig.* 26, which consists of a strong glass tube, containing a tightly-fitting piston. By forcing the piston downward, the air in the cylinder can be compressed. If a piece of cotton, moistened with bisulphide of carbon, be put into the syringe previously, a brilliant flash of light will follow the compression. Even tinder may be set on fire in the same way.

The variation in capacity of substances with variation of volume was explained on Black's doctrine, as follows: If a sponge soaked in water be compressed, a portion of the water ex-

---

What effect does change in volume of a gas have on its capacity for heat? Describe the experiment *Fig.* 25. Explain the appearance of fog in rarefying air. How may this be illustrated experimentally? Describe the fire-syringe and its method of action. How is the variation in capacity for heat explained on Black's theory?

udes, just as in the syringe the air allows heat to escape. On relaxing the sponge, it will take up more water, just as air, when dilated, has its capacity for heat increased. According to the dynamical theory of heat, the heat developed during compression would be regarded as a conversion of the muscular movement of the arm into molecular motion of the atoms of air.

The great degree of cold which prevails in the upper regions of the atmosphere, as is shown by the following table, is due, to a considerable extent, to the capacity of that dilated air for heat.

Altitude in feet.	Temperature.	Altitude in feet.	Temperature.	Altitude in feet.	Temperature.
0	80°	15,000	31°.4	25,000	− 7°.6
5,000	64°.4	20,000	12°.8	30,000	−30°.7
10,000	48°.4				

The formation of clouds is also explained on the same principle. A stratum of air resting on the sea or moist earth becomes saturated with moisture, and by the warmth of the sun begins to rise through the atmosphere. As it rises it expands, on account of the decreasing pressure, and its capacity for heat increases. A portion of the moisture is therefore deposited in the form of drops, and constitutes a cloud.

The small capacity of quicksilver for heat renders it a suitable liquid for thermometers, because it cools or warms rapidly, and follows variations of temperature more quickly than water and most other fluids.

There is a connection between the specific heat of bodies and their atomic weights. In most cases, if substances are compared together in the ratio of their combining proportions, it will be found that the same amount of heat will raise them an equal number of degrees; that is, *the specific heat of an elementary body is inversely as its combining proportion.* In the exceptional cases there is generally some simple multiple relation, as is shown in the following table:

How is the variation in the capacity for heat explained on the dynamical theory? What is the cause of the cold in the upper regions of the air? What is the temperature at 30,000 feet? Explain the formation of clouds. Why is quicksilver specially suitable for thermometers? What is the relation between specific heat and atomic weight?

Table of the Specific Heats of the Elementary Atoms.

Iron	3.1861	Sulphur	2.8416
Zinc	3.1054	Mercury	3.3320
Copper	3.0162	Silver	6.1570
Lead	3.2530	Arsenic	6.1050
Tin	3.3063	Antimony	6.1939
Nickel	3.2045	Gold	6.3777
Cobalt	3.1553	Iodine	6.8732
Platinum	3.1976	Bismuth	6.4764

From this table it appears that the first ten substances show a close approximation in their capacities for heat, if the quantities used be in proportion to the *atomic* weights instead of *equal* weights. The remainder have a double capacity. In compounds the specific heat may be calculated from the sum of the atomic heats of their components.

Latent Heat.

First Change of Form.—Heat of Fluidity.

When solid substances, which can resist a high temperature without decomposition, are exposed to an increasing heat, a point is eventually reached at which they assume the liquid state. This point, known as the point of fusion, or melting point, is fixed for each substance.

Table of the Melting Points of Bodies.

Mercury	−39	Bismuth	512
Oil of Vitriol	−30	Nitrate of Soda	591
Bromine	9.5	Lead	620
Ice	32	Nitrate of Potassa	642
Phosphorus	111.5	Zinc	773
Potassium	136	Silver	1773
Yellow Wax	143.6	Copper	1996
Sodium	207.7	Gold	2016
Iodine	224.6	Cast Iron	2786
Sulphur	239.	Wrought Iron	above 3280
Tin	451		

Some substances, perhaps all, to a greater or less extent, pass through a condition intervening between the solid and liquid state, assuming a pasty consistency.

What quantities of substances must be used instead of equal weights? What change in *form* occurs on heating solid substances? State the melting points of various bodies—mercury, oil of vitriol, etc. What intermediate condition may substances pass through?

The manufacture of glass depends on such a property. It is also shown strikingly by various oils and wax. Indeed, different liquids may be said to present different degrees of liquidity. This is well seen when sulphuric acid, a dense, sluggishly-moving body, is compared with sulphuric ether, a substance of remarkable mobility. The liquidity of the liquid state seems generally to be increased by elevation of temperature.

If we take a mass of ice, the temperature of which is at the zero point, and bring it into a warm room, examining the circumstances under which its temperature rises, they will be found as follows: the mass of ice, like any other solid body, warms with regularity until it reaches 32°; then, for a considerable period of time, no farther elevation is perceptible, but it undergoes a molecular change, assuming the liquid condition; when this is complete, the temperature again commences to rise.

That we may have precise views of these facts, let us suppose that the mass of ice and the warm room into which it is carried have such relations to each other that the temperature of the former can rise from the zero point one degree per minute; for thirty-two minutes the temperature of the ice will be found to increase, and at the end of that time, a thermometer, if applied, would stand at 32°. But now, although the heat is still entering the ice at the rate of a degree per minute, the process of warming ceases, and for 142 minutes no farther rise takes place, but the ice melts, and is completely liquefied at the end of that time. The temperature then again steadily rises, and continues to do so with regularity.

We know, from experiments like the foregoing, that about 142 degrees of heat are absorbed by ice in passing into the condition of water; and, as this heat is not discoverable by the thermometer, it is designated as latent heat.

A similar fact appears when any liquid, such as wa-

Give an example of the intermediate condition substances may pass through. Give an example of different degrees of liquidity. Describe what occurs on warming ice. At what degree does its temperature remain stationary for a time? How long is the pause? What change in form occurs at the same time? What is meant by latent heat?

ter, passes into the vaporous condition. If some water be exposed to a fire which can raise its temperature at the rate of one degree per minute, that effect will continue till 212° are reached; at that point, no matter how much the heat be increased, the temperature remains stationary. The water undergoes a change of form, assuming the condition of a vapor, and the change is completed in about 967 minutes. In this, as in the former instance, we infer that a large amount of heat has become latent, or undiscoverable by the thermometer, and that it is occupied in establishing the elastic form which the water has assumed.

Table of Latent Heat of Bodies.

	F.°	Water = 1.
Water	142.65	1.000
Nitrate of Soda	113.34	.794
Nitrate of Potash	85.26	.598
Zinc	50.63	.355
Silver	37.92	.265
Tin	25.65	.179
Cadmium	24.44	.171
Bismuth	22.75	.159
Sulphur	16.85	.118
Lead	9.65	.067
Phosphorus	9.05	.063
Mercury	5.11	.035

By the method of mixtures the same results may be established. Thus, if a pound of water at 32° be mixed with a pound at 174°, the mixture will have the mean temperature, that is, 103°; but if a pound of ice at 32° be mixed with a pound of water at 174°, the mixture still remains at 32°; and the reason is clear, from the foregoing considerations, that ice, in passing into the liquid state, requires 142° of latent heat.

At what point does water pause when warmed? How many degrees of heat are required to vaporize water? What has become of the heat? Do all substances have the same amount of latent heat? How can the doctrine of latent heat be proved by the method of mixture?

LECTURE IX.

LATENT HEAT.—*Heat evolved in Solidification.—Theory of Freezing Mixtures.—Expansion during Solidification.—Freezing and Melting Point of Water.—Its Latent Heat affects the Duration of Autumn and Spring.—Regelation.—Heat of Elasticity.—Boiling Points of various Liquids.—Nature of Vapor.*

WHEN a liquid assumes the solid form a considerable amount of heat is evolved. The cause is readily understood from what we have seen taking place during the reverse process, which has led us to the fact that the difference between any given solid and the liquid which arises from it by melting is in the large amount of latent heat which is found in the latter, and which is occupied in giving it its form.

A saturated solution of sulphate of soda may be cooled from its boiling point to common temperatures in a vessel tightly corked without solidification taking place, but when the cork is withdrawn, crystallization ensues, and heat is evolved. This may be proved by taking a bottle filled with such a solution, and, having introduced the bulb of an air-thermometer through the neck by means of an air-tight cork, the mouth of the bottle is to be carefully stopped. When the whole apparatus has reached the ordinary temperature of the air the stopper is withdrawn, and solidification at once takes place; or, if it should at first fail, the introduction of a crystal of sulphate of soda will bring it on. At that moment it will be perceived that not only does the thermometer indicate a rise of temperature, but, if the bottle be grasped, it will be found to be sensibly warm.

On these principles depends the action of freezing mixtures, of which the following is an example. If we take 8 parts of crystallized sulphate of soda and mix it

What becomes of the latent heat of a fluid when it solidifies? What is the difference between the solid and liquid forms of a substance? Describe the experiment with sulphate of soda. What change of temperature is seen? Describe the action of a freezing mixture.

in a thin tumbler with 5 parts of hydrochloric acid, the sulphate of soda, from being a solid, assumes the liquid form; and taking, in order to effect that change of form, heat from surrounding bodies, it reduces their temperature. This may be shown by placing 4 parts of water in a thin glass test-tube and stirring it about in the mixture; the water speedily freezes, even on a summer day.

Table of Freezing Mixtures.

Mixtures.	Reduction of Temperature.
Nitrate of Ammonia 1, Water 1...............	From 50° to 4°
Sulphate of Soda 8, Hydrochloric Acid 5..	" 50° to 0
Snow or Ice 2, Common Salt 1...............	" 50° to – 4°
Snow 3, Dilute Nitric Acid 2.................	" 0 to –46°

All these mixtures depend essentially on the principle under consideration, that latent heat must be furnished to a substance passing from the solid to the liquid state; a cubic yard of ice would require a bushel of coal to melt it. They consist of various solid substances, the liquefaction of which is brought about by the action of other bodies. Even during the liquefaction of an alloy by quicksilver the same thing is observed. An alloy of lead, tin, and bismuth, dissolved in mercury, will cause a thermometer to sink from 63° to 14°.

Many substances, when solidifying, expand. This is the case with water, in which the amount of expansion is about $\frac{1}{9}$th of the bulk. The force which is exerted under these circumstances is very great, and capable of tearing open the strongest vessels. This may be shown by filling a bottle with water, and fastening the cork down tightly with wire. On putting it into a freezing mixture, congelation promptly takes place, and the bottle is burst. An iron bottle filled with melted bismuth, and allowed to cool, is broken in the same manner.

The freezing point of water is usually spoken of as a fixed point, and is marked as such upon the scales of our thermometers, but if water be cooled without agi-

Why is cold produced? Give examples of other freezing mixtures. What is the essential principle of freezing mixtures? What is the amount of expansion of water in freezing? How may the force exerted be shown? Is the freezing of water a fixed point?

tation it may be brought as low as 15°. By causing the freezing to take place in very strong vessels of steel, Mousson found that under a pressure of 13,000 atmospheres water remained liquid at 0°, and that if ice were introduced into the apparatus, on applying pressure, it was unable to preserve the solid state; its bulk was reduced at 13,000 atmospheres by $\frac{13}{100}$ of the volume at 32°.

But though water will retain its liquid form far below its freezing point, ice can not be brought above 32° without melting. The melting of ice is, therefore, the fixed thermometric point.

We have seen that the possession of a point of maximum density by water exerts a great effect upon the duration of the seasons; a similar observation might be made as respects its latent heat. If ice, by the absorption of a single degree of heat when it passes from 32°, could turn into water, the great deposits of winter would suddenly melt, and inundations be frequent, just as when Etna pours a torrent of lava down its snow-clad sides, the flood is more destructive than the molten mass; or if water, by losing a single degree of heat, turned into ice, freezing would go on with great rapidity. To the melting of snow or the freezing of water time is necessary; the 142° degrees of latent heat have to be disposed of; this, therefore, serves to procrastinate the approach of winter, and causes the spring to come on more slowly.

If two pieces of ice at 32°, with moistened surfaces, are placed in contact, they freeze to one another. This phenomenon is called *regelation*, and will occur, though the air may be at 90°, or the ice be in water of that temperature. In the same way, flannel, hair, or cotton will freeze to ice, though the metals will not. By placing fragments of ice in wooden moulds, as in *Fig.* 27, C D, H P, and applying a severe pressure, spheres, as B, cups, lenses, etc., may be formed. The cause of the refreezing has not been ascertained up to the present. Regelation explains the original formation of glaciers from damp snow, and accounts for their motions and

Describe Mousson's experiment. Is the melting of ice a fixed point? What effect has the latent heat of water on the seasons? What is meant by regelation?

Fig. 27.

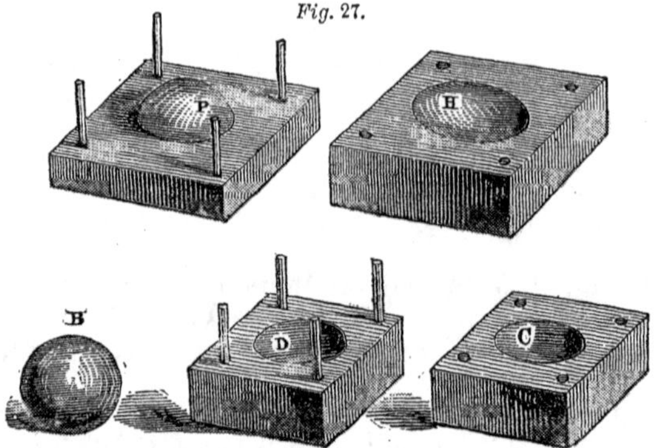

changes of form in descending to the valleys below. It avoids the necessity of supposing that the ice is a viscous body.

Second Change of Form.—Heat of Elasticity.

Exposed to a rise of temperature, liquid substances boil at a particular point, which varies with their nature.

Table of Boiling Points.

Sulphurous Acid........	17°	Acetic Acid..............	243°
Ether......................	95°	Nitric Acid..............	248°
Bisulphide of Carbon...	118°	Oil of Turpentine........	314°
Bromine..................	145°	Phosphorus..............	554°
Alcohol....................	173°	Sulphuric Acid..........	640°
Water......................	212°	Mercury..................	662°

A technical distinction is made between a gas and a vapor. By the latter we understand a gas which will readily take on the liquid form.

Fig. 28.

Some of the leading peculiarities in the constitution of vapors may be exhibited by the following experiment: take a glass tube, *a a*, *Fig.* 28, with a bulb, *b*, blown on its upper extremity; pour water into the bulb, filling the tube to within an inch or two of the end; this

What does *Fig.* 27 illustrate? Do all fluids boil at the same point? Give the boiling points of sulphurous acid, ether, etc. What is the distinction between a gas and a vapor? What fact is *Fig.* 28 intended to illustrate? Describe the instrument and its action.

vacant space fill with sulphuric ether; and now, closing the end of the tube with the finger, invert it in a glass of water, as is represented in the figure. The ether, being much lighter than water, at once rises to the upper part of the bulb, as is shown by the light space, the bulb being, of course, full of ether and water conjointly.

On the application of a spirit-lamp the ether vaporizes, and presses the water out of the bulb into the glass cup. Three important facts may now be established.

1st. Vapors occupy more space than the liquids from which they arise.

2d. They have not a misty or fog-like appearance, but are perfectly transparent.

3d. When their temperature is reduced, they collapse to the liquid state.

It has already been shown that a large amount of heat becomes latent, constituting the heat of elasticity of vapors. The temperature of steam is 212°, as is that of the water from which it rises, but it contains 966° of latent heat, which give it its new form. Different vapors contain different quantities of latent heat: ether, 163°; alcohol, 375°; bromine, 82°; iodine, 43°. It is the large amount of latent heat which makes steam so efficient in warming. The steam arising from one gallon of water will raise the temperature of five gallons from the freezing to the boiling point; its heat of elasticity is nearly sufficient, were the steam a solid body, to make

Fig. 29.

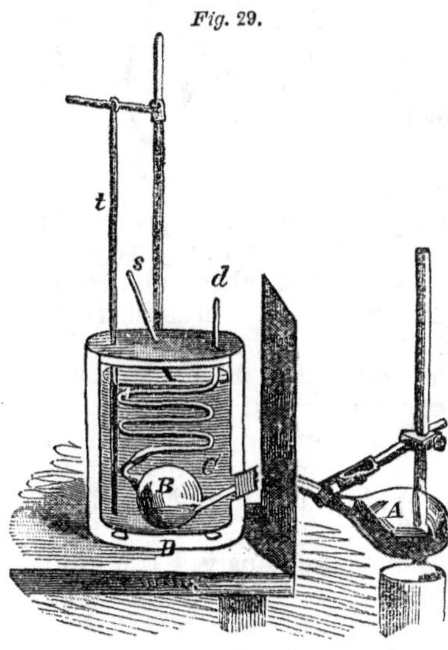

What facts are established by it? What is the difference between steam and water at 212°? Do all vapors have the same amount of latent heat?

it visibly red-hot in the daylight. In the warming of buildings by steam-pipes, each square foot of their surface will heat 200 cubic feet of the surrounding air to 75°, and will require about 170 cubic inches of boiler capacity for its supply.

The latent heat of vapors is determined by an apparatus arranged as in *Fig.* 29. *A* is a flask to contain the liquid; *B*, a receiver with a spiral tube, terminating at *d*. *C* contains a weighed quantity of water. The liquid is distilled into *B*, and the rise of temperature measured by the thermometer *t*; *s* is a tube for agitation. *R* is a tin-plate screen to shut off the influence of the lamp. The whole is inclosed in an outer tin vessel, *D*.

LECTURE X.

VAPORIZATION.—*Vapors form at all Temperatures.—Form instantly in a Vacuum.—Diminution of Pressure favors Evaporation.—Elastic Force of Vapors.—Cumulative Pressure of heated Vapor.—Restriction on Marriotte's Law.—Elasticity increases with Temperature.—Maximum density of Vapors.—Liquefaction of Gases.*

VAPORIZATION goes on at all temperatures. It is not necessary that the boiling point should be reached; even ice will evaporate away, and, if put in the vacuum of a barometer, will depress the mercurial column. The thin films of this substance often seen incrusting windows may disappear without melting, and a mass of ice freely exposed to the air on a dry frosty day loses weight. So camphor is constantly giving off a vapor, which crystallizes on the cold side of the bottle containing it. Steam, therefore, rises from water at all temperatures, but with more rapidity and a higher elastic force as the temperature is higher.

In a vacuum vapors form instantaneously. If a number of barometer tubes, *Fig.* 30, be taken and filled with mercury in the usual manner, the difference of volatility

---

Describe *Fig.* 29. What is its object? Is vaporization limited to any particular temperature? How can it be proved that ice may evaporate? What is the effect of a vacuum on vaporization?

Fig. 30.

of various substances may be shown. Let 1 be kept as a standard of reference. Pass a few drops of water into 2, of alcohol into 3, of bisulphide of carbon into 4, and of ether into 5. The mercury will be progressively more and more depressed in each case.

Diminution of pressure favors evaporation, and many liquids would be permanently gaseous if the pressure of the air were removed. Take a glass tube, A, *Fig.* 31, closed at one end, and, having filled it with water, invert it into a cup, B, and introduce into it a little sulphuric ether, which will rise to *a*. On covering the whole with an air-pump jar, and exhausting, the ether boils, and gives off a transparent vapor. On re-admitting the air, the ether goes back to the liquid condition. By increase of pressure, as well as by diminution of temperature, vapors may be condensed.

Fig. 31.

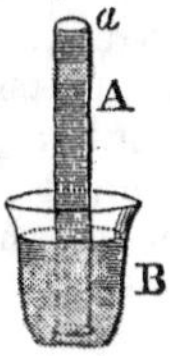

In the process of evaporation, vapor is supplied only from the surface, and consequently is more rapid as the surface is larger. In the salt-works of Salzburg the brine is made to trickle through brushwood, and the water evaporates with great quickness. By covering an evaporating surface of water with oil, evaporation is entirely suspended.

Though vapors occupy more space than the liquids from which they come, the increase of volume is by no

---

Describe the instrument *Fig.* 30. What effect has diminution of pressure on vaporization? Describe *Fig.* 31. What effect has surface on evaporation?

no means the same in all cases. At the ordinary pressure, a cubic inch of water produces 1696 cubic inches, nearly a cubic foot, of steam; alcohol, 528; ether, 298; oil of turpentine, 193. Hence, though the three last-named liquids take up less latent heat than water in vaporizing, they could not be economically used in a steam-engine, because the expansive force is proportionately less, a smaller bulk of vapor being given off.

The elastic force exerted by vapors within certain limits can be measured by the apparatus *Fig.* 30. The theory of the process is very simple. The height at which the barometer stands is determined by the pressure of the air. As long as there is nothing to counterbalance that pressure, the mercury is forced up by it in the tube to a height of 30 inches; but on introducing some ether, as at *a a*, *Fig.* 32, the vapor which forms, exerting an elastic force in the opposite direction, tends to push the mercury out of the tube. On the one hand, we have the pressure of the air; on the other, the elastic force of the ethereal vapor; they press in opposite directions, and the resulting altitude at which the mercury stands expresses, and indeed measures the elastic force of the vapor. Thus, at a temperature of 80°, water will depress the mercurial column about 1 inch, alcohol about 2 inches, and sulphuric ether about 20. These numbers, therefore, represent the elastic force of the vapors evolved.

Fig. 32.

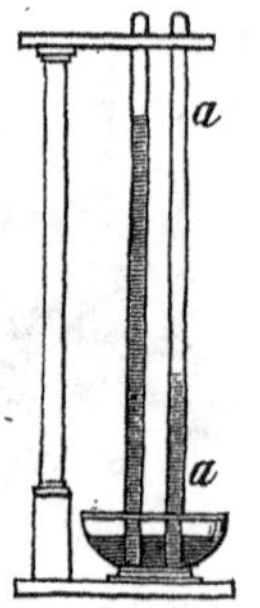

In close vessels, from which there is no escape, or from which the escape is greatly retarded, a constantly accumulated force is generated when the temperature is raised. Thus, if we place some water in a flask, *a*, *Fig.* 33, into which a tube, *b b*, is inserted air-tight by means of a cork, and bent in the form exhibited in the figure, and dip-

Fig. 33.

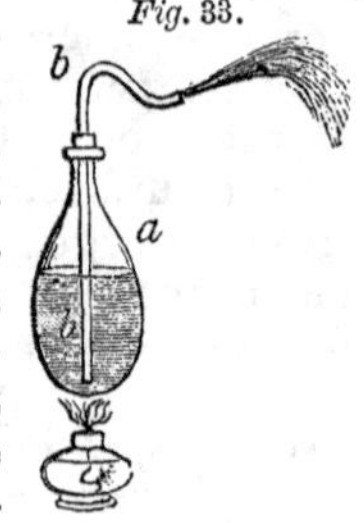

Do all liquids expand equally in assuming the vaporous state? How may the elastic force of vapors be measured by the barometer? What is the principle involved? What results from heating a liquid in a close vessel? Describe *Fig.* 33.

ping nearly to the bottom of the flask, on the application of a spirit-lamp, the vapor generated, having no passage of escape, accumulates in the upper part of the flask, and, exerting its elastic force, presses the liquid through the tube in a continuous stream. The mechanical force which thus arises when every avenue of escape is stopped, is strikingly exhibited by the little glass bulbs called candle-bombs. These are small globules of glass about as large as a pea, with a neck an inch long; into the interior a drop of water is introduced, and the termination of the neck hermetically sealed by melting the glass. When one of these is stuck in the wick of a candle or lamp, as in *Fig.* 34, the heat vaporizes a portion of the water, and, there being no passage through which the steam can escape, the bulb is burst to pieces with a loud explosion, a mechanical force which is wonderful, when we consider the amount of water employed. It is a miniature representation of what takes place on the large scale in the bursting of high-pressure steam-boilers.

Fig. 34.

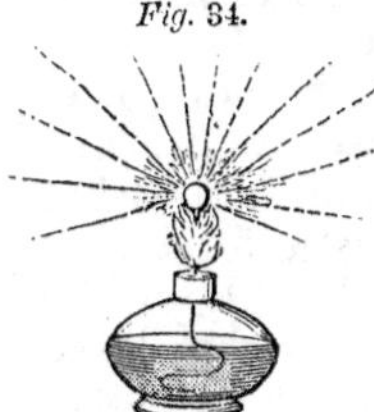

Marriotte's law, which assigns the volume of a gas under variations of pressure, applies, under certain restrictions, to the case of vapors. A permanently elastic gas, when the pressure is doubled, contracts to one half of its former volume; if the pressure be tripled, to one third, and so on; but not so with vapors. If upon steam, as it rises from water at 212°, any increase of pressure be exerted, this vapor at once loses its elastic form, and instantly condenses into water. But vapors, like atmospheric air, if the pressure upon them be diminished, follow Marriotte's law. Thus, if the pressure be reduced to one half, steam at once doubles its volume. For vapors, Marriotte's law, therefore, holds for diminutions of pressure; but when the pressure is increased it apparently fails, the vapors relapsing into the liquid form.

That the elasticity of a vapor increases with its tem-

What is the construction of a candle-bomb? What is Marriotte's law? Does it apply to vapors? What is the effect produced by increasing the pressure on a vapor?

Fig. 35.

perature may be readily proved by taking a tube one third of an inch in diameter, and 30 inches long, closed at one end, *a*, *Fig.* 35, with a jar, *b*, an inch or more in diameter, and 30 inches deep. Let the tube be filled with quicksilver, so as to leave a space of half an inch, into which ether may be poured; invert the tube in the deep jar, also containing quicksilver; the ether, of course, rises to the other closed extremity. If the tube be lifted in the jar as high as possible without admitting external air, a certain portion of the ether will vaporize, and, depressing the quicksilver, its elastic force may be measured by the length of the resulting column. If now the closed end of the tube be grasped in the hand, or if it be slightly warmed by the application of a lamp, the mercurial column is at once depressed, proving that the elastic force of the vapor is increasing. As soon as the tube is warmed to the boiling point of the ether, the column of mercury is depressed exactly to the level on the outside of the tube. At this point, therefore, it balances, or is equal to the pressure of the air.

Now let the tube be depressed in the jar, it will be seen with what facility the vapor reassumes the liquid condition. As the tube descends the vapor condenses, and the mercury keeps constantly at the same level. On raising the tube fresh portions of the ether vaporize, the mercury still retaining its height above the general level. There is, therefore, a point of maximum density for each temperature when the vapor is in con-

---

How may the relation between the elasticity of a vapor and the temperature be shown? What effect is seen on warming the tube, *Fig.* 35? When the tube is warmed to the boiling point of the ether what is seen?

tact with the liquid. It can not be surpassed by increasing the pressure, because a part of the evaporated liquid immediately condenses, while a decrease of pressure causes volatilization of a fresh portion. The point of maximum density rises with the temperature of the vapor. The density of air at 212° being taken as 1000, that of the vapor of water at its maximum density will be as follows:

Table of the Maximum Density of Water Vapor.

Temperature.	Density.	Weight of 100 cubic inches.
32°	5.690	.136 grains
50°	10.293	.247 "
60°	14.108	.338 "
100°	46.500	1.113 "
150°	170.293	4.076 "
212°	625.000	14.962 "

The tension or elasticity of all vapors is nearly the same, if compared at temperatures which represent differences of an equal number of degrees above or below the boiling points of the respective liquids.

By exerting severe pressure on various gases, as sulphurous acid, cyanogen, chlorine, carbonic acid, and protoxide of nitrogen, they have been made to assume the liquid condition; but other gases, as hydrogen, oxygen, and nitrogen, may be exposed to a pressure of 58 atmospheres, and cooled to −140° without liquefying. Air has been reduced by pressure to $\frac{1}{675}$ of its volume, and subjected to a bath of ether and solid carbonic acid (−166°) without changing its form.

What is meant by the maximum density of a vapor? What effect has severe pressure on certain gases? Give examples. Can all gases be liquefied?

LECTURE XI.

EBULLITION.—*Theory of Boiling.—In a close Vessel Water never Boils.—Effect of Air in Boiling Water.—Instantaneous Condensation of Water.—Variability of the Boiling Point.—Effect of Pressure.—Effect of the Nature of the Vessel.—Saline Solutions.—Boiling on Mountains.—Spheroidal State of Liquids.—Freezing in a red-hot Vessel.*

BY introducing different liquids into a tube arranged as represented in *Fig.* 35, we can prove that as soon as the boiling point of a liquid is reached, the elastic force of the vapor rising from it is equal to the pressure of the air. At a temperature of 80°, vapor of water will depress the mercurial column of a barometer about one inch; but if the temperature be raised to 212°, the mercury is at once depressed to the level in the cistern.

On these principles, the phenomena of boiling, or ebullition, are easily explained. When the temperature of a liquid is raised sufficiently high, vapor is rapidly generated from those portions of the mass which are hottest, and the violent motion characterized by the term boiling is the result. This is due to the fact that the elastic force of the generated vapor at that point is equal to the atmospheric pressure, and the vapor bubbles expanding can maintain themselves in the liquid without being crushed in; they rise to the surface, and there burst. But, just before ebullition takes place, a singing sound is often heard, due to the partial formation of bubbles, which, so long as they are in the neighborhood of the hottest part, have elasticity enough to maintain their form; but the moment they attempt to rise through the cooler portion of the liquid just above, their elasticity is diminished by their decline of temperature, and, the atmospheric pressure crushing them in, they resume the liquid condition. For a few moments,

---

What is the elastic force of a vapor at the boiling point? What is the elastic force of vapor of water at 80° and at 212°? Describe the phenomena of boiling. What is the singing sound previous to ebullition due to?

then, while the vapor has not gathered elastic force enough to maintain its condition perfectly, these bubbles are transiently formed and disappear, and the liquid is thrown into a vibratory movement, which gives rise to the singing sound.

Water, when heated in a vessel from which the steam can not escape, never boils. This takes place in the interior of Papin's Digester, a strong metallic vessel in which water is inclosed, and the orifice through which it was introduced fastened up. As the steam can not escape, the water can not boil, no matter what the temperature may be; but the vapor accumulating in the interior of the vessel exerts an enormous pressure. It is under the same conditions as were considered in the case of candle-bombs. Papin's Digester is used to effect the solution in water of bodies which are not acted on readily by that liquid at its common boiling point.

The presence of air assists the boiling of water; indeed, pure water can not boil at all. When the air is nearly expelled by heating in vacuo, the temperature will rise to 360° in an open vessel: a sudden burst of vapor will then often shatter it. Slowly-frozen ice, heated under oil, acts in the same manner. When a liquid has but little adhesion to air, as in the case of sulphuric acid, a bumping or irregular boiling is seen.

As a vapor rising from a vaporizing fluid will bear no increase of pressure, so neither will it bear any reduction of temperature without instantaneously condensing. This may be strikingly shown by an arrangement such as is represented in *Fig.* 36. Into the mouth of a flask, *a*, let there be fitted a tube, *b*, half an inch in diameter, bent as in the figure. Having introduced a little water into the flask, cause it to boil rapidly by the application of a spirit-lamp. The steam which forms soon

Fig. 36.

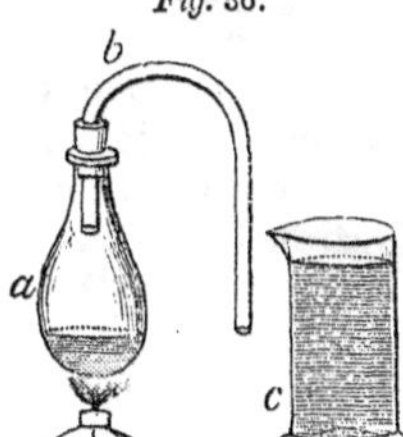

What results in heating water in a close vessel? Describe Papin's Digester. Why does not the water in it boil? What effect has the air in water on boiling? What is the effect on steam of reduction of temperature? How is this shown by the instrument represented in *Fig.* 36?

drives the atmospheric air from the flask and tube; and when this is entirely completed, and the vapor issues abundantly from the mouth of the tube, plunge the end of the tube beneath some cold water contained in the jar *c*, and take away the lamp. As soon as this is done, the cold water, condensing the steam in the tube, rises to occupy its place, and presently passing over the bend, introduces itself with surprising violence into the interior of the flask, filling it entirely full, or, which more commonly takes place, breaking it to pieces with the force of the shock. The low-pressure engine depends on this fact of the rapid condensibility of vapor; the high-pressure engine, on its elastic force.

Fig. 37.

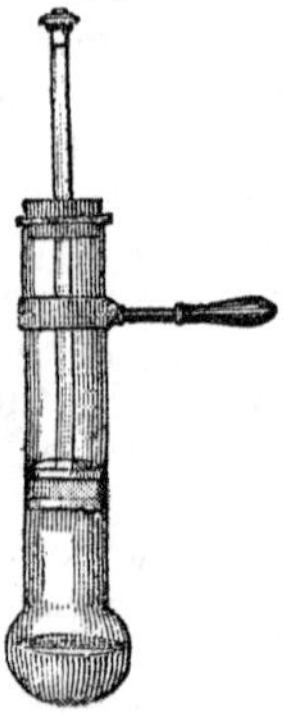

The principle involved in the action of the low-pressure engine, and more especially that form which was the invention of Newcomen, is well illustrated by the instrument *Fig.* 37. It consists of a glass tube, blown into a bulb at its lower extremity. In the bulb some water is placed, and a piston slides without leakage in the tube. On holding the bulb in the flame of a spirit-lamp, steam is generated, and the piston forced upward. On dipping it in a basin of cold water, the steam condenses, and the piston is depressed, and this action may be repeated at pleasure.

As the pressure of the atmosphere determines the boiling point of a liquid, and as that pressure is variable, the boiling point is not fixed, but is variable. If a glass of warm water be placed beneath the receiver of an air-pump, when the rarefaction has reached a certain point ebullition sets in, and the water continues to boil at a lower temperature as the exhaustion is more perfect. In a vacuum, water can be made to boil at 32°.

On this principle, that the boiling point depends on the existing pressure, we find an explanation of a cu-

---

What effect is produced by the rapidity of condensation? On what property of vapor does the low-pressure engine depend? On what the high-pressure? Describe the instrument *Fig.* 37. What effect has rarefaction of the air on the boiling point? At what temperature may water boil in a vacuum?

rious experiment, called the culinary paradox, in which ebullition is apparently brought about by the application of cold. Take a flask, *a*, *Fig.* 38, and, having filled it half full of water, cause the water to boil violently so as to expel the atmospheric air; introduce a cork which will fit the mouth of the flask air-tight a moment after it is moved from the lamp, and before any atmospheric air has been introduced. If the flask be now dipped into a jar, *b*, of cold water, its water begins to boil, and will continue to do so until the temperature is reduced quite low. This phenomenon is due to the fact, that the cold water condenses the steam in the flask, and a partial vacuum is the result. In this partial vacuum the water boils, and the steam, as fast as it is generated, is condensed by the cold sides of the flask.

Fig. 38.

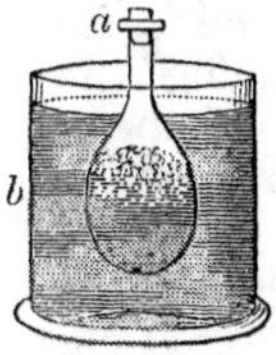

Besides this variation of the boiling point under variation of pressure, the nature of the vessel in which the process is carried forward exerts a certain action; thus, in a polished glass vessel the boiling point is 214°, but if a pinch of metallic filings be put in, it falls to 212°; in a rough metal vessel it is 212°. If the glass has been carefully cleaned with hot sulphuric acid, water may be heated to 221°, and then boiling takes place with bursts of steam, the temperature each time falling to 212°. The presence of oil elevates the boiling point three or four degrees.

The solution of a salt in water elevates the boiling point by the influence of adhesion, and more in proportion as the amount of salt is larger. A saturated solution of chloride of calcium boils at 355°, of acetate of potassa at 336°, of nitrate of lime at 304°, of common salt at 227°. In certain mountainous regions meat can not be cooked by the ordinary process of boiling. As we ascend to elevated regions in the air, the atmospheric pressure becomes less, because the column of air above is shorter, and therefore there is less air to press. Under such circumstances, the boiling point of water, of course, descends, and may become so low as not to bring

Describe the culinary paradox. What effect has the nature of the vessel on the boiling point? What effect has the presence of salt? Why is it impossible to cook meat on high mountains?

about the specific change required in the cooking of meat. An ascent of 596 feet lowers the boiling point one degree. Upon this principle we can determine the altitude of accessible elevations, by determining the thermometric point at which water boils upon them. A peculiar thermometer, the hypsometer, has been invented for this purpose.

When a drop of water is placed on a red-hot polished metallic surface, it does not, as might be expected, commence to boil rapidly, but remains perfectly quiescent, gathering itself up into a globule. In *Fig.* 39, *B* represents a heated metallic basin, turned upside down, and slightly indented at the bottom, so as to permit a drop of a mixture of ink and alcohol, *d*, to rest there. If the metal be now allowed to cool, by withdrawing the spirit-lamp as soon as its temperature has reached a certain point, the drop is suddenly dissipated in a burst of vapor. The temperature to which it is necessary to heat the metal varies with the liquid employed, being lower as the boiling point is lower, and as the latent heat is less. Water requires at least 340°; alcohol, 273°; ether, 142°. The liquid remains from 5° to 7° below its boiling point. The explanation of this phenomenon, which is called the spheroidal state of a liquid, is that, at the high temperature, the drop is not in contact with the red-hot surface, but a cushion of steam intervenes. By bringing the eye to the level of the bottom of the drop, a bright object beyond may be seen through the interval. The steam, being a bad conductor, prevents ebullition from occurring; but, as soon as the temperature declines and this steam no longer props up the drop, an explosive ebullition ensues, because of the contact which has taken place. The hand can be

Fig. 39.

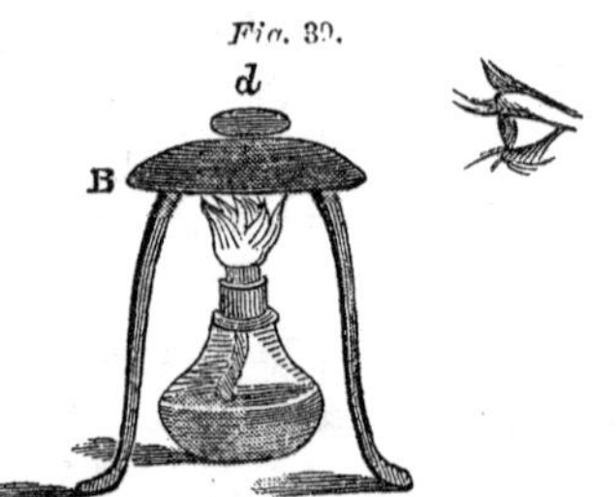

What number of feet must we ascend to lower the boiling point one degree? How may the heights of mountains be determined on this principle? Describe what is meant by the spheroidal state of liquids. What is the explanation of this condition?

passed through molten cast-iron, in consequence of the want of actual contact between the skin and metal, owing to the vaporization from its moist surface. If liquid sulphurous acid be brought into the spheroidal condition in a red-hot platinum capsule, and water dropped upon it, ice will be instantly formed.

LECTURE XII.

VAPORIZATION. — *The Boiling Point rises with the Pressure.—Relation between Insensible and Sensible Heat.—The Cryophorus.—Freezing Water in Vacuo.—Freezing by Evaporation.—Variability of Moisture in the Air.—Hygrometers: Saussure's, Daniell's, the Wet Bulb.—Drying of Gases.*

UNDER an increase of pressure the boiling point rises, and the elastic force of the steam evolved becomes greater. As we have seen, the elastic force of steam from water boiling at 212° is equal to the pressure of one atmosphere. If the pressure be doubled, the boiling point rises to 250°; if quadrupled, to 291°, and under a pressure of fifty atmospheres, it is 510°.

These results may be established by the aid of the boiler, *a*, represented in *Fig.* 40. It is a globular vessel of brass, about three inches in diameter. In its upper part are three perforations, into one of which the stopcock, *b*, is secured; through the second a tube, *c*, is inserted, deep enough to reach nearly to the bottom of the boiler, and through the third a thermometer, *d*, is introduced. Some quicksilver is poured in, sufficient to cover the end of the tube, *e*, half an inch or more deep, and upon it water is poured, the bulb of the thermometer being immersed in it. The stopcock, *b*, being open, a spirit-lamp is applied to bring the water to its boiling point, and as the steam can freely pass

Fig. 40.

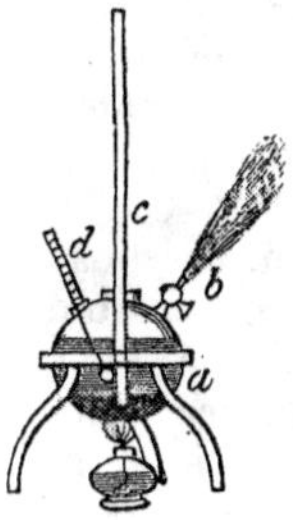

How may ice be formed in a red-hot vessel? What effect has increased pressure on the boiling point of water? Describe *Fig.* 40, and its method of action.

out, this, of course, takes place at 212°. On closing the stopcock the steam can no longer escape, but, exerting its elastic force on the surface of the boiling liquid, presses the mercury up the tube, *c*. The altitude of the mercurial column measures the amount of this pressure, and the thermometer indicates the corresponding change in the boiling point; as soon as the pressure is equal to two atmospheres the thermometer will be found to have risen to 250°.

It is immaterial at what temperature vaporization is carried on, a very large amount of heat must always be rendered latent; and, in point of fact, vapors generated at a low temperature contain more latent heat than those generated at a high one. The relation which exists in the amount of heat rendered latent at different temperatures is, according to Watt, very simple. The sum of the insensible and sensible heat is always the same; thus, water boiling at 212° absorbs 966°, the sum being 1178°; but vapor from water at 32° contains 1146°, the sum again being 1178°. But Regnault has shown that, although in practice the rule works well, yet that it is not strictly correct, the latent heat not decreasing as fast as the sensible heat increases. At 8 atmospheres the sum of the two is 1216°, instead of 1178°, the boiling point being 340°.

The increase of elasticity by equal additions of heat is greater at high than at low temperatures, and this renders the employment of high-pressure engines more economical. But it is only when in contact with water that this increase is seen; dry steam follows the general law of gases.

When vapors return to the liquid condition, the heat which has been latent in them reassumes the sensible form. They may thus be regarded as containing a great store of heat, of the effect of which many natural phenomena furnish us with examples. Thus, there is a remarkable difference between the climate of the eastern coast of America and the opposite European coasts in

---

What is the boiling point under a pressure of two atmospheres? What is the relation between the amounts of latent heat at different temperatures of vaporization? Does the elasticity of vapors increase with regularity? When vapors liquefy, what becomes of their latent heat?

the same latitude, and this arises from the action of the Gulf Stream—a great stream of warm water, which, issuing from the Gulf of Mexico, and passing the Atlantic States, stretches across toward the European continent. The vapors which arise from it give forth their latent heat to the air, and the southwest winds, which are therefore damp and warm, moderate the climates of those countries.

The cryophorus, or frost-bearer, invented by Wollaston, in which water may be frozen by the cold produced by its own evaporation, depends for its action on the laws relating to latent heat. It is represented in *Fig.* 41, and consists of a bent tube half an inch or more in

Fig. 41.

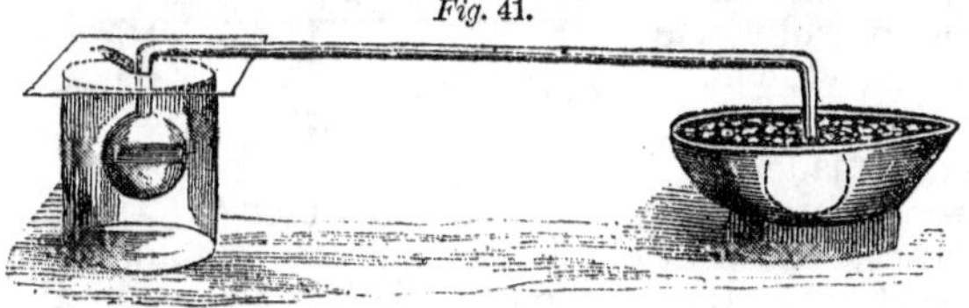

diameter, with a bulb at each extremity; the left-hand bulb is filled one half with water, and the rest of the space, with the tube and the other bulb, is filled with the vapor of water only. If now the right-hand bulb be immersed in a freezing mixture of nitric acid and snow, although the tube may be of considerable length, the water in the distant bulb presently freezes; hence the name of the instrument, frost-bearer, because cold applied at one point produces a freezing effect at another which is at a distance. The action is simple: in the cold bulb, which is in contact with the freezing mixture, the vapor is condensed, fresh quantities rise with rapidity from the water in the other bulb, to be in their turn condensed; a continual condensation therefore goes on in the one, and a continual evaporation in the other, but the vapor thus formed must have heat of elasticity; it obtains it from the water from which it is rising, the temperature of which, therefore, descends until solidification takes place.

What effect has the Gulf Stream on the climate of Europe? Describe the construction of the cryophorus. What is the principle of its action?

Leslie's process for freezing water in vacuo by its own evaporation is an example of the same kind. If some water in a watch-glass be placed in an exhausted receiver, with a large surface of sulphuric acid, as fast as vapor rises it is condensed by the acid, a rapid evaporation of the water therefore takes place, the temperature falls, and congelation ensues. In *Fig.* 42 this apparatus is represented. *T* is the watch-glass, containing water; *S*, a wide dish, filled with sulphuric acid, and *P*, a low bell-jar, in which the exhaustion is made.

Fig. 42.

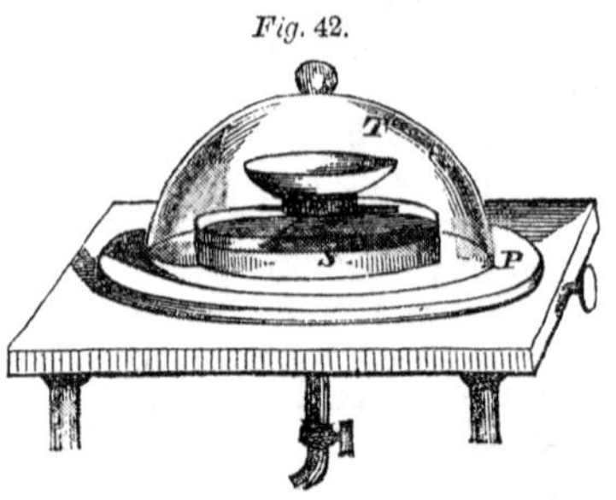

A drop of prussic acid held in the air on the tip of a rod solidifies, the portion that evaporates obtaining its latent heat from the portion left behind; and on the same principle liquid carbonic acid can also be solidified. If a drop of water be placed between two watch-glasses, and a little ether poured into the upper one, on putting it under an exhausted receiver the ether will boil, and the two glasses freeze together.

The pulse-glass is an instrument which serves to illustrate the fact that evaporation is a cooling process. It consists of a glass tube bent twice at right angles, and terminated by bulbs, as in *Fig.* 43. It is partially filled with alcohol, the rest being occupied by vapor of that substance. On grasping one of the bulbs in the hand, the warmth is sufficient to boil the liquid, and as it distils over into the other bulb an impression of cold is felt.

Fig. 43.

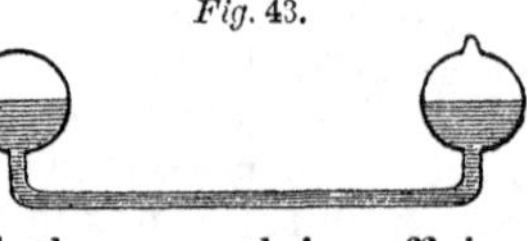

The amount of watery vapor contained in the air is very variable. Many common facts prove this. The swelling of wooden furniture takes place in consequence

Describe Leslie's process for freezing water. Why does liquid prussic acid solidify when exposed to the air? How may water be frozen by the aid of ether? Describe the pulse-glass.

of damp weather, and the opposite effect, or its shrinking, occurs during dry. Several instruments have been invented to determine what the amount is at any time; they are called hygrometers.

Saussure's hygrometer consists of a human hair 8 or 10 inches long, *b c*, *Fig.* 44, fastened at one end to a screw, *a*, and at the other passing over a pully, *c*, being strained tight by a silk thread and weight, *d*. From the pully there goes an index, which plays over the graduated scale, *e e'*, so that as the pully turns through the shortening or lengthening of the hair the index moves. The instrument is graduated to correspond with others by first placing it under a bell-jar, with a dish of sulphuric acid, or other substance having an affinity for water, which, absorbing all the moisture of the air of the bell, brings it to absolute dryness. The point at which the index then stands is marked 0°. The hygrometer is next placed in a jar, the interior of which is moistened with water; when the index has again become stationary the point is marked 100°, and the intervening space divided into 100 equal parts. The hair should have its oily matter removed by soaking in sulphuric ether. This renders it more sensitive.

Fig. 44.

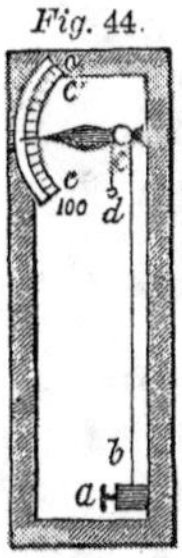

There is a simple and ingenious instrument, the movements of which depend on these principles: it is represented in *Fig.* 45. A thin slip of pine wood, *a a*, cut across the grain, a foot long and an inch wide, has inserted into its corners four needles, all pointing in one direction backward. If this instrument be set upon a floor or flat table, in the course of time it will crawl a considerable distance. During dry weather the thin board contracts, and the two fore legs taking hold of the table, the hind ones are drawn up a little space; when the weather turns damp the board expands, and now the hind legs, pressing against the table, cause the fore ones to advance. Every change from dry to damp, or the reverse, produces a walking

Fig. 45.

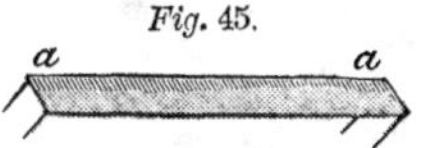

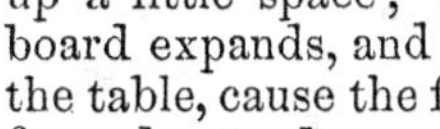

What is a hygrometer? Describe Saussure's hygrometer. How is Saussure's hygrometer graduated? What is the construction of the instrument *Fig.* 45? Describe its method of action.

motion in a continuous direction, and the distance passed over is a register of the sum total of all these changes.

But, of all hygrometric methods, the process known as "the determination of the dew point" is the most philosophical. This method consists in cooling the air until it begins to deposit moisture. When there is much moisture in the air, it obviously requires but a slight diminution of the temperature to cause a portion of the vapor to deposit as a dew; but when the air is dryer, the cooling must be carried to a greater extent. The precise thermometric point at which the moisture begins to deposit is called the dew point.

Daniell's hygrometer affords a ready and beautiful method of determining the dew point. It consists of a cryophorus, *a c b*, *Fig.* 46, the bulb *b* being made of black glass, and *a* covered over with muslin. The bulb *b* contains ether instead of water, and into it there dips a very delicate thermometer, *d*. Usually another thermometer is affixed to the stand of the instrument. When a little ether is poured on *a*, by its evaporation it cools that bulb, and ether distils over from *b*, which, of course, becomes cold. After a time the temperature of *b* sinks to the dew point, and that bulb becomes covered with a dew. The thermometer, *d*, then shows at what temperature this takes place, and of course gives the dew point. Knowing the temperature of the air, the dew point, and the barometric pressure, the absolute amount of vapor can be determined by calculation.

Fig. 46.

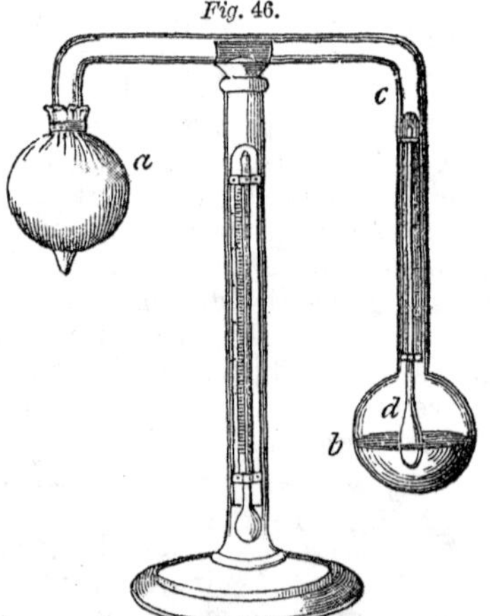

The amount of evaporation, and its great variation at

What is meant by determination of the dew point? Describe Daniell's hygrometer. How is it used?

different seasons of the year, is shown by the following table:

Mean Temperature.	Mean Dew Point.	Maximum Evaporation in 24 hours.
Annual...... 49°.88	44°.31	7,166 galls. per acre
Summer..... 62°.21	54°.56	15,048 " "
Winter...... 38°.95	35°.64	3,450 " "

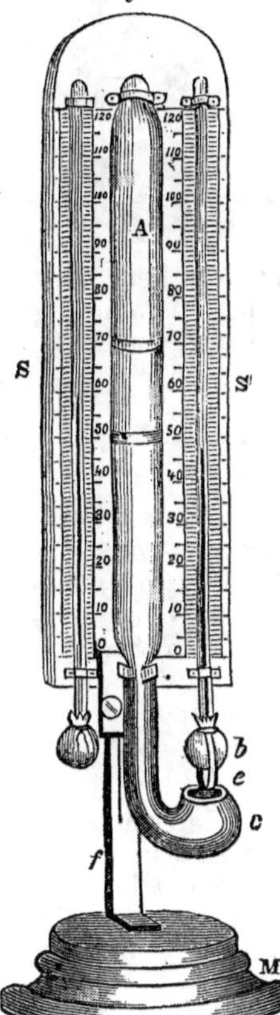

Fig. 47.

The wet bulb hygrometer consists of two mercurial thermometers, S S, which exactly correspond. The bulb of one of them, *b*, *Fig.* 47, is covered with muslin, and kept constantly wet by water supplied from a reservoir, A *c*, by capillary attraction along a thread, *e*. The other bulb is covered with dry muslin. Owing to the evaporation from the wet bulb, its temperature will be lower than the dry one, and this in proportion to the rate of evaporation or the dryness of the adjacent air. M *f* is the supporting stand. If both thermometers, the wet and the dry, coincide, the air contains moisture at its maximum density, and the greater the difference between them, the dryer the air. In practice it is found that this instrument requires a variety of corrections to ascertain the dew point, the difference between the two thermometers being multiplied by a number which varies with the actual temperature.

It is frequently necessary to remove moisture from air or gases. This may be done by conducting them through tubes containing

---

Describe the wet bulb hygrometer. What is its principle of action? How may gases be freed from moisture?

bodies having a strong attraction for water, such as chloride of calcium, sulphuric or phosphoric acids. Such an arrangement is shown in *Fig.* 48, in which A is the

Fig. 48.

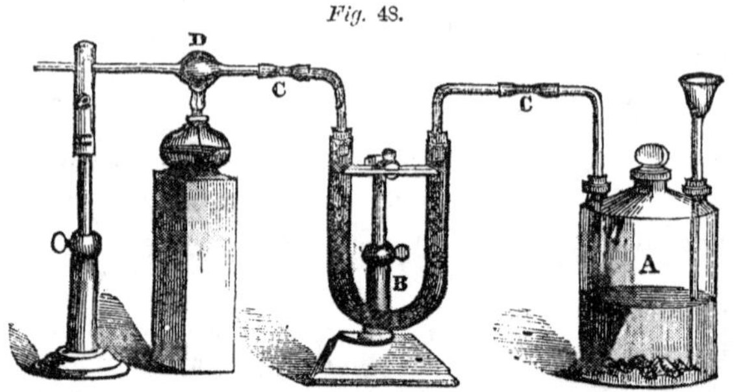

bottle in which the gas is generated; B, a bent tube filled with chloride of calcium; D, a bulb containing material on which the dried gas is to act; and C C, caoutchouc connecting pieces.

LECTURE XIII.

EVAPORATION AND INTERSTITIAL RADIATION.—*Dumas's Method of ascertaining the Specific Gravity of Vapors.—Rapidity of Evaporation.—Control of Temperature, Dryness, Stillness, Pressure, and Surface.—Limit of Evaporation.—Conduction of Solids.—Difference among Metals.—Effect of Wire-Gauze on Flame.—Explanation on the Dynamical Theory of Heat.—Rumford's Experiments on Clothing.*

THE specific gravity of vapors may be determined in several ways. The method of Dumas consists in weighing a glass globe filled with the vapor to be tried. A portion of the substance is to be introduced into the globe, A, *Fig.* 49, the weight of which is *first* determined, and this is then held, as shown in the figure, by a handle, C, in a bath of fusible metal, placed over a small furnace. The heat of the melted metal vaporizes

Describe the apparatus *Fig.* 48. Describe Dumas's method of determining the specific gravity of vapors.

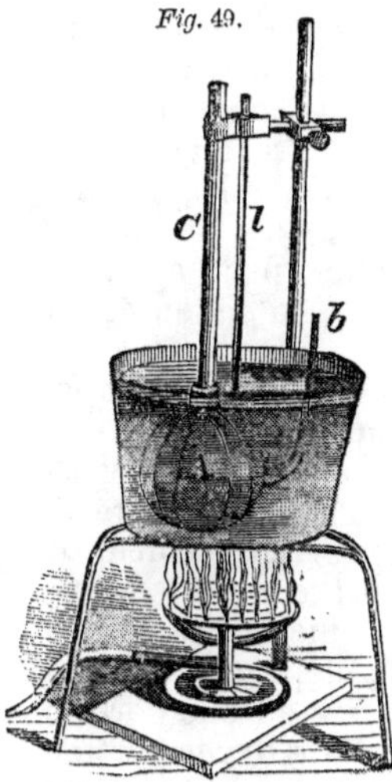

Fig. 49.

the substance, drives out the air, and occupies the whole cavity in a state of purity. When no more vapor escapes from the end of the tube, *b*, it is sealed by the blowpipe, and the temperature of the bath ascertained by the thermometer, *l*. The globe is now to be carefully weighed, when cold, a *second* time, and the point of the tube is then broken under quicksilver, which rises, and fills it completely; and this being subsequently emptied into a graduated jar, the volume of the globe is ascertained. Knowing the volume of the globe, we know the weight of the air it contains, and this, subtracted from the first weight, is the weight of the glass when empty. Subtracting this again from the second weighing gives us the weight of the vapor; and, as the air and vapor occupied the same volume, their densities are as their weights. But as their temperature was different, a farther calculation is required to bring them to the same standard.

There are several conditions which exert a control over the rapidity of evaporation. The amount of vapor which can exist in a given space depends entirely on the temperature. Thus, the air included in a glass jar which is standing over water contains at 32° a certain quantity of vapor, but if the temperature rises to 60° it contains more, and still more at 90°. Should the temperature descend, a part of the vapor is deposited as a mist.

It is the application of this principle which constitutes the most beautiful part of Watt's great invention, the low-pressure steam-engine. Taking advantage of the fact that the quantity of vapor which can exist in a given space is determined by the lowness of the tem-

What conditions determine the rate of evaporation? What improvement did Watt make in the steam-engine?

perature of any portion of it, he arranged a vessel maintained uniformly at a low temperature, in connection with the cylinder of the engine, and thus condensed the steam without cooling the cylinder. Previous to his time a jet of water had been thrown directly into the cylinder, and its whole mass cooled—a slow operation, and one which involved raising the whole mass again to the heat of the incoming steam before the piston could be forced up once more.

Among other causes exerting a control over evaporation into the air is the dry or damp state of that medium. As is well known, evaporation goes on with rapidity when the weather is dry, and is greatly retarded when the weather is damp. So, too, a movement or current exerts a great effect. When the wind is blowing water will evaporate much more quickly than when the air is quite calm. This obviously depends on a constant renewal of surface, so that as fast as one portion of air becomes moist it is removed, and a dryer portion takes its place. Extent of surface operates in the same way. The same quantity of water will evaporate much more rapidly if exposed in a plate than if exposed in a cup. Pressure also exerts a great control; for, as we have seen, evaporation takes place instantaneously in a vacuum. But while there are several circumstances which control the *rate* of evaporation, it is temperature alone which regulates the absolute and final *amount*.

At one time it was supposed that evaporation was due to a solvent power in the air, a kind of attraction between that medium and the water with which it was in contact; but such an opinion is wholly untenable, for the process goes on with the greatest rapidity in a vacuum when the air is totally removed.

Although the evaporation of liquids, such as water, will take place at very low temperatures, there is reason to believe that the process has a limit; thus, a minute quantity of vapor will rise from quicksilver at a temperature of 60°, but at 40° not a trace can be discovered.

What was the disadvantage of the method previously used? What effect has dryness of the air on evaporation? What effect have currents? What other circumstances influence the rate of evaporation? What was evaporation formerly supposed to be due to? Is there a temperature below which evaporation ceases?

Sulphuric acid does not evaporate at all at ordinary temperatures. The cohesion of the liquid overcomes the evaporating tendency.

Interstitial Radiation or Conduction.

It is commonly said that heat is transmitted through bodies by conduction, a term which involves the idea that the particles are in contact, whereas it has been proved that they are separated by interspaces. The passage of heat across these interstices is called interstitial radiation. From the convenience of the expression, the term conduction will be frequently used.

Different solids conduct heat with different degrees of facility. If we take a cylindrical mass of metal, and hold tightly against its surface a piece of white writing paper, the paper may be placed in the flame of a spirit-lamp for a considerable time without scorching; but if we take a cylindrical piece of wood of the same dimensions, and, wrapping the paper round it, expose it to the flame, it rapidly scorches. The metal, therefore, keeps the paper cool by carrying off the heat, but the wood, being a bad conductor, suffers the paper to burn.

By the aid of the apparatus of Ingenhausz, *Fig.* 50, the same fact can be proved in a more general way. It consists of a trough of brass, 6 inches or more long, 3 wide, and 3 deep; from the front of it project cylinders of metallic and other substances, of the same length and diameter. They may be of silver, copper, brass, iron, porcelain, wood, etc., in succession. The surface of each cylinder is smeared with beeswax. On pouring boiling water into the trough, the heat passes along these cylinders with a rapidity corresponding to their conducting power, and the wax correspondingly melts. On the silver bar the wax melts most rapidly, and on the wood most slowly; on the others intermediately: thus affording a clear proof that different solids conduct heat with different degrees of facility. The actual conducting power is in this experiment com-

Fig. 50.

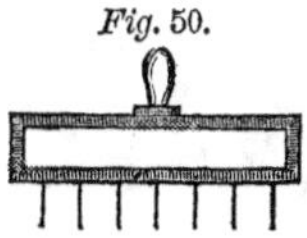

Why is the term conduction erroneous? Do all solids conduct alike? Describe the experiment with a cylinder of wood, and one of metal. Describe the apparatus of Ingenhausz. What does it prove?

plicated with the specific heat of the substance, that having a low specific heat seeming to conduct more rapidly. It is only necessary, in order to correct this source of error, to wait a certain length of time, and observe the distance to which the melting takes place.

The apparatus *Fig.* 51 demonstrates the same fact.

Fig. 51.

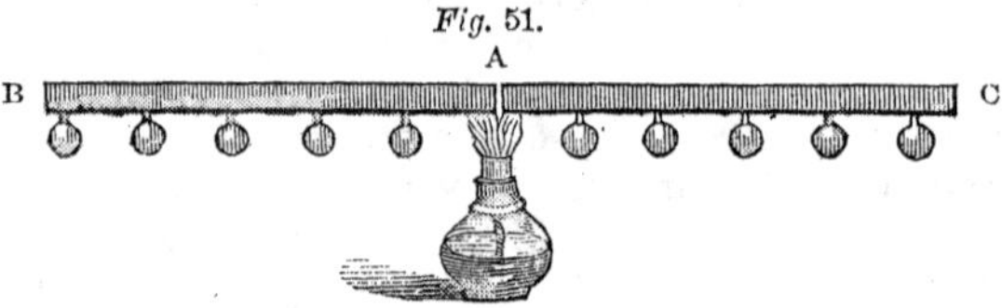

It consists of two similar bars of metal, B C, placed end to end; one may be of copper, the other of iron. To their under sides a number of balls of wood are attached by wax. On heating the junction A, the power of conduction is seen to differ, from the fact that, on the copper, the balls drop off to a greater distance from the source of heat than on the iron.

The instrument *Fig.* 52 can be made to exhibit to a number of persons differences in conducting power. From a central ring of brass a number of arms of various substances, as copper, brass, iron, porcelain, etc., project radially. On the tip of each arm is a piece of phosphorus. When the ring is held by the aid of the handle in the flame of a spirit-lamp, so that the flame passes through the central aperture, it will be found that the pieces of phosphorus inflame one after another, and not simultaneously. That on the copper takes fire first, the brass next, etc., the order being the same as that shown by Ingenhausz's apparatus.

Fig. 52.

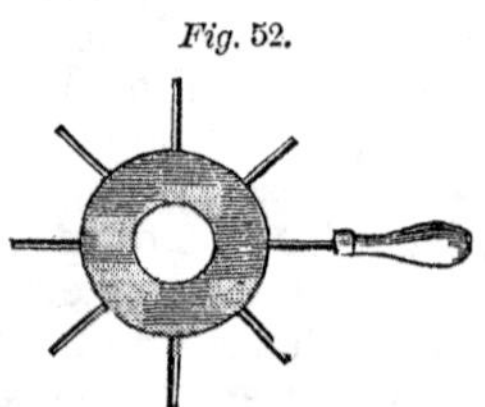

Describe *Fig.* 51. What is the construction and mode of action of *Fig.* 52?

Table of the Conducting Power of Solids.

	For Electricity.	For Heat.
Silver	100	100
Copper	73	74
Gold	59	53
Brass	22	24
Tin	23	15
Iron	13	12
Lead	11	9
Platinum	10	8
German Silver	6	6
Bismuth	2	2

The purity of the metals is of importance. Gold, when alloyed with 1 per cent. of silver, loses 20 per cent. of its conducting power. Some alloys conduct in the ratio of the mean of the two metals; some no better than the inferior metal. Carbon diminishes the conductivity of iron; malleable iron, steel, and cast iron being as 44 to 40 to 36.

If a piece of wire gauze be held over the flame of a candle or gas-jet, *Fig.* 53, the flame fails to pass through; but the gaseous matter of which the flame consists freely escapes through the meshes of the gauze, and may be set on fire above it, as in *Fig.* 54. Flame is either gaseous matter or solid matter in a state of minute subdivision, temporarily suspended in gas brought to a very high temperature. It can not, therefore, pass through a piece of wire gauze, because the metallic threads, exerting a high conducting power, abstract its heat from the incandescent gas, and bring its temperature down to a point at which it ceases to be luminous. The safety-lamp of Davy is an application of this principle; by it combustion is prevented from

Fig. 53.

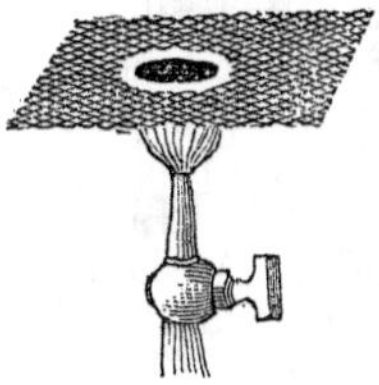

Fig. 54.

Give the order of conducting power among metals. What effect has the purity of a metal on its conducting power? What effect is seen on holding wire gauze over the flame of a lamp? How may we show that the combustible matter passes through?

Fig. 55.

spreading through masses of explosive gas by calling into action the conducting power of metallic gauze with which the lamp-flame is surrounded, as in *Fig.* 55. Hemmings's safety-tube, used to prevent explosions in the oxy-hydrogen blowpipe, acts on the same principle. The action of gauze is explained on the dynamical theory of heat, as follows: in a flame, the molecular movement is very intense, but the weight of the moving particles is but small. If their motion be communicated to a heavy body, the intensity of the motion must fall, just as a light bullet shot from a rifle could communicate to a 100-pound cannon-ball but a low velocity of motion. In placing a gauze over a flame, the intensity of motion is so much reduced that it is unable to propagate the combustion to the opposite side of the gauze.

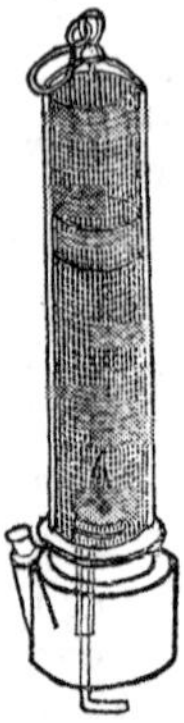

Count Rumford made several experiments to determine the conducting power of the materials that are used for clothing. He placed the bulb of a thermometer in the centre of a glass globe of large diameter, and filled the interspace with the substance to be tried. Having immersed the apparatus in boiling water until it was at 212°, he transferred it to a freezing mixture, and ascertained how many seconds it took to cool 135°. Linen and cotton were found to be better conductors than wool and the various furs, and hence the reason that they are preferred as articles of summer clothing. The greatest impediment to the transmission of heat was offered by hare's fur and eider down. The state of compression also influences the result, raw silk taking 1264 seconds to cool the specified amount, while twisted silk only took 917. Such bodies act not so much by their own bad conducting power as by calling into action the non-conducting quality of air.

Only those crystalline bodies which belong to the regular system conduct equally in all directions. If

How is Davy's lamp constructed? How is the action of gauze explained on the dynamical theory of heat? What was the object of Rumford's experiments? How were they conducted? What are the worst conductors of heat among solid bodies? What is the reason of their resistance to conduction?

Fig. 56.

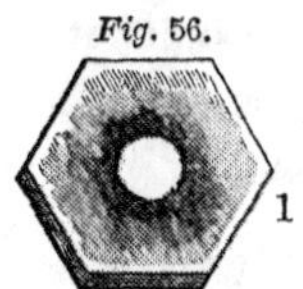

1

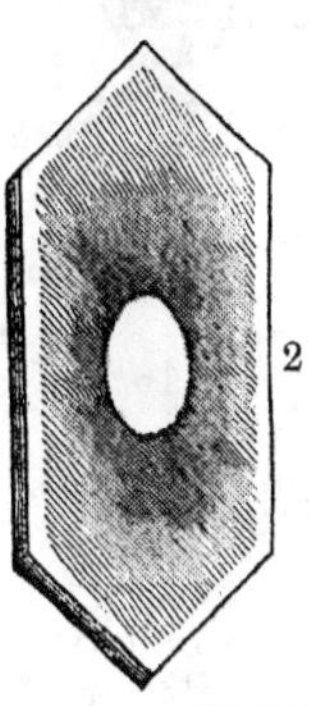

2

two plates be cut from a rhombohedral crystal, as a prism of quartz, one parallel to the axis, the other at right angles to it, and these be warmed from a point at the centre by the aid of a silver wire, if the plates have been coated previously with wax, it will be seen that the first plate, 2, *Fig.* 56, shows an elliptical spot of melting, while the other, 1, shows a circle, and demonstrates an equal rapidity of conduction in all directions. When a substance is altered by unequal tension, as in the case of a plate of unannealed glass, heat is conducted most slowly in the line of greatest density or pressure. Wood also shows similar differences of conduction in different directions.

LECTURE XIV.

Conduction.—*Conduction of Liquids.*—*Convection, or Heating by Circulation.*—*Conduction of Gases.*—*The Trade-Winds.*—*Land and Sea Breezes.*—*Applications of the Non-conducting Power of Air.*

The conducting power of most liquids, such as water, is very low; a thin stratum is sufficient almost entirely to cut off the passage of heat. This may be shown by the apparatus *Fig.* 57, consisting of a funnel partly filled with water, with an air-thermometer included in such a manner that the bulb is within a short distance of the surface, a depth of a quarter of an inch or less intervening. The tube of the thermometer may be passed through the lower mouth of the funnel, water-tight by means of a cork, and the position at which the index-liquid stands having been marked, some ether is poured on the surface of the water, upon which it readily floats, and set on fire. A voluminous flame is the result, and

---

Do solids conduct equally in all directions? Describe *Fig.* 56. What effect has pressure on conduction? How does the conducting power of liquids compare with that of solids? Describe *Fig.* 57.

a great deal of heat is evolved; and since the bulb of the thermometer is apparently separated from the flame by a thin film of water only, if the heat traversed that film the thermometer should rapidly move, but it does not; we therefore conclude that water is a very bad conductor of heat.

Fig. 57.

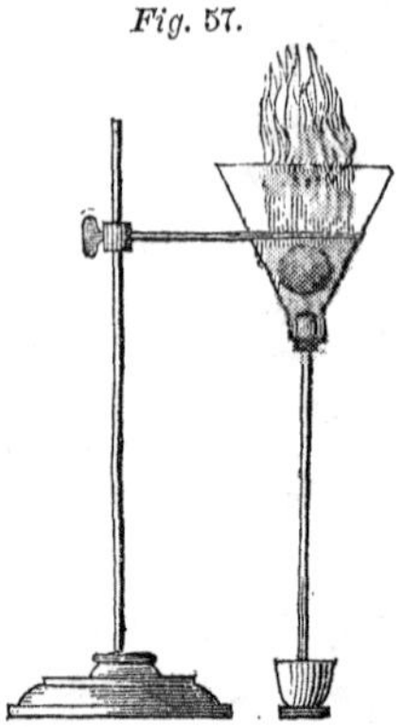

But the experiment is very deceptive; for as the flame is hollow, and only incandescent on its surface, it is really a great distance from the thermometer bulb, and, in addition, the evaporation of the ether is a cooling operation. The conclusion is nevertheless true.

To a certain extent all liquids conduct. Mercury is a good conductor; but, in those of which water is the type, the dissemination is chiefly effected by a process called convection or circulation, which depends on the free mobility of their particles.

The apparatus *Fig.* 58 illustrates this process. It consists of a wide tube, into which the water may be poured; the lower portion as high as *a* being colored blue by the addition of some coloring substance, the intermediate portion from *a* to *b* being colorless, and the upper portion, from *b* to *c*, being tinged yellow. By the application of a red-hot ring, *d*, of such a diameter that it can surround the jar, a space of an inch or more intervening all round, the upper yellow portion may be made even to boil, without showing any disposition to intermix with the portions beneath; but if the red-hot ring is lowered so as to surround the blue portion, as it becomes warm it will be found to ascend first through the colorless stratum, and finally that tinged yellow on the top. When the lower portion of the liquid is warmed, currents are established, which, rising through the

Fig. 58.

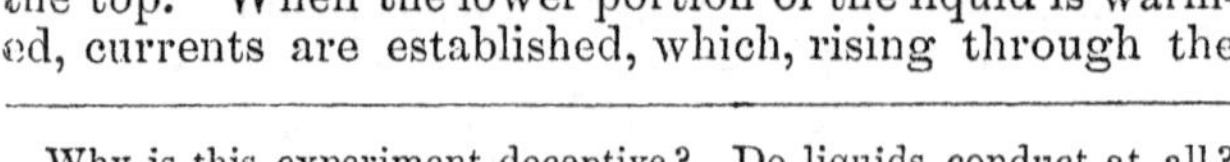

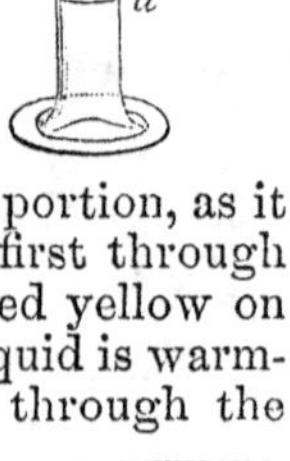

Why is this experiment deceptive? Do liquids conduct at all? How is heat disseminated in liquids? Describe *Fig.* 58.

strata above, bring about a rapid dissemination of the heat.

This may also be shown by taking a flask, *Fig.* 59, and filling it with water, in which some light substance, such as bran, is suspended. On applying a lamp to the bottom of the jar, currents are established in the water, rising up the centre and descending down the sides of the liquid; and in this manner, new portions constantly presenting themselves on the surface exposed to the flame, the whole mass becomes uniformly hot.

Fig. 59.

The cause of this movement is due to the fact that when water is heated it expands. Those portions, therefore, which rest on the bottom of the vessel, and to which the heat is applied, as soon as they become warm dilate, and, being lighter than before, rise to the top of the liquid, while colder, and therefore heavier ones, occupy their place.

In the vegetable world, advantage is taken of the non-conducting power of water in a very beautiful way. Soon after sunset, the leaves and delicate parts of plants become covered with little drops of dew, which invest them on all sides Under these circumstances, the process of convection, or the establishment of currents, is entirely cut off, for each of the drops is isolated and has no communication with those around. The cold air does not suddenly affect the delicate organs, as it would do were not this non-conducting film spread over them; their action is therefore less liable to be deranged.

All liquids possess true, though in most cases extremely feeble powers of conduction, as compared with solids. They also vary among themselves.

If the conducting power of liquids is small, that of gaseous bodies is so much less, that it is doubtful whether they can be proved to conduct at all. In these the

---

How may the currents in liquids be shown? What is the cause of the movement? Describe the formation of dew on plants at nightfall. Of what use is the covering of water? Do gases conduct heat?

mobility of particles is so great that heat is readily diffused through them. On burning sulphur in oxygen gas, the establishment of currents is well shown. The ventilation of buildings and mines, and the proper construction of furnaces and chimneys, depend on gaseous convection. The trade-winds are an illustration, on a grand scale, of the movements of a gas caused by heat. The temperature of the earth's surface is greatest in the tropics, and the air there expands, rises, and is replaced by cooler air flowing in from the polar regions. The heated air, after ascending a certain distance, flows over and tends to go toward the poles, to take the place of the air that has gone to the tropics. If the earth were at rest, there would be a steady breeze toward the poles in the upper parts of the atmosphere, and one in the opposite direction below; but, as the globe is revolving from west to east at the rate of 1000 miles an hour at the equator, and with less rapidity as the poles are approached, where the motion vanishes, the direction of the flowing streams of air is changed. The equatorial air, which has been moving at the same rate as the earth below, when it travels toward the poles, moves faster than the ground, and the wind has a westerly direction. On the contrary, the polar air, going toward the equator, drags against the surface of the earth until it has acquired the same velocity, and has apparently an easterly direction, which is less and less marked as the equator is approached.

Land and sea breezes are to be explained on the same principles. During the day the surface of an island will warm faster than the sea around, the air over it will dilate, and an inward current, the sea breeze, will be established. At night it cools more quickly, and the air over becoming denser, flows out toward the sea, causing the land breeze.

By taking advantage of the non-conducting power of air, rooms may be kept warm with a small consumption of fuel by furnishing them with double windows. A stratum of air an inch thick cuts off the loss of heat

Why does heat diffuse easily through them? Describe the trade-winds. What is the apparent direction of the trade-winds? Explain the cause of land and sea breezes. Upon what principle do double windows act?

through the windows to a great extent. The same fact accounts for the difference of conduction in Rumford's experiment, when the materials were closely and loosely packed. In the latter case, the slow conduction is due to the entrapped air, in which convection is almost entirely prevented by the fibres of the materials. The same explanation affords the reason for the warmth of the down and fur of animals.

LECTURE XV.

RADIATION.—*Preliminary Ideas on Radiant Heat.—Analogies with Light.—Effect of Surface on Radiation.—Relations of Radiation, Reflection, and Absorption.—The Florentine Experiment.—The Cold-Ray Experiment.—Opacity of Glass to Heat of Low Refrangibility.—Diathermacy of Different Solids, Liquids, Gases, and Vapors.—Action of Perfumes.—Heat is an Ethereal Vibration.—Melloni's Apparatus.*

THOUGH gases are bad conductors of heat, they freely allow its transmission by radiation, the solar rays, in traversing 6000 feet of air, only losing one fifth of their heat. A person who stands at one side of a fire receives the heat of it, although no currents of warm air can reach him. In a vacuum a piece of red-hot metal rapidly cools.

The heat which under these circumstances escapes from bodies is entirely invisible to the eye: it moves in straight lines, exhibiting many of the phenomena of the rays of light. Thus, if we interpose between a fire and thermometer an opaque screen, the moment the rays of light are stopped the heat is simultaneously intercepted.

The rays of heat, like the rays of light, are capable of being reflected by polished metallic surfaces. If a piece of planished tin be held before a fire in such a position as to reflect the light of it upon the face, the heat also is similarly reflected, and gives rise to a sensation of warmth.

Do gases transmit radiant heat? How may it be proved that radiant heat moves in straight lines? Is it capable of reflection?

The analogy between light and heat is still farther observed when the rays of the latter fall upon bodies of a different physical constitution from the metals. As glass is transparent to light, there are many bodies transparent to the rays of heat, though, as we shall find, these are not the same in both instances; and as there are substances, like lampblack, which will absorb all the light which impinges on them, there are many which perfectly absorb heat. Reflection, transmission, and absorption are therefore common to both these agents.

If we take two metallic vessels of the same size and shape, and, having blackened one of them all over with the smoke of a candle, fill them both with hot water and notice their rate of cooling, it will be seen that the blackened one cools faster: the same thing may be observed if, instead of blackening the vessel, it is covered with layers of varnish. These results may be proved by the aid of Leslie's Canister, which consists of a cubical brass vessel, *a*, *Fig.* 60, set upon a vertical stem, upon which it can rotate. At a little distance is placed the blackened bulb of a differential thermometer, *d*; a mirror, *M*, receives the rays of the canister and reflects them on the thermometer. One of the vertical sides of the cube is left with a clear metallic surface, a second washed over with one coat of varnish, a third with two, and the fourth with three coats. If these sides be presented in succession to the thermometer, they will be found to radiate heat with very different degrees of speed, more heat escaping from them as the number of coats is increased. Melloni found that the maximum was not attained until sixteen coats were applied.

Fig. 60.

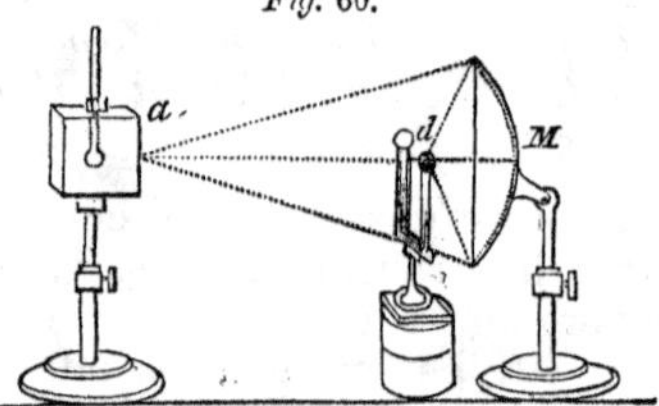

These results can only be explained on the principle that radiation does not take place from the surface of

---

What farther analogies are there between light and heat? In what manner can the radiating power of a surface be increased? Describe the experiment, *Fig.* 60. Does radiation take place from the surface alone?

bodies merely, but from a certain depth in their interior.

A highly polished metal is a bad radiator, but on roughening it this quality is improved. As a general rule, good radiators are bad reflectors, and good reflectors are bad radiators. This latter statement is exemplified in the case of a silver teapot, which retains its heat much longer than an unpolished vessel.

It is important to bear in mind that *the absorbing and radiating powers of a substance are directly proportioned to one another.*

When rays of light diverging from the focus of a concave parabolic mirror impinge on the surface, they are reflected in parallel lines; when parallel rays fall on such a surface, they are reflected to its focus. Thus, if from the point *a*, *Fig.* 61, the focus of a parabolic concave, *c f*, rays diverge, they will be reflected in parallel lines, *c g*, *d h*, *e i*, *f k;* and if at these points they be intercepted by the mirror, *g k*, they will be reflected to its focus, *b*.

Now, as the laws of the reflection of radiant heat are the same as the laws of the reflection of light, it is plain that if we place any incandescent body, such as a red-hot cannon-ball, in the focus *a*, heat which radiates from it will be found at the other focus, *b*.

This is beautifully illustrated by an experiment known under the name of the experiment with the conjugate mirrors. In the focus *a*, *Fig.* 61, of a parabolic mirror, *c f*, place a red-hot cannon-ball, and in the focus *b* of a second mirror, *g k*, set opposite, but twenty or thirty feet off, place a piece of phosphorus, a screen intervening between. As soon as the arrangements are completed, remove the screen, and in a moment the phosphorus takes fire. That this effect is due to the reflecting action of the mirrors, as has been described, may be proved by removing the mirror *c f*, when it will be found that the phosphorus can not be lighted, even though the ball be brought within a very short distance of it. If the

---

What effect on the radiating power is observed on roughening a polished surface? What relation is there between absorbing and radiating power? What is the action of a parabolic mirror on parallel rays? What is its action on rays diverging from its focus? Describe the experiment, *Fig.* 61.

mirrors are well adjusted, the cannon-ball may be replaced by an ordinary gas flame.

Fig. 61.

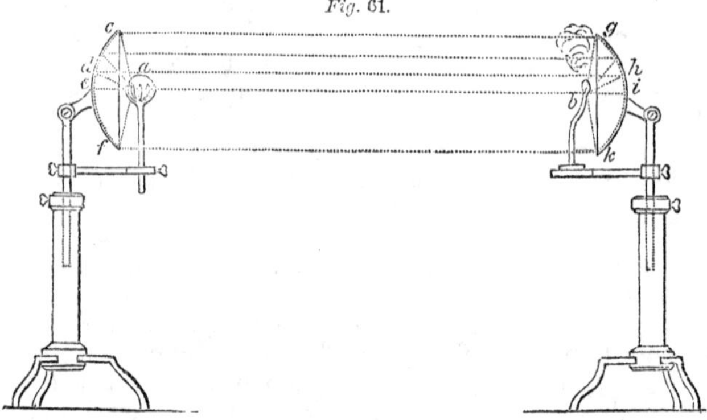

This striking experiment proves, first, that the rays of heat move in straight lines, like those of light; and, second, that in the same manner they are subject to the ordinary laws of reflection, for the apparatus is adjusted by the flame of a candle.

A variation of the experiment may be made by using a snowball, *C*, *Fig.* 62, instead of the heated body, in the focus of *N*, in which case a thermometer, *B*, in the focus of the opposite mirror, *M*, will exhibit a reduction of temperature, even though shaded by a screen, *A*, from direct radiation. From this it was at one time supposed that there exist rays of cold precisely analogous to rays of heat, and that they observe the same

Fig. 62.

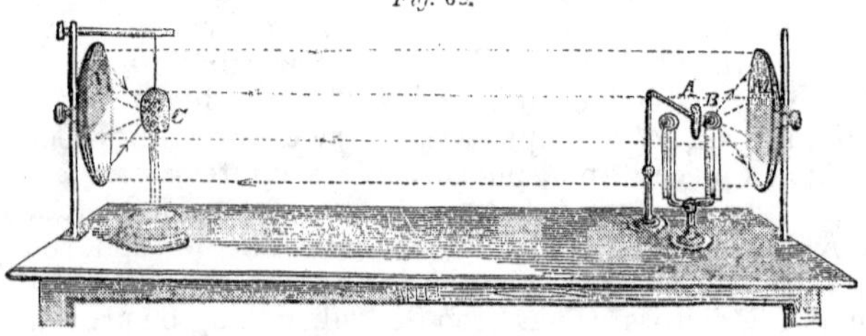

What does the experiment of the conjugate mirrors prove? Describe the experiment, *Fig.* 62. What was it supposed to prove?

laws; but, as we shall see in speaking of the Theory of the Exchanges of Heat, a simple explanation can be given without implying the existence of a principle of cold.

Let it be supposed that in the focus of the mirror *g k*, *Fig.* 61, the bulb of a delicate thermometer is placed, and in the focus of the other mirror, *c f*, a metalline mass, *a*, the temperature of which can be varied at pleasure. Between the mirrors let there be interposed a screen of transparent plate glass, and let us farther suppose that the temperature of *a* is 212°, or considerably below the point at which it is visibly red-hot. Under these circumstances the thermometer exhibits no rise of temperature so long as the glass intervenes, but the moment it is removed the heat passes.

A piece of transparent glass is therefore opaque to the rays of heat which come from a non-luminous source.

Let us now suppose that the temperature of the metalline mass *a* continually rises. When it has reached a red-heat, a certain proportion of the rays emitted by it begins to pass through the glass, as is shown by the effect upon the thermometer. When the mass is visibly red-hot in the daylight, the rays go through the glass more readily; and when it has become white-hot, or has reached the highest temperature we can give it, the glass transmits the rays with facility.

These facts are of the utmost importance. They show that bodies transparent to light are not necessarily transparent to heat, and, therefore, that there is a distinction between the two agents. They farther show that, as respects glass, its transparency for heat differs with the temperature of the source from which the rays come.

There is only one known solid that approaches perfect transparency for heat, rock salt. If, in the preceding experiment, a plate of it were substituted for the glass, no matter what the temperature of the mass *a*, the rays would pass through it with equal facility. It is the glass for heat, and stands at the head of *diather-*

What effect has glass on heat of low intensity? What changes occur in the transmissive power of glass as the temperature rises? What facts are shown by these experiments? What substance is transparent to heat?

mic bodies, as those transparent to heat are called. An *athermic* substance is one which does not permit heat to pass.

Diathermacy of different Solids.

Rock Salt........	92.3 per cent.	Limpid Quartz ...	38 per cent.
Sulphur..........	74 "	Smoky Quartz....	37 "
Iceland Spar....	39 "	Alum...............	9 "
Plate Glass......	39 "	Ice...................	6 "

The plates of the substances used in determining the above percentages were one tenth of an inch thick, and the source of heat was a naked flame. When copper at 212° was used, only the rock-salt, sulphur, and quartz permitted any rays to pass, the rest being perfectly athermic.

Diathermacy of Liquids.

Bisulphide of Carbon........................	63 per cent.
Chloride of Sulphur..........................	63 "
Spirit of Turpentine..........................	31 "
Olive Oil..	30 "
Ether...	21 "
Sulphuric Acid..................................	17 "
Alcohol..	15 "
Distilled Water.................................	11 "

Absorptive Power of Gases.

Air..............................	1	Carbonic Acid............	90
Oxygen........................	1	Nitrous Oxide............	355
Nitrogen......................	1	Sulphureted Hydrogen ..	390
Hydrogen.....................	1	Sulphurous Acid..........	710
Chlorine........................	39	Olefiant Gas...............	970
Carbonic Oxide.............	90	Ammonia....................	1195

The last table exhibits the relative absorbing powers for heat at 212°. The absorbing power of many vapors is quite as remarkable, as is also that of the perfume of flowers for rays of obscure heat. Aqueous vapor has also a powerful absorbent action on heat of low refrangibility. Among gaseous bodies, the same rule holds good as for solids—the best absorbents are the best radiators.

We see, therefore, that there are different varieties of radiant heat. The difference is due to the same

What is meant by diathermic and athermic bodies? What is the absorptive power of ammonia compared with air? What is the cause of the varieties of radiant heat?

cause which gives different colors to light, namely, that heat, being an undulatory motion, has waves of various lengths, and varying velocity of vibration. In the prismatic spectrum the maximum of heat is found below the red end, and it steadily declines on passing toward the violet. But in the interference spectrum, produced by a ruled grating, as discovered by Prof. J. W. Draper, the maxima of heat and light coincide, and are found in the centre of the yellow. The prismatic spectrum induces the supposition that heat is produced by slower and longer vibrations than those of light, but the interference spectrum corrects this hypothesis, and shows that they coexist in the same place at the same time, and are probably one and the same force. Obscure heat is invisible light, and light, when extinguished, produces heat.

The apparatus by the aid of which Melloni prosecuted his extensive researches on heat is shown in *Fig.* 63.

Fig. 63.

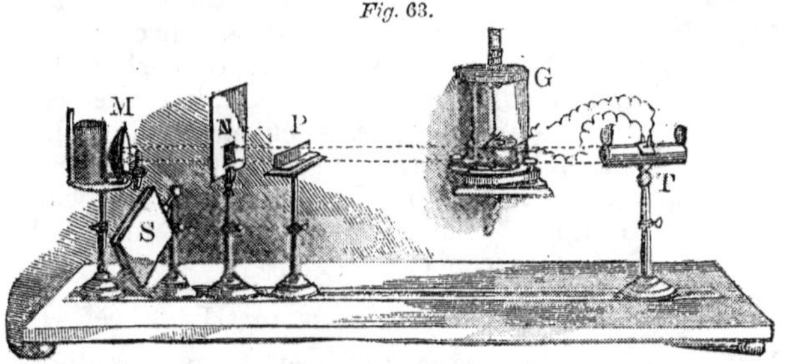

At M is a stand for the source of heat, with a concave mirror for concentrating the rays; at N, a perforated screen; at P, a plate of the substance to be examined; at T, a thermo-electric pile; at G, a galvanometer. S is an unperforated screen that can be turned down out of the way.

What is the distribution of heat in the prismatic spectrum? What is the distribution in the interference spectrum? What is the relation between light and heat?

LECTURE XVI.

THEORY OF THE EXCHANGES OF HEAT.—*Relations of Light and Heat.—Theory of Exchanges.—Explanation of the Cold Ray Experiment.—Wells's Theory of the Dew.—Cold on Mountain Tops.—Temperature of the Sun.—Polarized Heat.—Heat of Chemical Combination.*

THE facts which militate against the doctrine of the unity of light and heat are, 1st, that the relations of transparency for the two are not the same, smoky quartz or dark-colored mica allowing heat, but not light, to pass; and, 2d, that the radiations from such a source as a vessel of hot water are not visible to the eye, and can not be made to assume the luminous condition. But it has been shown by Tyndall that if the heat coming from an electric lamp, which has passed through an opaque solution of iodine in bisulphide of carbon, be concentrated upon a thin strip of platinum, it will cause the strip to glow, though the eye, when screened by the solution, may be directed to the sun without perceiving the faintest trace of light. In that case heat has been converted into light. Such phenomena are included under the term Calorescence.

THEORY OF THE EXCHANGES OF HEAT.

The theory of the exchanges of heat, comprehending an explanation of a great many of the phenomena we ordinarily witness, depends on the following principles. It assumes, 1st, that all bodies, no matter what their temperature may be, are constantly radiating heat at all times; 2d, that the ratio of radiation depends on the temperature, increasing as the temperature rises, and diminishing as it declines.

Thus the various objects around us are constantly emitting heat, the warm bodies to the cold, and the cold ones to the warm. A mass of snow and a red-hot

What facts militate against the unity of light and heat? Describe Tyndall's experiment. What is Calorescence? On what does the theory of the exchanges of heat depend?

cannon-ball respectively give off heat, the ball emitting it in greater quantities and the snow in less. And, even when the adjacent bodies have reached the same thermometric point, they still continue to exchange heat with one another.

Upon these principles we can readily account for the fact that bodies of different temperatures at first, finally come to an equilibrium. If an ignited cannon-shot be placed in the middle of a large room, it radiates heat to the ceiling, the walls, the floor, and the various objects around; they also radiate back upon it. But, from its elevated temperature, it emits its heat faster than they, and therefore gives out more than it receives. Its temperature constantly descends, and continues to do so until it receives just as much as it gives, which takes place when it has reached the same degree as the objects around; for, other things being equal, bodies at the same temperature radiate with equal speed.

The process must, however, stop as soon as that equality of temperature is attained; for, if we suppose the shot to cool below that point, it would evidently begin to receive more heat from the objects around than it gave forth; and the excess accumulating in it, the temperature would at once rise.

When an equilibrium is obtained, the process of radiation still continues, but the exchanges are equal. Two lighted candles placed together do not extinguish each other or cease to exchange light with each other, nor do two bodies equally warm cease for that reason to exchange heat. In a room, therefore, in which every thing has the same temperature, rays are continually exchanging, but each object maintains its own temperature, because it receives as much as it gives.

If a red-hot ball and a thermometer bulb be placed near one another, the bulb receives more heat from the ball than it gives to it, and its temperature therefore rises; but if a thermometer bulb and a snowball be placed in presence of one another, the bulb, being the hotter body, gives more than it receives, and its temper-

---

How do bodies of different temperatures come to an equilibrium? When does the descent of temperature cease? What is the reason that it ceases? Does radiation continue when an equilibrium is reached? Give the explanation of the cold ray experiment.

ature therefore descends. This is the explanation of the cold ray experiment with the conjugate mirrors. That experiment, as was observed, affords no proof that there are rays of cold; the effect is due to the fact that a mutual exchange is going forward between the two bodies, and the temperature of the hotter descends. The mirrors of course take no part in this phenomenon; their office is merely to direct the path of the rays, as has been explained.

On the principles of the radiation of heat is founded Wells's theory of the dew. After the sun goes down of an evening drops of water condense on the leaves, grass, stones, and other objects exposed to the air. It was once a question whether this dew descended in the form of a light shower, or ascended from the ground. There are also certain circumstances, apparently very mysterious, attending its formation. The dew rarely falls on a cloudy night; it also apparently possesses a selecting power, depositing itself on some bodies in preference to others. The theory of Dr. Wells furnishes a beautiful explanation of these curious facts. During the day the various bodies on the surface of the earth, receiving the rays of the sun, become warm, but at nightfall, when the sky is unclouded, they begin to cool, for, the process of radiation continuing without any source of supply, their temperature must descend. While the sun shone they received as much heat from him as they gave forth to the sky, but, when he is set, the supply is cut off, and they therefore cool, and, as there is moisture always in the air, their temperature descending, by-and-by the dew point is reached, and they become cold enough to condense water from the surrounding air. This is the dew. And as different bodies, according to the roughness or physical condition of their surfaces, radiate with different degrees of speed, as Leslie's canister proves, some of the objects exposed to the sky cool rapidly and are covered with dew, but with others the dew point is never reached; hence the apparent selecting power. When there is a canopy of clouds over the sky dew can

When does the dew form? What circumstances attending it seem difficult of explanation? What is Wells's theory of the formation of dew? Why does dew settle on some bodies and not on others? What is the effect of clouds?

not form, for the cloud radiates to the earth as much as the earth radiates to it—the exchanges are equal, and the equilibrium is maintained; but, if the cloud disappears, the heat of the surface of the ground escapes away into the regions of space. Hence cloudy nights are warm, and a clear is often a frosty night.

For similar reasons mountain tops are always colder than valleys. In a valley the radiation is obstructed by the sides of the adjacent hills, but on the top of a mountain the free exposure to the sky permits of unchecked radiation.

An interesting conclusion may be drawn from the conditions of the passage of radiant heat through glass. We have seen it is necessary that the heat should come from a source of very high temperature to pass this medium with facility. Now the heat of the sun passes it with the greatest freedom, as is shown when we stand before a window through which the sun shines. In the focus of a convex lens bodies may be readily set on fire. We infer, therefore, that the temperature of the sun is very high, a result corroborated by proofs from other sources.

Radiant heat is susceptible of polarization by tourmalines, and transmission through bundles of mica set at a proper angle to the incident ray. It also exhibits the phenomena of diffraction and interference.

The quantity of heat produced by chemical combination is definite, though the precise determination of its amount is difficult, owing to the complication with latent heat and changes of volume. By the aid of delicate calorimeters, tables of the heat evolved during combustion and combination have been constructed, and it has been proved that, for example, the union of a given amount of oxygen with the various elements does not in all cases produce the same quantity of heat. Substances presenting different allotropic conditions evolve in combustion different quantities of heat. In many decompositions a large development of heat is seen, as when chloride of nitrogen or gun cotton is exploded. In precipitations,

---

Why is a mountain top colder than a valley? What reason is there for supposing the temperature of the sun to be high? How may radiant heat be polarized? Is the amount of heat produced in chemical combination definite?

as well as in the reaction of acids upon bases, and solution of salts and gases, disturbances of temperature occur. It is stated that, during the combination of equivalent quantities of the different acids with a given base, nearly the same amount of heat is produced. But these inquiries are as yet very incomplete.

The great source, however, of all the heat, and indeed all the force exhibited in the various manifestations on the face of the globe, is the sun; that from other sources is but insignificant in comparison. He is the prime mover, and his extinction would make the earth a desolate waste like the visible side of the moon.

LECTURE XVII.

OF LIGHT.—*Sources of Light.—The Sun.—Incandescence.—Combustion.—Colors of Lights.—Shadows.—Conditions of the Intensity of Light.—Photometers: Rumford's, Ritchie's, Extinction of Shadows, Chlorine and Hydrogen.—Velocity of Light.—Reflection and Refraction.—Burning-glasses.*

LIGHT may be artificially produced by many different processes, such as the ignition of solids, combustion, and phosphorescence. Any solid, if sufficiently heated, becomes luminous, combustible gases take fire at a certain temperature in the air, and the diamond will emit a phosphorescent glow in a dark place after it has been exposed to the day. It is, however, as has just been remarked in the case of heat, to the sun that we are chiefly indebted. The quantity of light furnished by him far exceeds that of all natural or artificial sources, and its brilliancy is so great that the electric spark alone rivals it.

When the temperature of solid substances is raised to 1000° they begin to be luminous in the daylight, or, as it is termed, are visibly red-hot. It requires a far higher temperature to render a gas incandescent. This may be shown by holding a piece of thin platinum wire in the current of hot air which rises from the apex of the

---

What is the principal source of heat? How may light be produced artificially? What is the point of incandescence of solids? How may we prove that that of gases is higher?

flame of a lamp. The air is not visibly ignited, but the platinum wire instantly becomes red-hot, showing a great difference in this respect between this metal and a gas.

Different vapors and gases evolve different quantities of light when ignited. The flame of burning hydrogen is scarcely visible in the daylight; that of alcohol is but little brighter, but, under the same circumstances, sulphuric ether emits much light. If we take a glass of the form *Fig.* 64, consisting of a bulb, *a*, and a curved tube, *b*, and, having filled the bulb with ether, cause it to boil by the application of a lamp, *c*, the ether may be set on fire as it is forced out of the vessel by the pressure of its vapor. It burns in a beautiful arch of great brilliancy; but if we substitute alcohol for ether the light becomes quite insignificant.

Fig. 64.

The light which is emitted by lamps and candles is, however, in reality due to the disengagement of solid matter. The constituents of the gas which produce the flame are carbon and hydrogen chiefly; of these the latter is the more combustible, and is first burned; for a moment the carbon exists in a solid form in a state of extreme subdivision, and at a high temperature, but, being in contact with the external air, it is immediately consumed.

Artificial lights differ in color. If alcohol be mixed with common salt and set on fire, the flame is of a yellow tint; if with boracic acid, it is green; if with nitrate of strontian, it is red. It is upon these principles that the art of pyrotechny depends.

From whatever source light may come, it exhibits the same physical properties. It moves in straight lines. When it impinges on polished metallic surfaces it is reflected, on dark surfaces it is absorbed, on transparent surfaces, as glass, it is transmitted. In the last case it is frequently forced into a new path, as we shall presently see, and then the phenomenon takes the name of

Do all substances evolve the same amount of light in burning? What renders the flame of a candle luminous? How may flames be colored? Mention some of the properties of light. What is refraction?

refraction, because the ray is *broken* from its primitive course.

There are two different kinds of opacity, black and white. Charcoal is a black opaque substance, earthenware is opaque white.

The shadows formed by opaque bodies arise from the interception of light in its rectilinear progress. They may be of two different kinds, the common and the geometrical. The former arises from a luminous surface, the latter from a lucid point. The former consists of two portions, the *umbra* and *penumbra;* in the latter case the passage from total darkness to light on the side of the *shadow* is abrupt, and without the intervention of any *shade.*

The illuminating power of a light depends upon several conditions. As the distance increases it becomes less, the effect being inversely as the square of the distance; that is, at two feet it gives only one fourth of what it would do at one; at three feet, only one ninth. The absolute intensity of the light also determines the result; thus, there are flames that are very brilliant, and others that are paler. The magnitude of the luminous surface is another of these conditions; the absorbent effect exerted on the passing rays by the air, or medium traversed, another; as is also the direct or oblique manner in which the rays are received on the illuminated surface.

Of Photometers and the Measurement of Light.

The methods resorted to for the measurement of light are quite numerous, the most common being Rumford's method, Ritchie's method, the method of the extinction of shadows, and the chlorine-hydrogen photometer. The precipitation of gold from its chloride by the aid of peroxalate of iron, a process invented by Dr. J. C. Draper, and various photographic operations, are more or less used for special purposes.

Rumford's depends on the principle that, of two lights, the most brilliant will cast the deepest shadow.

What is the cause of shadows? What is the difference between common and geometrical shadows? What does the illuminating power of a light depend upon? What are the principal methods of photometry? Describe Rumford's method.

If, therefore, the lights to be compared are made to cast shadows of the same opaque body side by side on a piece of paper, the eye can without difficulty determine which of the shadows is darkest, and the light which casts it, being moved to a greater distance, or the other brought nearer, when the two shadows are of precisely the same depth, the distances of the lights from the paper will indicate their relative illuminating power. Thus, if one is twice as far off as the other, its intensity is four times as great.

Ritchie's photometer depends on the equal illumination of surfaces. It consists of a box, *a b*, six or eight inches long, and one broad and deep, *Fig.* 65, in the middle of which a wedge of wood, *f e g*, is placed, with its angle, *e*, upward. This wedge is covered with white paper neatly doubled to a sharp line at *e*. In the top of the box there is a conical tube, with an aperture, *d*, at its upper end, to which the eye is applied, and the whole may be raised to any suitable height by means of the stand *c*. On looking down through *d*, having previously placed two lights, *m n*, the intensity of which we desire to determine, on opposite sides of the box, they illuminate the paper surfaces exposed to them, and the eye sees both those surfaces at once. By changing the position of the lights, we eventually make them illuminate the surfaces equally, and then, measuring their distances from *e*, their illuminating powers are as the squares of those distances.

Fig. 65.

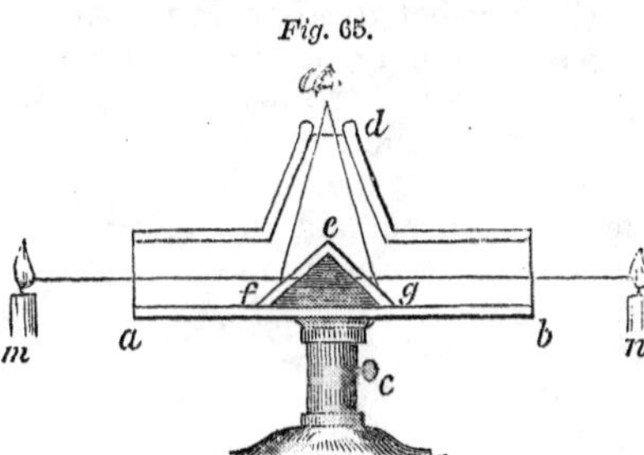

In both this and the preceding method a difficulty arises when the lights to be compared are of different tints. To some extent this may be avoided in Ritchie's instrument by placing a colored glass at *d*.

The method of extinction of shadows is much more

Describe Ritchie's photometer. How is it used? What difficulty is there with these two methods?

exact, differences in the colors of the lights even serving to give greater accuracy. It depends on the following principle. If a light is made to throw the shadow of an opaque object upon a white screen, there is a certain distance at which, if a second light be brought, its rays illuminating the screen, will totally obliterate all traces of the shadow. It has been found that eyes of average sensitiveness fail to distinguish the effect of a light when it is in presence of another sixty-four times as intense. The precise number varies with different eyes, but to the same eye it is always the same. If there be any doubt as to the perfect disappearance of the shadow, the receiving screen may be agitated or moved a little: this brings the shadow, to a certain extent, into view again. Its place can then be traced, and, on ceasing the motion, the disappearance verified.

When, therefore, we desire to ascertain the relative intensities of lights, we have only to determine at what distance they will extinguish a given shadow. Their intensities are as the squares of those distances.

Fig. 66.

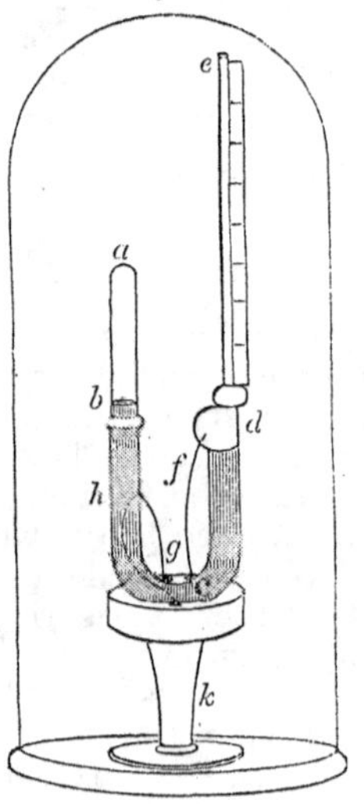

The chlorine-hydrogen photometer, invented by Professor J. W. Draper, depends on the fact that chlorine and hydrogen, if mixed in the proper proportion to form hydrochloric acid, do not unite in the dark, but, if exposed to even a feeble light, combine, the gases contracting more quickly as the light is stronger, and turning into the acid. The contrivance consists of a glass tube, *a b c d*, *Fig.* 66, with three platinum wires, *f g h*, fused into it. From one end, *d*, a fine tube, *e*, projects, provided with a scale; the other end, *a*, is closed. The stand *k* supports the whole. When the instrument is to be used, it is filled with hydrochloric acid, and the wires *f* and *g* connected with a voltaic bat-

What is the principle of the method of extinction? Describe the chlorine-hydrogen photometer. How is it made ready for use?

tery, so that chlorine may be evolved from *g* and hydrogen from *f*. As soon as the acid in the limb *a b h* is saturated with chlorine, the pole of the voltaic battery that had been connected with *f* is dipped into the mercury cup of *h*. Chlorine and hydrogen then accumulate together at *a b*, which part is to be covered with an opaque screen. The acid at the same time rises in the fine tube *e*.

On taking off the screen over *a b* and exposing the photometer to light, the chlorine and hydrogen at once commence to unite, and the acid descends in the tube *e*. The amount of action can be quantitatively ascertained by examining the scale. If exposed to sunlight, a violent explosion will result, and the instrument be destroyed. This photometer was extensively used by Professor Draper in his Researches on Light, published in the Philosophical Magazine.

Light does not move from point to point instantaneously, but at a rate which is measurable. From observations on the eclipses of Jupiter's satellites, and the experiments of Foucault and Fizeau, it appears that this velocity is about 192,000 miles per second.

When a ray falls upon a polished surface it suffers reflection, and when it falls on a transparent medium it undergoes refraction. It is in consequence of this that convex lenses, as in *Fig.* 67, converge the rays of the sun, and so produce a high temperature. In this application they are called burning-glasses, and, until the invention of the voltaic pile and oxyhydrogen blowpipe, presented the most energetic means for elevation of temperature. If made of a diameter from one to three feet, they will effect the fusion of most earthy and metallic bodies. Even gold and silver volatilize at the focus.

Fig. 67.

How is the amount of light known? What is the velocity of light? What are burning-glasses? How high a temperature will they produce?

LECTURE XVIII.

THE CONSTITUTION OF SOLAR LIGHT.—*Newton's Discoveries.—The Solar Spectrum.—Order of Intensity of Light.—Fluorescence.—Chemical Effects of Light.—Fraunhofer's Lines.—Spectral Analysis.—Chemical Composition of the Sun.—The Spectroscope.—Photography of the Spectrum.—Electric Spectra.*

SIR ISAAC NEWTON first succeeded in proving the compound nature of light by the aid of a glass prism. If the shutters of a room be closed, and through an aperture in one of them a beam of the sun enters, *a*, *Fig.* 68, it pursues a straight path, following the dotted line *a e*. Now let the prism interpose in the position *b c*, so as to intercept completely the beam. It goes no longer to *e*, but is bent out of its course and moves in the direction *d*.

Fig. 68.

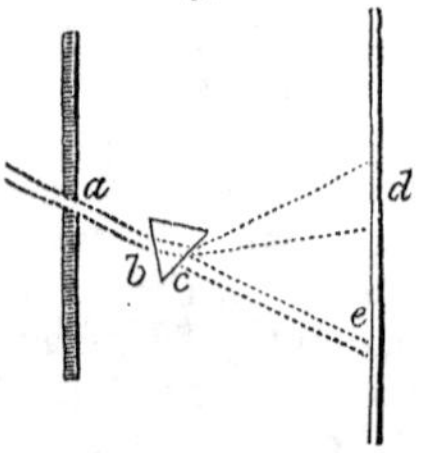

Two striking facts are now to be remarked; first, the ray *a* is broken or refracted from its path; and, second, instead of forming on the surface *d*, upon which it falls, a white spot, an elongated and beautifully colored image is produced. These colors are said to be seven in number, though they shade imperceptibly into one another—red, orange, yellow, green, blue, indigo, violet. This separation of the colors from one another is designated by the term *dispersion*.

Newton has shown that white light consists of these various colored rays blended together, and their separation in the case before us is due to the fact that the prism refracts them unequally. On examining the position of the colors in relation to the point *e*, to which they would all have gone had not the prism intervened, it is ascertained that the red is least disturbed or refracted from its original path, and the violet most. For

How may the compound nature of light be proved? What facts are remarked on passing light through a prism? What is dispersion? What is meant by least and most refrangible rays?

this reason we call the red the least refrangible ray, the violet the most refrangible, and the yellow intermediately.

That the mixture of these colored rays reproduces white light, may be proved by resorting to any optical contrivance which will reassemble them, as, for instance, another prism, set with its back in the opposite direction to the first one, or a color-blender, *Fig.* 69. This consists of a light disc mounted upon an axis, around which it may be caused to revolve with rapidity. Upon its face are painted the seven principal colors. On causing it to rotate, the colors blend together and the disc seems nearly white.

Fig. 69.

Fig. 70.

Let *v r*, *Fig.* 70, represent the spectrum which is given by a sunbeam after its passage through a prism, and *e* the point to which it would have gone had not the prism intervened. The order of colors, commencing with that which is least disturbed from its path or nearest to *e*, is as follows:

Red,	Blue,
Orange,	Indigo,
Yellow,	Violet.
Green,	

Besides this difference in color, the light differs in intrinsic brilliancy in the different spaces. Thus, if we receive the spectrum on a piece of finely-printed paper, we can read the letters in each color at very different distances. In the yellow region the light is most brilliant, and there we can read farthest. From this point the light declines in brilliancy to the two ends of the spectrum, its intensity in the colored spaces being in the following order: yellow, green, orange, red, blue, indigo, violet.

It was discovered, however, by Stokes that if the paper be impregnated with an acid solution of sulphate

How may the recomposition of white light be effected? Describe the color-blender. What is the order of the colors? How can we show the difference of brilliancy in different parts of the spectrum? What is the order of brilliancy of the colors?

of quinine, the visibility of the rays at the violet end of the spectrum is much increased. He attributed it to the lowering of the refrangibility of those extreme rays so as to bring them into the visible group. Decoctions of horse-chestnut bark shows the phenomenon particularly well: it is termed *fluorescence*.

The alchemists found, centuries ago, that the chloride of silver, a substance of snowy whiteness, turns black on exposure to light. More recently a great number of such bodies have been found—bodies which change with greater or less rapidity under the influence of this agent. The iodide and bromide of silver, which form the basis of photography on glass, are such. In the same manner, changes take place in a great variety of organic compounds; the most delicate vegetable hues are soon bleached, and, indeed, a ray of light can scarcely fall on a surface of any kind without leaving traces of its action.

If a piece of paper spread over with chloride of silver be placed in the solar spectrum, it soon begins to blacken; but it does not blacken with equal promptitude in each of the colored spaces, the effect taking place most rapidly among the more refrangible colors, and especially in the indigo region. As in the case of heat, the effect extends far beyond the limits of the spectrum, and where the eye can not discover a trace of light.

By placing a chlorine-hydrogen photometer successively in the various colored spaces, we can readily determine the place of maximum action, and the distribution of chemical influence throughout the spectrum. The greatest effect is found among the more refrangible colors, and from that point diminishes toward each end of the spectrum.

When the aperture which admits a ray of light into a dark room, *Fig.* 68, is a narrow fissure or slit not more than one thirtieth of an inch in width, the spectrum which is formed by the action of a prism is crossed by great numbers of black lines. These, called Fraunhofer's lines, are always found in the same position as re-

What is fluorescence? What changes occur in chloride of silver in the light? Mention other substances affected by light. How does chloride of silver act in the solar spectrum? How may the point of maximum action be found? How can Fraunhofer's lines be shown?

spects the colored spaces, and, from the invariability of that position, are used as boundary marks. The more prominent ones are designated by the letters of the alphabet, and their relative magnitude and position is given in *Fig.* 71.

Fig. 71.

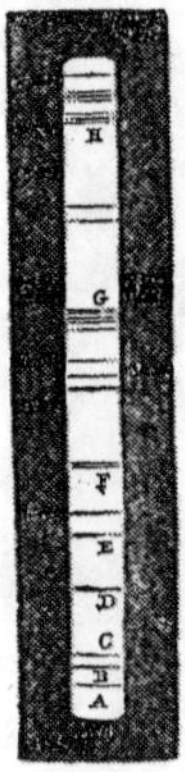

When, instead of solar light, light from other sources, as a candle, the stars, the electric spark, etc., is used, the number and position of the lines is entirely changed. On causing various bodies to volatilize in the flame of a spirit-lamp or gas-burner, a characteristic bright group appears for each, and by the aid of a prism properly arranged, we are enabled to find out the chemical composition of many bodies, and even ascertain the presence of traces too minute to be detected in any other manner. This branch of chemical inquiry, called spectroscope analysis, received a great impulse from the discovery, by Kirchhoff and Bunsen, of two new metals, caesium and rubidium, in certain mineral waters. The former gives characteristic blue lines in the spectrum, the latter red ones.

They also made the discovery that on passing a beam of light from an ignited solid, which was shown by Professor Draper to have no fixed lines, through a flame in which such a substance as sodium, for example, was volatilizing, instead of the bright yellow lines usually given by that substance appearing, there came, in their place, black lines, the flame having absorbed the rays it would otherwise have yielded. They were hence led to conclude that the lines in the solar spectrum are due to the absorptive action of the atmosphere of that body, and that we could infer his chemical composition by comparing the solar spectrum with that of flames in which the various elements were burning. In the same manner the composition of the stars has been examined. The atmosphere of the sun contains sodium, potassium, magnesium, calcium, iron, chromium, and nickel, besides many other substances. The only difficulty in the way of this hypothesis is, that it is known that a change in temperature may vary the position of the spectral lines.

How are they designated? Are they the same in all kinds of light? What is spectroscope analysis? What were Bunsen and Kirchhoff's discoveries? What metals are found in the sun?

The instrument, the spectroscope, which is in common use in laboratories, is constructed as follows: *P*, *Fig.* 72,

Fig. 72.

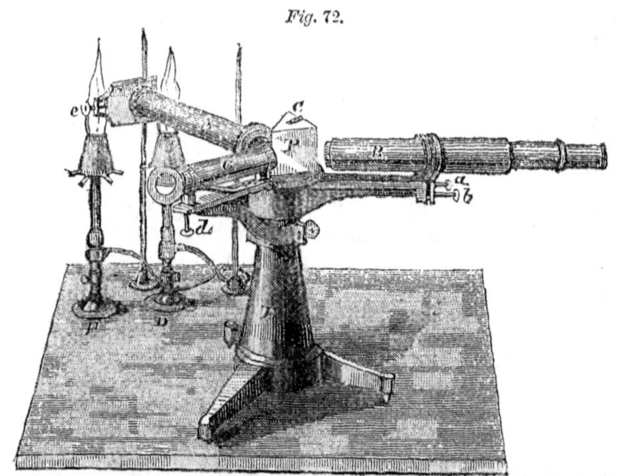

represents a flint glass prism, supported on a firm stand, *F*. The tube *A* has a lens at the end next the prism, and a vertical slit at the other, the tube being of such a length that the focus for parallel rays falls on the slit. At *B* is an observing telescope, having a lens at the end next the prism, and an eye-piece at the other: it is adjustable by the two screws *a b*. The telescope *C* has also a lens toward the prism; but at its other end, and in the focus of the lens, is a scale on an opaque ground, which is to be illuminated by a feeble light; *d* is an adjusting screw.

In the use of the instrument, the substance to be examined is placed in the flame, *e*, of a gas burner, *F*. The rays pass down the tube *A*, are rendered parallel by its lens, pass through the prism, are refracted and suffer dispersion, and the resulting spectrum is examined by the telescope *B*. As a reflection of the scale *C* is seen at the same time in the field of *B*, the position of the different lines may be measured. By the aid of the flame *D* and a reflector, two spectra may be seen in *B* at the same time.

Describe the construction of the spectroscope. How is it used?

The delicacy of the reactions is shown by the statement, that by the aid of the spectroscope $\frac{1}{2500000}$ of a grain of sodium can be detected.

The position of many of these fixed lines of the solar spectrum, including those at the more refrangible end, has been determined by the aid of photography, Professor Draper having first photographed them in 1842, and discovered four great groups, containing many hundred lines, which he named M, N, O, P, and which are invisible to the eye.

Fig. 73.

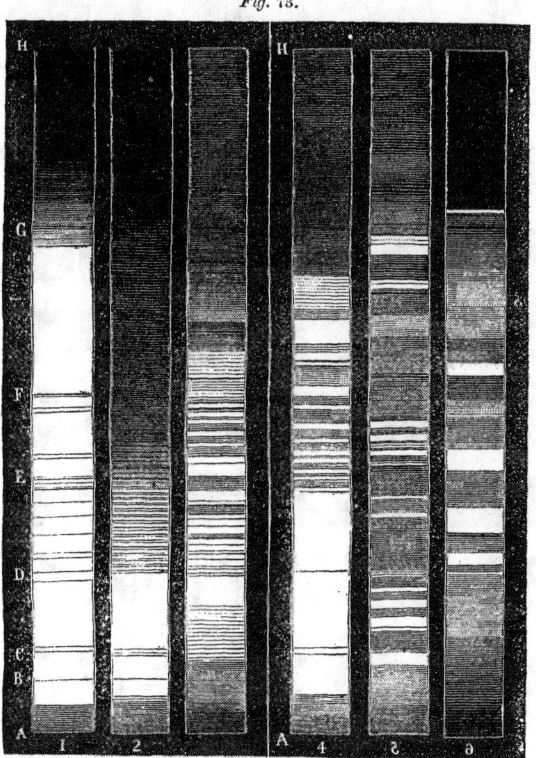

What is the delicacy of spectral reactions? Of what use is photography in spectral analysis?

The spectra of many substances can only be seen by having recourse to the aid of Ruhmkorff's coil, the terminal wires being coated with the body to be tried, or else the spark passed through a vessel or tube containing it.

In *Fig.* 73, No. 1 shows the principal dark lines of the solar spectrum; 2 represents the spectral lines or bands produced by the absorbing action of bromine; 3, those

Fig. 74.

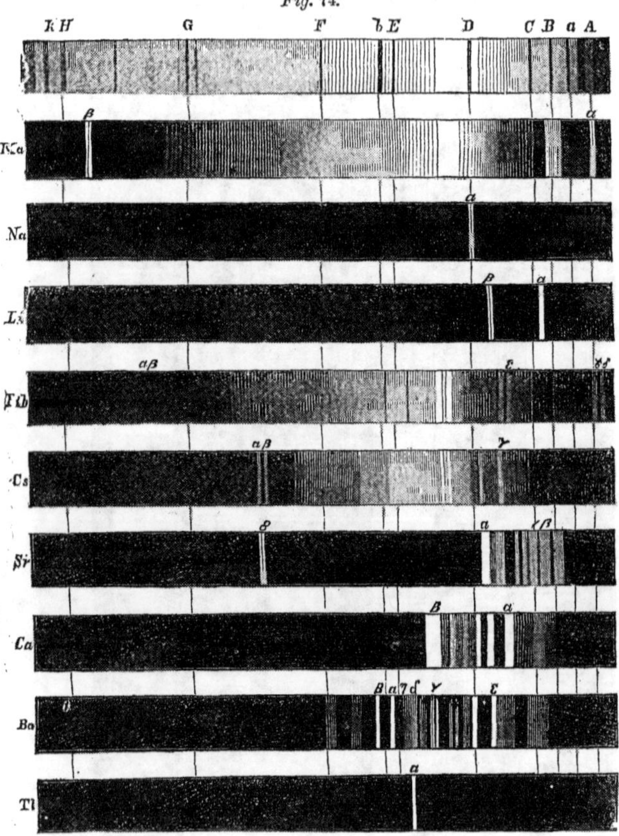

What does *Fig.* 73 represent?

of nitrous acid; 4, those of perchloride of manganese; 5 is the spectrum obtained by burning chloride of copper in alcohol; and 6, boracic acid in alcohol.

In *Fig.* 74 the solar spectrum, with its principal lines, is placed above, and below, in order, are the spectra of potassium, sodium, lithium, rubidium, caesium, strontium, calcium, barium, and thallium.

LECTURE XIX.

WAVE THEORY OF LIGHT.—*Proofs of the Existence of the Ether.—Light consists of Waves in it.—The Ethereal Particles move but little.—Distinction between Vibration and Undulation.—Fresnel's Theory of Transverse Vibrations.—Transverse and Normal Waves.—Brilliancy of Light depends on Amplitude of Vibration.*

THE cause of light is an undulatory movement taking place in the ethereal medium. That such a medium exists throughout all space, seems to be proved by a number of astronomical facts. It exerts a resisting agency on bodies moving in it. From its tenuity, we should scarcely expect that it would impress a marked disturbance on the great planetary masses, but on cometary bodies that are light and gaseous it produces a perceptible action. The comet of Encke, with a period of about 1200 days, is accelerated in each revolution by about two days, and that of Biela, with a period of 2460 days, is accelerated by about one day. As there is no other obvious cause for these results, astronomers have very generally looked upon them as corroborative proofs of the existence of a resisting medium—that universal ether to which so many other facts point.

In this elastic medium undulatory movements can be propagated in the same manner as waves of sound in the air. It is to be clearly understood that the ether and light are distinct things; the latter is merely the effect of movements in the former. Atmospheric air is one thing, and the sound which traverses it another.

What does *Fig.* 74 represent? What is the cause of light? What proofs have we of the existence of an ether? What distinction is to be made between ether and light?

The air is not made up of the notes of the gamut, nor is the ether composed of the seven colors of light.

Across the ether undulatory movements, resembling in many respects the waves of sound in the atmosphere, traverse with prodigious velocity. From the eclipses of Jupiter's satellites and other astronomical phenomena, it appears that the rate of the propagation of light, or the velocity with which these waves advance, is about 192,000 miles in a second. We are not, however, to understand by this that the ethereal particles rush forward in a rectilinear course at that rate: those particles, far from advancing, remain stationary.

If we take a long cord, *Fig.* 75, and, having fastened it by the extremity, *b*, to a fixed obstacle, commence agitating the end *a* up and down, the cord will be thrown into wavelike motions passing rapidly from one end to the other. This may afford us a rude idea of the nature of ethereal movements. The particles of which the cord is composed do not advance or retreat, though the undulations are rapidly passing.

Fig. 75.

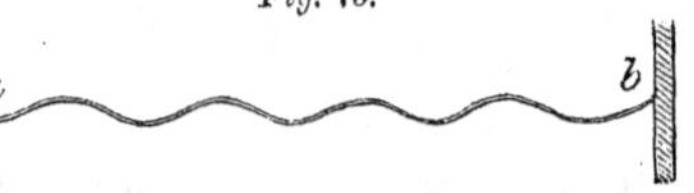

A distinction is to be made between the words vibration and undulation. In the case of the cord, *Fig.* 75, the vibration is represented by the movement exerted by the hand at the free extremity, *a;* the undulation is the wave-like motion that passes along the cord. We see, therefore, that these stand in the relation of cause and effect: the vibration is the cause, and the undulation the effect. Throughout the ethereal medium, each particle vibrates, and transmits the undulatory effect to the particles next beyond it.

In the same way as a vibrating cord agitates the surrounding air and makes waves of sound pass through it, so does an incandescent or shining particle, vibrating with prodigious rapidity, impress a wave-like motion on the ether, and the movement eventually impinging on the eye is what we call light.

To refer again to the simple illustration given in

What is the velocity of light? What does *Fig.* 75 illustrate? What is the distinction between vibration and undulation? How does a shining particle communicate an impression to the eye?

Fig. 75, it is obvious that there is an infinite variety of directions in which we may vibrate that cord, or throw it into undulations. We may move it up and down, or horizontally right and left, and also in an infinite number of intermediate directions, every one of which is transverse, or at right angles to the length of the cord, as *a a*, *b b*, *c c*, etc., *Fig.* 76. This is the peculiarity of the movement of light. Its vibrations are transverse to the course of the ray, and in this it differs from the movement of sound, in which the vibrations are normal; that is to say, in the direction of the resulting wave, and not at right angles to it.

Fig. 76.

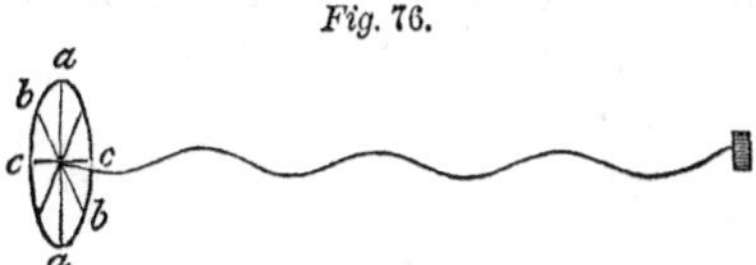

This great discovery of the transverse vibrations of light was made by M. Fresnel.

It may, however, be remarked, that though light consists of rays originating in these transverse motions, it is not impossible that there may be other phenomena which correspond to movements in other directions. To those movements our eyes are totally blind, and hence we can not speak of them as light. In the same way there may be motions in the air due to transverse vibrations, but to them our ear is perfectly deaf.

Lights differ from each other in two striking particulars—brilliancy and color. These are determined by certain qualities in the waves. On the surface of water we may have a wave not an inch in altitude, or a wave, as the phrase is, "mountains high." Under these circumstances waves are said to differ in amplitude; and transferring this illustration to the case of light, a wave, the amplitude of which is great, impresses us with a sense of intensity or brilliancy; but a wave, the amplitude of which is less, is less bright. The brilliancy of light depends on the magnitude of the excursions of the vibrating particles.

In how many ways may a cord be vibrated? Are other motions of the ether possible? In what particulars do lights differ? What is the cause of the difference in brilliancy?

LECTURE XX.

WAVE THEORY OF LIGHT.—*Colors of Light depend upon Wave Lengths.—Interference of Sounds.—Young's Theory of the Interference of Light.—Conditions of Interference.—Explanation of Lights and Shades in Shadows.*

BY the length of a wave upon water, we mean the distance that intervenes from the crest of one wave to that of the next, or from depression to depression. Thus, in *Fig.* 77, from *a* to *c* constitutes the wave length, and from *a* to *b* its altitude or height.

Fig. 77.

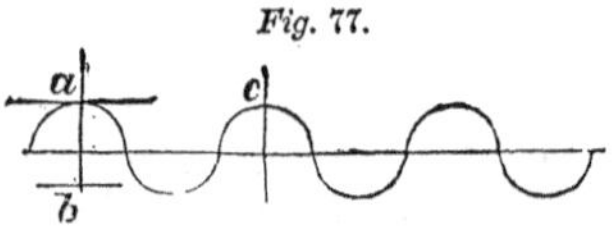

In the ether the length of waves determines the phenomenon of color. This may be rigorously proved, as we shall soon see when we come to the methods by which philosophers have determined the absolute lengths of undulations. It has been found that the longer waves give rise to red light, the shorter ones to violet, and those of intermediate magnitudes, the other colors in the order of their refrangibility.

Two rays of light, no matter how brilliant they are separately, may be brought under such relations to one another as to destroy each other's effect and produce darkness; two sounds may bear such a relation to each other that they shall produce silence; and two waves on the surface of water may so interfere with one another that the water shall retain its horizontal position.

Take two tuning-forks of the same note, and fasten by a little sealing-wax on one prong of each a disc of cardboard half an inch in diameter, as seen in *Fig.* 78, *a*. Make one of the forks a little heavier than the other by putting on the end of it a drop of the wax.

Fig. 78.

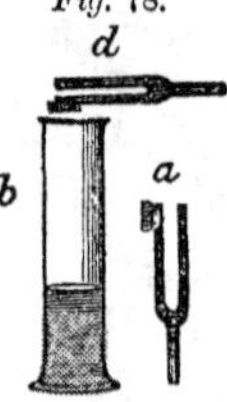

Then take a glass jar, *b*, about two inches in diameter and eight or ten long, and hav-

---

What is meant by wave length? What effect has the length of undulation on the color of light? Describe the experiment *Fig.* 78.

ing made one of the forks vibrate, hold it over the mouth of the jar, as seen at *d*, its piece of cardboard being downward; commence pouring water into the jar, and the sound will be greatly re-enforced. It is the column of air in the jar vibrating in unison with the fork, and we adjust its length by pouring in the water. When the sound is loudest we cease to pour in any more water, the jar is adjusted, and we can now prove that two sounds added together may produce silence.

It matters not which fork is taken, whether it be the light or the loaded, on making it vibrate and holding it over the mouth of the resonant jar, we hear a uniform and clear sound, without any pause, stop, or cessation. But if we make both vibrate over the jar together, a remarkable phenomenon arises—a series of sounds alternating with a series of silences; for a moment the sound increases, then dies away and ceases, then swells forth again, and again declines; and so it continues until the forks cease vibrating. The length of these pauses may be varied by putting more or less wax on the loaded fork; and as we can see that, even during the periods of silence, both forks are rapidly vibrating, the experiment proves that two sounds taken together may produce silence.

Under these circumstances, waves of sound are said to interfere with each other, and in like manner interference takes place among the waves of light. We can gather an idea of the mechanism by considering this case in waves upon water, in which, if two undulations encounter each other under such circumstances that the concavity of the one corresponds with the convexity of the other, they mutually destroy each other's effect.

If two systems of waves of the same length encounter each other after having come through paths of *equal* length, they will not interfere; nor will they interfere, even though there be a difference in the length of these paths, provided that the difference be equal to one whole wave, or two, or three, etc.

But if two systems of waves of equal length encoun-

What is the use of the jar of water? How may silence be produced by the vibrating tuning-forks? What is meant by interference of sound? How may it be illustrated by waves on water? Under what circumstances do undulations re-enforce one another?

ter each other after having come through paths of *unequal* length, they will interfere; and that interference will be complete when the difference of the paths through which they have come is half a wave, or $1\frac{1}{2}$, $2\frac{1}{2}$, $3\frac{1}{2}$, etc.

These cases are respectively shown at *a b* and *c d*, *Fig.* 79, at the point of encounter, *x*. In the first instance, the two sets of waves are in the same phase, that is, their concavities and convexities respectively correspond, and there is no interference; but in the second case, at the point of encounter, *x*, the two systems are in opposite phases, the convexity of the one corresponding with the concavity of the other, and interference takes place.

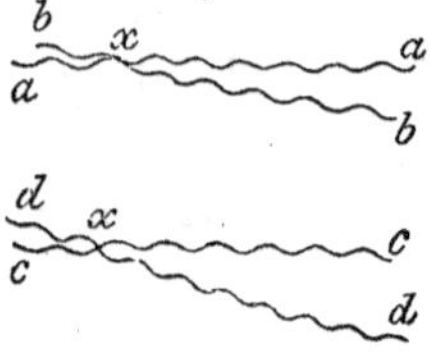

Fig. 79.

Upon these principles we can account for the remarkable results of the following experiment: From a lucid point, *s*, *Fig.* 80, which may be formed by the rays of the sun converged by a double convex lens of short focus, or by passing a sunbeam through a pinhole, let rays emanate, and in them place an opaque obstacle, *a b*, which we will suppose to be a cylindrical body, seen endwise in the figure. At some distance beyond place a screen of white paper, *c d*, to receive the shadow. It might be supposed that this shadow should be of a magnitude included between *x y*, because the rays *s a*, *s b*, which pass the sides of the obstacle, impinge on the paper at those points. It might farther be supposed that within the space *x y* the shadow should be uniformly dusky or dark, but on examining it such will not be found to be the case. The shadow will be found to consist of a series of light and dark stripes, as represented in *Fig.* 81. In its middle, at *e*, *Figs.* 80 and 81, there is a white stripe; this is succeeded on each side by a dark one, this again by a bright one, and so on alternately.

Fig. 80.

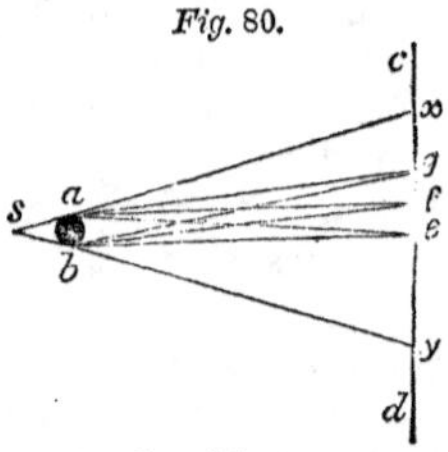

Fig. 81.

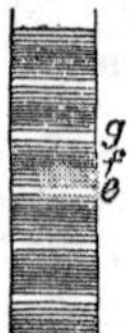

Under what circumstances do they interfere? Describe *Fig.* 79. Describe the experiment *Fig.* 80. What is seen on the screen? How is this explained on the undulatory theory?

Upon the undulatory theory all this is readily explained. Sounds easily double round a corner, and are heard though an obstacle intervene. Waves upon water pass round to the back of an object on which they impinge, and the undulations of light in the same manner flow round at the back of the cylinder *a b*, *Fig.* 80. And now it is plain that two series of waves which have passed from the sides of the obstacle to the middle of the shadow, that is, along the lines *a e*, *b e*, have gone through paths of equal length, and, therefore, when they encounter at the point *e*, they will not interfere, but exalt each other's effect.

But, leaving this central point, *e*, and passing to *f*, it is plain that the systems of waves which have come through the paths *a f*, *b f*, have come through different distances, for *b f* is longer than *a f*; and if this difference be equal to the length of half a wave, they will, when they encounter at the point *f*, interfere and destroy each other, and a dark stripe results.

Beyond this, at the point *g*, the waves from each side of the obstacle *a g*, *b g*, again have come through unequal paths; but, if the difference is equal to the length of one whole wave, they will not interfere, and a white stripe results.

Reasoning in this manner, we can see that the interior of such a shadow consists of illuminated and dark spaces alternately—illuminated spaces when the light has come through paths that are equal, or that differ from each other by 1, 2, 3, 4, etc., waves, and dark when the difference between them is equal to $\frac{1}{2}$, $1\frac{1}{2}$, $2\frac{1}{2}$, $3\frac{1}{2}$, etc., waves.

That it is the interference of light coming from the opposite sides of the opaque object which is the cause of these phenomena, is proved by the circumstance that if we place an opaque screen on one side of the obstacle, so as to prevent the light passing, the fringes all disappear.

What is the cause of the white stripe in the centre? What is the cause of the dark stripes on each side, and of the white stripes which succeed them? How can it be proved that these stripes result from interference?

LECTURE XXI.

WAVE THEORY OF LIGHT.—*Measurement of the Length of a Wave of Light.—Lengths differ for different Colors.—Measurement of the Rapidity of Vibration.—Nature of Polarized Light.—Plane, Circular, and Elliptical Polarized Light.—Reflection, Refraction, and Absorption of Light.—Cause of the Color of Objects.*

THE experiment, *Fig.* 80, may enable us to determine the length of a wave of light. This may be done by measuring the distances *a f* and *b f*, or from the sides of the obstacle to the first bright stripe from the central one, for at that point the difference between those two lines, *a f* and *b f*, is equal to the length of one wave. We might employ the second bright stripe; the difference would then be equal to two waves.

Farther: if, instead of using ordinary white light, radiating from the lucid point *s*, we use colored lights, such as red, yellow, blue, etc., in succession, we shall find that the wave length determined by the process just explained differs in each case; that it is greatest in red, and smallest in violet light. By exact experiments made upon methods more complicated than the elementary one here given, it has been found that the different colored rays of light have waves of the following length:

Wave Lengths of the Different Colors of Light.

The English inch is supposed to be divided into 10 million equal parts, and of those parts the wave lengths are,

For red light............	256	For blue light............	196
" orange light........	240	" indigo "	185
" yellow "	227	" violet "	174
" green "	211		

In this manner it is proved that the different colors of light arise in the ether from its being thrown into waves of different lengths.

Knowing the rate at which light is propagated in a

How can the length of a wave of light be measured? What results on using various colored lights? Give the wave lengths of the different colors.

second, and the wave length for a particular color, we can readily tell the number of vibrations executed in a second, for they plainly are obtained by dividing 192,000 miles, the rate of propagation, by the wave length. From this it appears that if a single second of time be divided into one million of equal parts, a wave of red light trembles or pulsates 458 millions of times in that inconceivably short interval, and a wave of violet light 727 millions of times.

Common light, as has been said, originates in vibratory motions taking place in every direction transverse to the ray. With polarized light it is different. To gather an idea of the nature of polarized light, we must refer once more to the cord, *Fig.* 76, which, as has been said, serves to imitate common light when its extremity is vibrated vertically, horizontally, and in all intermediate positions in rapid succession; but if we simply vibrate it up and down or right and left, then it imitates polarized light. Polarized light is therefore caused by vibrations transverse to the ray, but which are executed in one direction only.

There is a certain gem, the tourmaline, which serves to exhibit the properties of polarized light. If we take a thin plate of this substance, *c d*, properly cut and polished, and allow a ray of light, *a b*, *Fig.* 82, to fall upon it, that ray will be freely transmitted through a second plate if it be held symmetrically to the first, as shown at *e f*; but if we turn the second plate a quarter round, as seen at *g h*, then the light can not pass through. The rays of the meridian sun can not pass through a pair of crossed tourmalines.

Fig. 82.

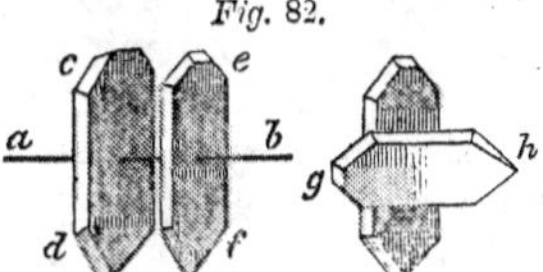

Fig. 83.

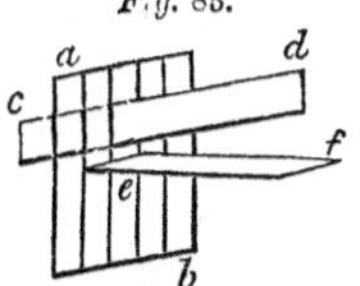

The cause of this is obvious. If we take a thin lath or strip of pasteboard, *c d*, *Fig.* 83, and hold it before a cage or grate, *a b*, it will readily slip through when its plane coincides with the bars;

How can we determine the number of vibrations of light in a second? What is the difference between common and polarized light? What are the properties of the tourmaline? How are they illustrated by *Fig.* 83?

but if we turn it a quarter way round, as at *e f*, then, of course, it can not pass the bars. Now the plate of tourmaline, *Fig.* 82, *c d*, polarizes the light, *a b*, which falls upon it; that is, the waves that pass through it are all vibrating in one plane. They pass, therefore, readily through a second plate of the same kind so long as it is held in such a way that its structure coincides with that motion, but if it be turned round so as to cross the waves, then they are unable to pass through it.

There are many ways in which light can be polarized —by reflection, refraction, double refraction, etc. The resulting motion impressed on the ether is the same in all cases.

Light modified as just described is designated plane polarized light, but there are other varieties of polarization. If the end of the rope, *Fig.* 76, be moved in a circle, circular waves will be produced, imitating circularly polarized light, and if it be moved in an ellipse, elliptical polarized light.

When a beam of polarized light is transmitted in certain directions through plates of double-refracting bodies, splendid phenomena of coloration are produced. Plates of mica, selenite, or quartz, cut in a direction parallel to that of the optic axis, cause the image to assume a tint which may be made to vary by rotating the analyzing plate. If, instead of an analyzing plate in the reflecting apparatus, a crystal of calcareous spar be substituted, two images are seen, which are tinged of complementary hues at all parts of the revolution. By using a plate cut from an uniaxial crystal perpendicular to the optic axis, a series of colored rings, intersected by a cross, will be observed. With biaxial crystals, a double system of colored rings is formed. In uncrystallized media, such as glass, which are submitted to pressure or not well annealed, the same phenomena are shown, and can be varied by altering the shape of the piece used.

The undulatory theory of light gives a clear account of the ordinary phenomena of optics. The general law under which light is reflected from polished surfaces is

By what means can light be polarized? What is circular polarized light? What is elliptical polarized light? How is coloration produced in polarized light?

Fig. 84.

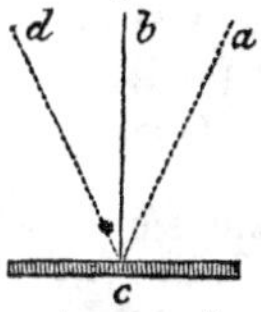

d b a

c

a direct consequence of it. That law is, that the angle *d c b*, *Fig.* 84, made by the reflected ray *d c*, with a perpendicular, *c b*, drawn to the point *c*, at which the light impinges, is equal to the angle *a c b*, which the incident ray makes with the same perpendicular, or, as it is briefly expressed, "the angles of incidence and reflection are equal to each other, and on opposite sides of the perpendicular."

By the aid of this law we can show the action of reflecting surfaces of any kind, and discover the properties of plane and curved mirrors, whether they be concave or convex, spherical, elliptical, paraboloidal, or any other figure.

From the undulatory theory, the law of the refraction of light also follows as a necessary consequence. It is, "In every transparent substance, the sines of the angles of incidence and refraction are to each other in a constant ratio," and by the aid of this law we can determine the action of media bounded by surfaces of any kind, plane or spherical, concave or convex. It explains the action of lenses, and the construction of refracting telescopes and microscopes.

Sir Isaac Newton's discovery that white light arises from the mixture of the different colored rays in certain proportions explains the cause of the colors which transparent media often exhibit. Thus, if glass be stained with the oxide of cobalt, it allows a blue light to pass it; and upon such principles the art of painting on glass depends, different colors being communicated by different metallic oxides. The cause of this effect is readily discovered; for, if we make the light which enters a dark room, as in *Fig.* 68, pass through such a piece of stained glass before it goes through the prism, and examine the resulting spectrum, we find that several rays are wanting in it; that the glass has absorbed or detained some, and allowed others to traverse it. A piece of blue thus suffers most of the blue light to pass, but stops the green, the yellow, etc. But it is also to be observed that the light which is transmitted by any of

---

What is the law of the reflection of light? What is the law of refraction of light? What is the cause of the color of transparent media? Is the light transmitted pure?

these colored media is not pure; it is contaminated with other tints. The blue glass, for instance, does not stop all the rays except the blue; it allows a large portion of the red to pass, and hence the light it transmits is more or less compound.

The color of an opaque object is due to the same cause, the substance absorbing the rays of one or more colors and reflecting the rest. It appears to be of the color that it reflects.

LECTURE XXII.

PRODUCTION OF LIGHT.—*Professor J. W. Draper's Researches.—By Incandescence.—Point at which Bodies become Red-hot.—All Solids shine at the same Degree.—Colors emitted.—Rate of Increase of Intensity.—Production by Combustion.—Nature of Flames.—By Phosphorescence.—Control of Temperature.—Application of the Theory of Undulations.*

A THEORETICAL explanation of the chemical action of light must depend on the views entertained of the nature of that agent. In a series of memoirs published in the London and Edinburg Philosophical Magazine between 1847 and 1851, Professor Draper investigated the circumstances under which light arises by artificial processes. The chief results obtained are as follows.

There are three general processes by which light is obtained artificially: 1st. By the ignition of bodies; 2d. By their combustion or burning; 3d. By phosphorescence.

1st. *Of the Production of Light by Ignition.*—All solid substances shine when their temperature is raised to a certain degree. The point at which this occurs has been variously estimated. Sir Isaac Newton places it at 635°, Davy at 812°, Wedgewood at 947°, Daniell at 980°. By inventing improved means, Dr. Draper found that for platinum it is 977°, or, if Laplace's coefficient of dilatation be used in the calculation, 1006°.

What is the cause of the color of opaque objects? How may light be obtained artificially? What is the temperature to which a solid must be raised before it shines?

By inclosing a number of different substances with a mass of platinum in a gun-barrel, the temperature of which was gradually raised, it was found, on looking down the barrel, that they all commenced to shine at the same moment, and this even though, as in the case of lead, the melted condition had been assumed. Prof. Draper therefore inferred that all solids and liquids begin to shine at the same degree of the thermometer.

The color of the light which the ignited substance emits depends upon the degree of heat to which it is exposed. Making due allowance for the physiological imperfections of the eye, there can be no doubt that the first rays which appear are the red, and, as the temperature is made gradually to ascend, the yellow, orange, green, blue, indigo, and violet are emitted in succession. At 2130° all these colors are exhibited, and from their commixture the substance appears *white-hot.*

As the temperature of an incandescent body rises, therefore, it emits rays of light of increasing refrangibility.

By the aid of the method of extinction of shadows, it was proved that, as the temperature of an ignited solid rises, the intensity of the light increases very rapidly. For example, platinum at 2600° emits almost forty times as much light as at 1900°, as the following table shows.

Intensity of Light emitted by Platinum at different Temperatures.

Temperature of the Platinum.	Intensity of its Light.
980°	0.00
1900°	0.34
2015°	0.62
2130°	1.73
2245°	2.92
2360°	4.40
2475°	7.24
2590°	12.34

From a parallel series of experiments, in which the

Do all solids shine at the same temperature? What does the color of light emitted by an ignited solid depend upon? What is the first ray that appears? At what temperature are all the colors emitted? What is the relation between the intensity of the light and the temperature?

heat radiated by the ignited platinum was measured, a striking analogy between the two agents appears. Thus, if the quantity of heat radiated by platinum at 980° be taken as unity, it will have increased at 1440° to 2.5, at 1900° to 7.8, at 2360° to 17.8 nearly. The rate of increase is, therefore, very rapid, as in the preceding case.

2d. *Of the Production of Light by Combustion.*—It has been long known that all common flames are incandescent shells, the interior of which is dark, and it has been supposed that there are certain flames which emit particular rays only, but an examination by the prism showed that in every flame every prismatic color is found. The red which burning cyanogen, and the blue which burning sulphur emits, are compound colors.

By burning solid carbon in oxygen gas, it appeared that there is a connection between the refrangibility of the light which a burning body yields and the intensity of the chemical action going on, and that the refrangibility always increases as the chemical action increases.

From this it appears that flames, such as those of lamps and candles, consist of a series of concentric and differently-colored shells, the most interior one being red, and having a temperature of 977°. Upon this, in succession, are placed orange, yellow, green, blue, indigo, and violet shells. The flame, looked at directly, appears to yield white light because of the commixture of these rays; but, on being submitted to the action of a prism, they are separated from each other, and their individual existence proved. If, therefore, we could isolate a horizontal section of such a flame, it would have the aspect of an iris, or rainbow ring.

Upon the principle that the more energetic the chemical action the higher the refrangibility of the light emitted, we may explain without difficulty the colors which different flames present. The red tints predominate in the flame of burning cyanogen, because in that gas there is an element wholly incombustible, the nitro-

---

What analogy was shown between heat and light? What is the construction of a flame? What relation is there between chemical action and refrangibility? What is the color of the interior of a flame? What colors surround it? Why is the flame of cyanogen red?

gen. This, as it is set free, cuts off the free access of the air, and the burning goes on tardily, very much in the same manner as in an oil-lamp to which the air is imperfectly supplied. On the other hand, carbonic oxide burns blue, because of the small quantity of air required to carry it to its maximum of oxidation. The color of flames depends, therefore, on the completeness or incompleteness of the combustion, this principle readily accounting for those cases in which means are used for retarding or promoting the rate of burning, as where an atmosphere of oxygen is used, or air introduced into the interior of a flame by means of a blowpipe, the bright blue cone arising in this latter instance being a striking indication of the increased rapidity of combustion.

There is, therefore, a direct connection between the vehemence with which chemical affinity is satisfied and the refrangibility of the resulting light. If, as there are many reasons for supposing, all chemical changes are attended by vibratory movements of the particles of the bodies engaged, it might well be anticipated that these vibrations should increase in frequency as the action becomes more violent; but it is to be remembered that an increased frequency of vibration is the same thing as an increased refrangibility.

3d. *Of the Production of Light by Phosphorescence.*—All solid substances, except the metals, possess the property of shining after they have been exposed to the sun. Even the hand, after being dipped in the sunshine, emits subsequently light enough to be visible in a dark place. In some the effect lasts but for a moment, in others it is of longer duration and considerable splendor. Among the best phosphori may be mentioned the sulphide of barium, the sulphide of calcium, certain varieties of fluor spar and of diamond. Phosphorescence has generally been regarded as unattended by the emission of heat. It does not require that the exposure to light should be protracted; the flash of an electric spark is sufficient; but the phosphorogenic rays can not pass through glass.

Why is the flame of carbonic oxide blue? What does the color of a flame depend on? Why is a blowpipe flame blue? What substances are phosphorescent? Mention some of the best phosphori. Is phosphorescence attended by heat?

By suitable experimental arrangements Professor Draper ascertained that the best phosphori, when at their maximum of glow, do not increase in volume by so much as the $\frac{1}{12000}$ part; but that there is a minute expansion can not be doubted, since, when means sufficiently delicate are resorted to, a feeble rise of temperature can be detected. The intensity of the light disengaged is to some extent deceptive; for, by resorting to the method of the extinction of shadows, it was shown that a fine specimen of chlorophane, at its maximum of brightness, yielded a light three thousand times less intense than the flame of a very small oil lamp.

The quantity of light a substance can receive, when exposed to the sun, depends on the temperature. The colder the phosphorus is, the more brightly will it subsequently shine; if kept hot during its exposure, it will not shine at all. If a diamond placed upon ice be submitted to the sun and then brought into a dark room the temperature of which is 60°, for a time there is a glow, but presently the light declines and dies out. Let the diamond now be put in water at 100°; again it shines, and again its light dies away. If it next be removed from that water and suffered to cool, and then be re immersed, it will not shine again; but if the water be heated to 200° and the diamond be dropped into it, again it glows, and again its light dies away. There is therefore a correspondence between the light disengaged and the temperature applied.

The phenomena of phosphorescence may all be explained on the principles of the theory of undulations; for, from a shining body undulations are propagated in the ether, and these, impinging on a phosphorescent surface, throw its molecules into a vibratory movement. These, in their turn, impress on the ether undulations; but, by reason of the difference of its density compared with that of the molecules, they do not lose their motion at once, but it continues for a time, gradually declining away, and ceasing when the *vis viva* of the molecules is exhausted.

We may therefore abandon expressions derived from

What effect has temperature on phosphorescence? Describe the experiment with a phosphorescent diamond. Explain phosphorescence on the undulatory hypothesis.

the material theory of light, such as *absorption* and subsequent *emission* of the luminous agent, and conclude that, whenever a radiation falls upon a surface of any kind, it throws the particles thereof into a state of vibration, as when a stretched string is made to vibrate in sympathy with a distant musical sound. This view includes at once all the facts of the radiation of heat, on the dynamical hypothesis and the theory of calorific exchanges; it also offers an explanation of the connection of the atomic weights of bodies and their specific heats. It suggests that all cases of the decomposition of compound molecules under the influence of light are owing to a want of consentaneousness in the vibrations of the impinging ray and those of the molecular group, which, unable to maintain itself, is broken down, under the periodic impulses it is receiving, into other groups, which can vibrate in unison with the ray.

LECTURE XXIII.

CHEMICAL ACTION OF LIGHT.—*Action of Natural and Artificial Lights.—Preliminary Absorption.—Change in the Ray.—Necessity of Absorption.—Spectral Impressions.—Light causes Vegetation by its Yellow Ray.—The Indigo Ray determines the Direction of Growth.—Effect of Amplitude, Frequency, and Direction of Undulations.*

FROM whatever source it may be derived, light exerts chemical action. The moonbeams are sufficiently intense to give copies of that satellite on sensitive surfaces, as my father found in 1841. Recently I have succeeded, by the aid of a reflecting telescope of 15½ inches aperture which I constructed, in producing photographs of that body 50 inches in diameter. Lamplight and other artificial lights are often peculiarly energetic. These decomposing effects take place on those portions of the substance only on which the rays actually fall. There is no lateral spreading, nothing analogous to conduction.

When a sensitive substance receives light for a short

What is the cause of the decomposition of substances by light? Does the moonlight exert chemical action?

space of time no change takes place, the rays are being actively absorbed; but as soon as that preliminary absorption is over they act in a manner which is perfectly definite: if, for instance, it be a decomposition that they are bringing about, the amount of decomposing effect will be precisely proportional to the quantity of rays absorbed.

When a beam from any shining source causes a decomposing effect, it is always itself disturbed; the medium which is changing impresses a change in the ray. Thus, a mixture of chlorine and hydrogen unites under the influence of a ray, but that portion of the ray which passes through the mixture has lost the quality of ever bringing about a like change again.

When a beam from any shining source falls on a changeable medium, a portion of it is absorbed for the purpose of effecting the change, and the residue is either reflected or transmitted, and is perfectly inert as respects the medium itself.

No chemical effect can therefore be produced by rays, except they be absorbed. It is for this reason that water is never decomposed by the sunshine, nor oxygen and hydrogen made to unite; for these substances are all transparent, and allow the rays to pass without any absorption, and absorption is absolutely necessary before chemical action can ensue.

But with chlorine the case is very different. This substance exerts a powerful absorbent action on light: the effect takes place on the most refrangible rays. When mixed with hydrogen and set in the light, it unites with a violent explosion.

As connected with the minute changes of surface which are effected when the different radiant principles fall upon bodies, as in the case of photography, we may allude to the formation of *spectral impressions*, which, though invisible, may be brought out by proper processes. Professor Draper described the following instance many years ago: Take a piece of polished metal, glass, or japanned tin, the temperature of which is low,

---

What is meant by preliminary absorption? What change is impressed on the ray? Does the ray undergo absorption? Why do not oxygen and hydrogen unite in the sunshine? Why do chlorine and hydrogen unite? What are spectral impressions?

and, having laid upon it a wafer, coin, or any other such object, breathe upon the surface and allow the breath entirely to disappear; then toss the object off the surface and examine it minutely. No trace of any thing is visible, yet a spectral impression exists on that surface, which may be evoked by breathing upon it. A form resembling the object at once appears; and what is very remarkable, it may be called forth many times in succession, and even at the end of many months. Other instances of the kind have subsequently been described by M. Moser.

Light exerts no ordinary control over the phenomena of the natural world. Thus it is to its influence that the vegetable world owes its existence; for plants can only obtain carbon from the air while the sun is shining on them, and it is of that carbon that their solid structures are chiefly formed. It has been a question to which ray this effect is due, but in 1843 Professor Draper proved that it is the yellow light which is involved. Dr. Priestley discovered that the leaves of plants will effect the decomposition of carbonic acid gas under water; and, on immersing tubes filled with water holding this gas in solution, and containing a few green leaves, Professor Draper found that at the blue extremity of the spectrum no effect whatever took place, while decomposition went on rapidly in the yellow ray. Dr. Gardner showed, however, that the more refrangible rays have the power of influencing the direction of growth of plants; turnip-seeds which had germinated in the dark, when placed in the solar spectrum, leaning toward the indigo region.

On the Chemical Action of Light.

In considering the action of a ray of light upon a decomposable body, there are three different points to be discussed, so far as the ray itself is concerned: 1st. To what extent and in what manner is the result affected by the *intensity* of the ray—that is, by the amplitude of the vibrating excursions; 2d. How is it affected by the frequency of the pulsatory impressions? 3d. How

What effect has light in the natural world? How did Professor Draper show the function of the yellow ray? What influence has the indigo ray on plants? What points have to be considered in examining the action of a ray?

by the direction in which the vibrations are made, as involved in the idea of polarization?

1st. By means of burning lenses Professor Draper found that it is not the intensity of a beam which determines its decomposing power, and that we can not produce greater effects by concentrated light, than we can by the application of the simple sunbeam continued for an equivalent period of time. Nor can such optical contrivances effect the decomposition of substances on which a feeble beam has no action.

2d. Rays of the highest refrangibility, and therefore of the most frequent vibration, commonly have the greatest activity. On the number of impulses a ray can communicate in a given period of time depends its power of destroying the constitution of any group of atoms; and the phenomena of interference arising from the superposition of wave motions occur exactly as might have been predicted.

3d. The direction of wave motion, as involved in the idea of polarization, whether plane or circular, seems to exert no effect.

LECTURE XXIV.

PHOTOGRAPHY.—*Origin of.—Daguerre and Talbot.—The Daguerreotype.—Prof. Draper first takes Portraits from Life.—The Collodion Process.—Negatives and Positives.—The Albumen Paper Process.—Toning and Permanency.—Dry Collodion Processes.—Formulas for Collodion, the Nitrate Bath, Developing Solution, Fixing Solution.*

PHOTOGRAPHY, literally, writing by light, is a creation of the present century, although many facts that might have led to its cultivation were known in the Middle Ages. The first germ is to be found in observations on the darkening of the chloride of silver by the sunlight; and the first photographs, truly speaking,

What effect has the intensity of the ray? Which rays have the greatest decomposing power? What is the effect of polarization in decompositions? What is the meaning of the term Photography? What are some of the early facts in photography?

were those made by Wedgewood, who exposed leather soaked in nitrate of silver to the sun under the slide of a magic lantern. In 1835 Prof. Draper used bromide of silver and other sensitive compounds for investigating the solar spectrum.

In 1839 Daguerre announced the discovery of a means of fixing the images of the camera obscura, and about the same time Talbot, in England, did the same. The process of the former, known as the daguerreotype, involves the use of a silver-plated sheet of copper, which, having been carefully cleaned with rotten-stone and a buff, is exposed to the vapors of iodine. It becomes tinged yellow, red, blue, gray, depending on the time of the exposure to the vapor, but is most sensitive when yellow. In that condition it is exposed in the camera obscura to the image of some object, being carefully preserved from extraneous light. The latent image has then to be developed by the aid of vapor of mercury rising from a portion of that metal heated to 170°. In about four minutes the image is fully brought out, if the camera exposure has been sufficient, and it is then only necessary to dissolve away the remainder of the film of iodide of silver to prevent farther change.

The process of Daguerre was very sluggish, and was generally regarded as only suitable to the reproduction of still objects; but soon after its publication Prof. Draper succeeded in taking portraits from life, and arrived at such a degree of excellence at once that those pictures have scarcely been surpassed since.

The great subsequent improvements of the process were the discovery of the use of bromine as an accelerating agent, and that of the process of gilding, for fixing the image by covering it with an extremely thin film of gold, deposited from a solution of the chloride of gold in hyposulphite of soda.

The daguerreotype image consists in the whites of an amalgam of silver, and in the dark parts of the unchanged silver. Prof. Draper showed that this amalgam presents a dotted or stippled arrangement, and

---

Describe the daguerreotype process. Who took the first portraits from life? State some of the subsequent improvements in the daguerreotype process. What is the nature of the daguerreotype image?

may be copied by isinglass, and the original in this way reproduced.

Talbot's process has never come into general use, except for enlarged pictures on paper, being entirely superseded by the *Collodion* process. This process, invented by Scott Archer, depends on the fact that gun-cotton, or pyroxyline, is soluble in alcohol and ether, and affords, on drying, a clear and tenacious film. Such collodion, containing some soluble iodide or bromide, or both, is poured on a clean glass plate, and, when set, is immersed in a bath of nitrate of silver, which changes the iodide and bromide into iodide and bromide of silver. In this state it is exposed to light, and, after a sufficient lapse of time, the invisible image is developed by pouring over the plate a solution of pyrogallic acid, or protosulphate of iron, slightly acidified. This causes a reduction of metallic silver in the black state on all those parts which have received the impression of light, the amount of deposit being greater as the light was greater.

After washing in water, the plate is soaked in hyposulphite of soda, or cyanide of potassium, to get rid of the iodide and bromide of silver, and is then again washed and dried. On being varnished, the *negative* thus produced may be used for printing on paper.

If the exposure to light be made shorter, and the development not pushed so far, on placing behind the resulting picture a blackened surface, a *positive* is seen, in which the lights and shades are not reversed, as in the *negative*. Very many forms of these pictures are produced, and many names, as ambrotype, melainotype, have been introduced to distinguish them.

The great superiority of the collodion process results from the fact that negatives can be used to multiply indefinitely copies of the original. The albumen process on paper, by means of which this multiplication is effected, is conducted as follows.

A sheet of paper, which is free from metallic grains or other impurities, is floated on the surface of a bath consisting of albumen that has been thoroughly beaten

Describe the collodion process. Why is the collodion plate soaked in hyposulphite of soda? What is the difference between a *positive* and a *negative?* Describe the albumen process on paper.

up and filtered, containing chloride of ammonium. After the paper has dried, it is floated in a strong solution of nitrate of silver, which causes the production of chloride of silver, and coagulates the albumen. As soon as dry, the paper is exposed behind a negative to either sunlight or daylight. The operator judges of the time required by the darkness of tint assumed by the chloride. The paper has next to be soaked in water, and then *toned.*

The object of the toning operation is to replace the reduced silver by gold, and thereby change the reddish color of the print to a shade either of purple, brown, or black. The toning-bath may be composed of many various substances; one of the best, however, is a solution of chloride of gold, rendered neutral by carbonate of lime, and containing a small quantity of chloride of lime. This rapidly produces a beautiful color, and is the one used by the author in taking his photographs of the moon.

The next step is to remove the superfluous chloride of silver by the aid of hyposulphite of soda, and afterward to submit the proof to a current of water to effect the removal of the hyposulphite, which, if allowed to remain, would infallibly cause it to turn yellow and fade away.

The permanency of paper positives depends on the thoroughness of the toning operation, the gold substituted for silver not being liable to sulphurization. The early photographs, which have, in the last ten years, so nearly passed away altogether, were toned by a bath of hyposulphite of soda and chloride of gold; and although their tints were very beautiful in the first instance, yet, owing to the sulphide of silver that entered into them, they could not be permanent.

The collodion and paper processes have undergone a great number of modifications, in order to adapt them to special purposes. Dry plate photography, for example, involves a method of retaining the sensitiveness of a collodion film, though the excess of nitrate of silver that in the wet process is upon the plate be washed off. The desired result is usually accomplished by flowing

What is toning? What influences the permanence of paper pictures? Describe the process of dry plate photography.

upon the collodion film, after it has been taken from the nitrate of silver bath and washed, a preservative solution of tannin, malt, or some similar substance, which permits the film to be subsequently permeated by water during its development. These dry processes are of great convenience to the landscape photographer, and have led to many ingenious forms of apparatus for carrying the plates and introducing them into the camera obscura.

The following formulas are in common use:

Composition of Iodized Collodion.

Ether	4 ounces (fluid).
Alcohol	4 ounces.
Pyroxyline	30 to 60 grains.
Iodide of Cadmium	50 grains.
Bromide of Cadmium	6 grains.

The nitrate of silver bath for negatives is merely a solution in water of 40 grains to the ounce.

Developing Solution.

Protosulphate of Iron	3 ounces.
Acetic Acid, No. 8	4 ounces.
Alcohol	3 ounces.
Water	40 ounces.

Or,

Pyrogallic Acid	1½ grains.
Alcohol	1 drachm.
Acetic Acid, No. 8	1 drachm.
Water	1 ounce.

Fixing Solution.

Cyanide of Potassium	1 drachm.
Water	4 ounces.

Or,

Hyposulphite of Soda	6 ounces.
Water	16 ounces.

What is the use of the tannin or malt? What are the advantages of this process? Give the formulas for iodized collodion, the developing solution, the fixing solution.

LECTURE XXV.

ELECTRICITY.—*First Discoveries in Electricity.—Leading Phenomena.—Conductors, Non-Conductors, and Insulators.—Two Kinds of Electricity, Vitreous and Resinous, or Positive and Negative.—Law of Electrical Attraction and Repulsion.—Plate Machine.—Cylinder Machine.—Method of Using.—Miscellaneous Experiments.*

MORE than 2000 years ago it was discovered that when amber is rubbed it acquires the property of attracting light bodies. This incident has served to give a name to the agent whose operations we have now to explain, which has been called electricity, from ηλεκτρον, amber.

The catalogue of substances in which electrical development can be produced was greatly increased by Gilbert, who showed that glass, resin, wax, and many other bodies are equally effective as amber. They, too, when rubbed, can attract light substances, and, when the excitement is vigorous, emit sparks like those which are seen when the back of a cat is rubbed on a frosty night.

If a piece of brown paper be thoroughly dried at the fire until it begins to smoke, and then rubbed between woolen surfaces, it will emit sparks on the approach of the finger, attract pieces of paper and then repel them. This latter phenomenon is not, however, peculiar to it, but is noticed in the case of all highly-excited bodies. Electrified bodies therefore exhibit repulsions as well as attractions.

Let there be taken a glass tube, *a b*, *Fig.* 85, an inch in diameter and a foot or more long, closed at the end *b* by means of a cork, into which there is inserted a wire with a round ball, *c*.

Fig. 85.

If the tube be excited by rubbing with a piece of dry

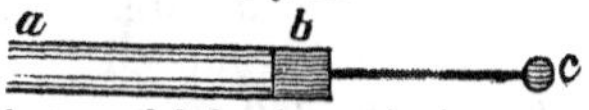

From what fact does the term electricity originate? What occurs when glass, resin, etc., are rubbed? What phenomena may be exhibited by brown paper? Describe *Fig.* 85.

silk, it may be shown that not only does the space rubbed possess the powers of attraction and repulsion, but also the cork and the ball. Nor does it matter how long the wire may be; the electric power is transmitted through the whole of the metal. A metal, therefore, can conduct electricity.

But if, instead of a piece of metal, we terminate the glass tube with a rod of glass or sealing-wax, or hang a ball to it by a thread of silk, in all these cases the electric power can not pass. Such substances are therefore non-conductors of electricity. The important fact, that all substances may be divided into two classes, conductors and non-conductors, was first accidentally discovered by Mr. Grey, who found that all metals and moist bodies are conductors, and that glass, resins, wax, sulphur, atmospheric air, are non-conductors. The following table exhibits the relation of bodies in this respect. The nearer the substance is to the bottom of the table, the better is its conducting power.

Non-Conductors.

Dry Gases and Dry Steam.
Shellac.
Sulphur.
Amber.
Resins.
Gutta Percha and Caoutchouc.
Diamond.
Silk.
Dry Fur.
Glass.
Ice.
Spermaceti.
Turpentine and Volatile Oils.
Fixed Oils.
String and Vegetable Fibres.
Moist Animal Substances.
Water.
Saline Solutions.
Flame.
Melted Salts.
Plumbago.
Charcoal.
All the Metals.

Conductors.

When electricity is communicated to a body which is supported on any of these non-conducting substances, its escape is cut off, and the body is said to be *insulated.*

To Otto Guericke, who was also the inventor of the air-pump, we owe another of the most important discoveries in electricity, that bodies which have touched an excited substance are subsequently repelled by it. Thus, if we rub a glass tube, *a*, *Fig.* 86, until it becomes electrified, and then present to it a feather, *b*, suspended

What is a non-conductor of electricity? What is a conductor? Give examples of each. When is a body said to be insulated?

Fig. 86.

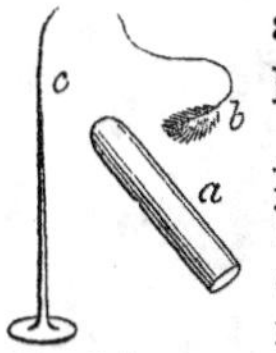

by a silk thread to a stand, *c*, the feather is at first attracted, and then immediately repelled.

A very celebrated French electrician, Dufay, having caused a light, downy feather to be repelled by an excited glass tube, intended to amuse himself by chasing it round the room with a piece of excited sealing-wax. To his surprise, instead of being repelled, the feather was at once attracted. On examining the cause of this more minutely, he arrived at the conclusion that there are two species of electricity, the one originating when glass is excited, and the other from resin or wax. To these he gave the names of vitreous and resinous electricity, thus pointing out their origin. They are also called positive and negative electricities.

He found that these different electricities possess the same general physical qualities: they are self-repulsive, but the one is attractive of the other. This is readily proved by hanging a feather by a linen thread to the prime conductor of the electrical machine, and, when it is excited, bringing near to it an excited glass tube. The feather is already vitreously electrified, and the tube, being in the same condition, at once repels it. But a stick of excited sealing-wax, being resinously electrified, that is to say, in the opposite condition to the feather, at once attracts it.

These various results may all be grouped under the following general law, which includes the explanation of a great many electrical phenomena. Bodies electrified dissimilarly attract, and bodies electrified similarly repel. *Like electricities repel, unlike ones attract.*

For the sake of observing electrical phenomena more readily, instruments have been invented called electrical machines. They are of two kinds, the plate machine and the cylinder. They derive their names from the shape of the glass employed to yield the electricity. The plate machine, *Fig.* 87, consists of a circular plate of glass, *a a*, which can be turned upon an axis, *b*, by

---

Describe *Fig.* 86. Describe Dufay's discovery. What conclusion did Dufay arrive at? What is the difference between vitreous and resinous electricity? What is the general law of attraction and repulsion? Describe the plate machine.

Fig. 87.

means of a winch, *c*; at *d* is a pair of rubbers, which compress the glass between them, and a piece of oiled silk extends over the glass plate, as shown at *e*. In the same manner, on the opposite side of the plate, there is another pair of rubbers, *d*, and an oiled silk, *e*; *f* is the prime conductor, which gathers the electricity as the plate revolves. It must be supported on an insulated stem.

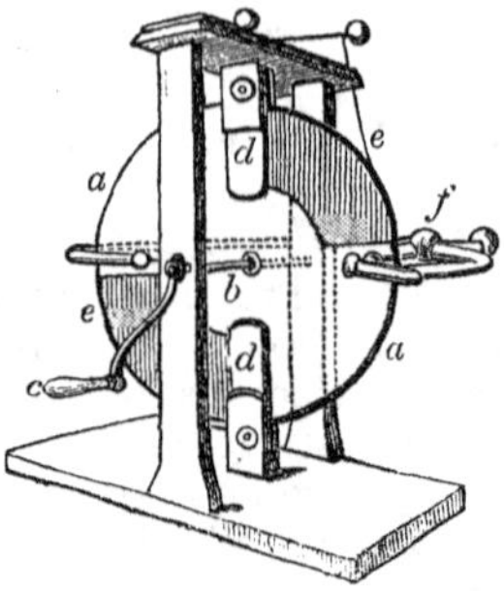

The cylinder machine is represented in *Fig.* 88. It consists of a glass cylinder, *a a*, so arranged that it can be turned on its axis by the multiplying wheel *b b*. The rubber bears against the glass on the opposite side to that seen in the figure, and the oiled silk is shown at *c*; *d* is the prime conductor, usually a cylinder with rounded ends, made of thin brass, and *e* its insulating support.

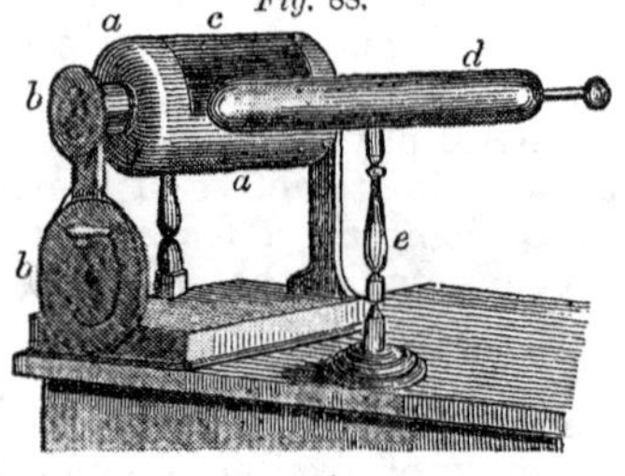

Fig. 88.

Of these machines the plate is commonly the most powerful. It is more liable to be broken than the cylinder, from the disadvantageous way in which the power to turn it round is applied.

To bring an electrical machine into activity, it must be thoroughly dried, but a plate machine should never be set before the fire to warm, or it will almost certainly crack. The rubbers are to be spread over with a little Mosaic gold or amalgam of zinc, and the stem of the conductor made dry. If the rubbers of the machine are not in connection with the ground, there must be a chain hung from them to reach the table. Then, when the instrument is in activity, on presenting the finger

---

Describe the cylinder machine. Which is the best? How is an electrical machine prepared for action? What occurs on presenting the finger to the prime conductor?

to the prime conductor, a succession of *sparks* is emitted, attended by a crackling sound.

A great many beautiful experiments may be made by the aid of this machine. They are, for the most part, illustrations of the luminous effect of the spark, attractions and repulsions, and certain physiological results, as the electrical shock.

If small pieces of tin-foil be pasted round a glass tube in a spiral form, as shown in *Fig.* 89, *a b c*, distances of the twentieth of an inch intervening between, and the ends of the tube terminated by balls, on presenting one of these balls to the prime conductor, and holding the other in the hand, as the spark passes it has to leap over each interstice between the spangles of tin-foil, and exhibits a beautiful spiral line of light.

Fig. 89.

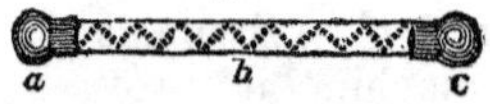

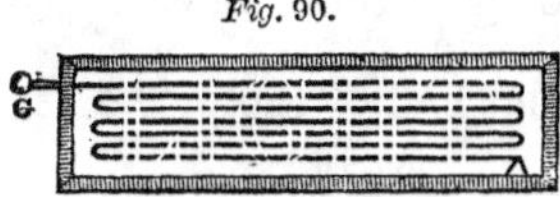

Fig. 90.

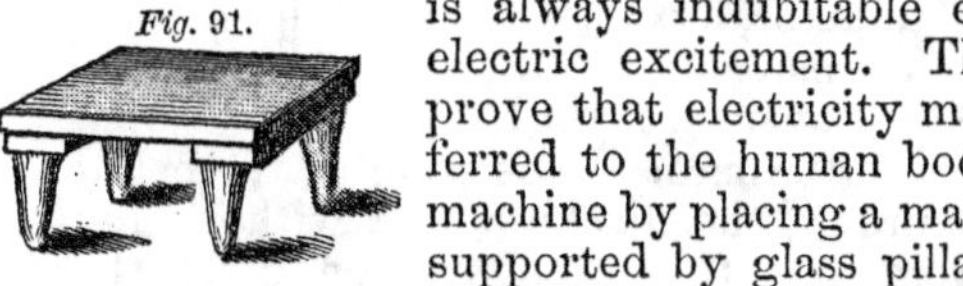

By pasting the tin-foil on a pane of glass in such a way as to direct the spark properly, words may be written in electric light, as shown in *Fig.* 90.

As the electric spark can not be confounded with any other physical phenomenon whatever, its presence is always indubitable evidence of electric excitement. Thus we can prove that electricity may be transferred to the human body from the machine by placing a man on a stool supported by glass pillars, *Fig.* 91. If he touches the prime conductor with one hand, sparks may be drawn from any part of his clothing or body.

Fig. 91.

Describe *Figs.* 89 and 90. What does the electrical stool, *Fig.* 91, demonstrate?

LECTURE XXVI.

THEORY OF ELECTRICAL INDUCTION. — *Phenomena of Induction.—Independent of Shape.—Permanent Excitement by Induction.—Takes place through Glass.—Experiments illustrative of Attraction and Repulsion, and Induction.—Medicated Tubes.*

THERE are many ways in which electrical excitement can be developed. In the common machine it is by friction; in the tourmaline, a crystallized gem, by heat; and in other cases by chemical action, and by conduction. Electrical disturbance also very often arises from induction.

By the term *electrical induction*, we mean that a body which is already excited tends to disturb the condition of others in its neighborhood, inducing in them an electrical condition.

Thus, let *a*, *Fig.* 92, be the terminal ball of the prime conductor, and a few inches off let there be placed a secondary conductor, *b c*, of brass, supported on a glass stand, and at each extremity, *b* and *c*, of the conductor, let there be arranged a pair of cork balls, suspended by linen threads, as shown in the figure. As soon as the ball *a* is electrified by turning the machine, and without any spark passing from it to the secondary conductor, the balls will begin to diverge, showing that the condition of that conductor is disturbed by the neighborhood of the excited ball, *a*.

Fig. 92.

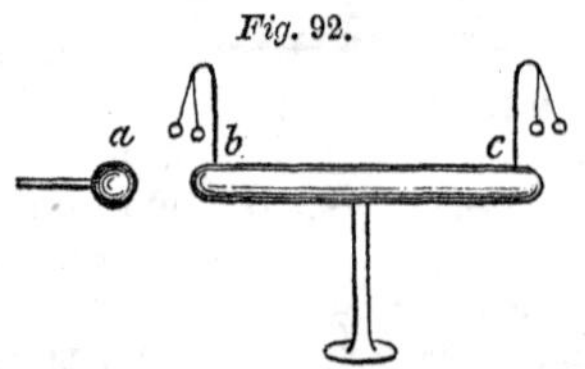

It will farther be found, on presenting an excited piece of sealing-wax to the pairs of cork balls, that one set is attracted and the other repelled. They are therefore in opposite electrical states. The disturbing ball is vitreously electrified, and that end of the secondary conductor nearest it is resinous, the farther end being vitreous.

How may electrical excitement be produced? What is meant by electrical induction? Describe *Fig.* 92. What occurs to the balls on the secondary conductor?

If the disturbing ball *a* be now removed, the electric disturbance ceases and the corks no longer diverge.

The phenomena of electrical induction are not dependent on the shape of bodies. Let there be two flat circular plates, *a b*, *Fig.* 93, supported on glass stands, and set a few inches apart, looking face to face. Let one of them, *a*, be electrified positively by contact with the prime conductor, as indicated by the sign +. It immediately induces a change in the opposite plate, the nearest face of which becomes negative, —, and the more distant positive. It is evident that this disturbance is a consequence of the law that "like electricities repel, and unlike ones attract." In the plate *b* both species of electricity exist; and *a*, being made positive, even though at a distance, exerts its attractive and repulsive agencies on the electric fluid of *b*, the negative electricity of which it attracts and draws near to it, the positive it repels and drives to the farthest side; so that the disturbed condition of the body *b* is a result of the fact that *a*, being electrified positively, will repel positive electricity and attract negative.

Fig. 93.

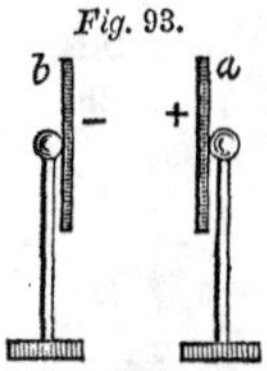

Now let the plate *b* be touched by the finger, or a channel of communication opened with the earth; the positive electricity of *a*, still exerting its repulsive agency on that of *b*, will drive it into the ground, and *b* will now become negative all over.

Let *b* be once more insulated by breaking its communication with the ground, and let *a* be removed; it will now be found that *b* is permanently electrified, and in the opposite condition to *a*.

Fig. 94.

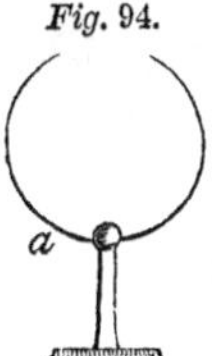

By manipulating in this manner, we can therefore effect a permanent disturbance in the condition of an insulated body by bringing an excited one in its neighborhood.

In these changes the intervention of a piece of glass makes no difference. Let a circular plate of glass, *a*, *Fig.* 94, be set so as to intervene between the metallic plates *a* and *b*,

What is their electrical state respectively? Does the shape of a body influence induction? Describe *Fig.* 93, and its method of action. How may the plate *b* be permanently electrified?

and still all the phenomena occur as before. Electrical induction, therefore, can take place through glass.

On the principles of induction, and of electrical attraction and repulsion, many interesting experiments may be explained. The following may serve as examples: To the ball of the prime conductor let there be suspended a circular plate of brass, *a*, *Fig.* 95, six inches in diameter; horizontally and beneath it let there be another plate, *b*, supported on a conducting foot parallel and at a distance of three or four inches. On the lower plate *b* place slips of paper or other light substance, cut into the figure of men or animals. On setting the machine in motion so as to electrify the upper plate, the objects move up and down with a dancing motion. The cause is obvious: the plate *a*, being positive, repels by induction the positive electricity of the figures through the conducting stand into the earth, and they thus, being rendered negative, are attracted by the upper plate. On touching it they become electrified positively like it, and then are repelled, and fall down to discharge their electricity into the ground, and this motion is continually repeated.

Fig. 95.

Upon a horizontal brass bar, *a b*, *Fig.* 96, three bells are suspended, the outer ones at *a* and *b* by chains, the middle one at *c* by a silk thread. Between the bells the metallic clappers *d* and *e* are suspended by silk, and from the centre bell the chain *f* extends to the table. On hanging the arrangement, by the hook at *g*, to the excited prime conductor, the bells ring, the clappers moving from the outer to the central bell and back, alternately striking them.

Fig. 96.

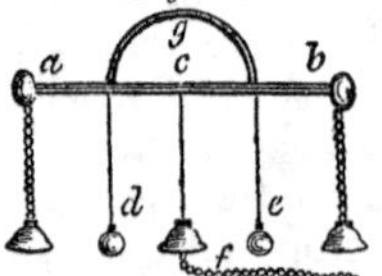

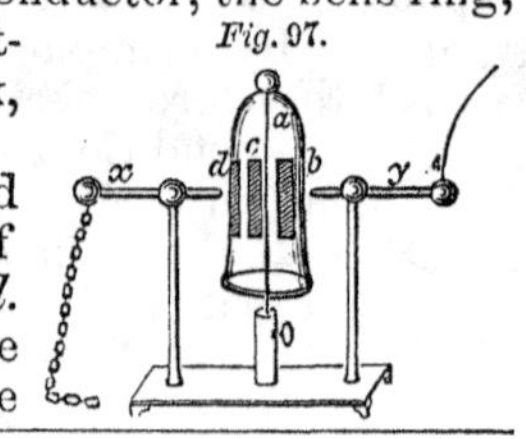

Fig. 97.

On a pivot, *a*, *Fig.* 97, suspend a bell jar having four pieces of tin-foil pasted on its sides *b c d.* Connect the jar, by means of the insulated wire *y*, with the prime

What effect has the intervention of glass? Describe the dancing figures experiment. What is the cause of the motion? Describe *Fig.* 96. Describe the rotating jar, *Fig.* 97

conductor, so that the pieces of tin-foil may receive sparks. On the opposite side arrange a conductor, x, in connection with the ground by a chain. On putting the machine into activity, the jar will commence rotating on its pivot.

Take a cake of sealing-wax or shellac eight or ten inches in diameter, electrify it by friction with a piece of flannel, and receive on its surface a few sparks from the prime conductor by bringing it near the ball. Then blow upon its surface, from a small pair of bellows, a mixture of flowers of sulphur and red lead which have been intimately ground together in a mortar. This mixture is of an orange color; but, the moment it impinges on the cake, it is, as it were, decomposed, the yellow sulphur settling on one portion and the red lead on another, giving rise to very curious and fantastical figures, called Lichtenberg's figures.

Soon after electricity became a subject of popular attention, it was currently believed that if medicines of various kinds were sealed up in glass tubes and the tubes electrically excited, their peculiar virtues would be exhaled in such a manner as to impress the patient with their specific purgative, emetic, or other powers. Like many of the popular delusions of our times, this imposture was supported by the most cogent evidence, and maladies cured publicly all over Europe. Like them, these "medicated tubes" have served to prove the worthlessness of human testimony when derived from the prejudiced and ignorant.

It should be remarked that, in their action upon material bodies, electricity and heat differ greatly. The former has no kind of influence in determining magnitude, whereas the size of any object depends on its temperature.

How are Lichtenberg's figures produced? What are medicated tubes? What is the difference between heat and electricity in their action on bodies?

LECTURE XXVII.

Laws of the Distribution of Electricity, and Theories of Electricity.—*Distribution of Electricity.—On a Sphere and Ellipsoid.—Action of Points.—Franklin's Discovery of the Identity of Electricity and Lightning.—Lightning Rods.—The Two Electrical Theories, Franklin's and Dufay's.—Electricity is a Compound Force.—The Leyden Jar.—Discharging Rod.—Electric Battery.*

When electricity is communicated to a conducting body, it does not distribute itself uniformly through the whole mass, but exclusively upon the surface. Thus, if to the spherical ball *a*, *Fig.* 98, supported on an insulating foot, *b*, there be adjusted two hemispherical caps, *c c*, also on insulating handles, it may be proved that any electricity communicated to *a* distributes itself entirely upon the surface; for if we place upon *a* the caps *c c*, and then remove them, it will be found that every trace of electricity has disappeared from *a*, and has accumulated upon the caps, which, while they were upon the ball, formed its superficies.

Fig. 98.

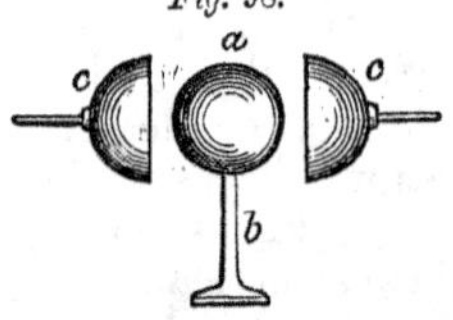

Again, if we take a large brass ball, *a*, *Fig.* 99, supported on an insulating stand, and having on its upper portion an aperture, *b*, through which we may have access to the interior, it will be found, on examination, that the most delicate electrometers can discover no electricity within the ball, the whole of it being on the external superficies.

Fig. 99.

In the case of a spherical body, not only is the distribution entirely superficial, but it is also uniform. Each portion of the sphere is electrified alike. But where, instead of a spherical, we have an

Upon what part of a body does electricity distribute itself? How may this be proved? What does *Fig.* 99 prove? What is the distribution on a sphere?

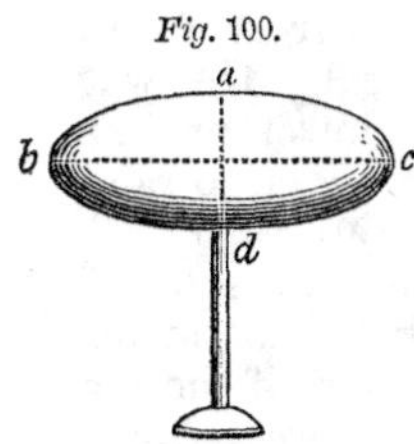

Fig. 100.

ellipsoidal body, it is different. Thus, if we examine the condition of such a conductor, *Fig.* 100, the quantity of electricity in its middle portion, as at *a d*, will be the smallest, and it will increase as we advance toward the ends, *b c*. In differen ellipsoids, as the length become greater, so the amount of electricit found on the extremities is greater. When, therefore a conductor of an oblong spheroidal shape is used, t intensity of the electricity at the extremity of the tw axes, *a d* and *b c*, *Fig.* 100, is exactly in the proporti of the length of those axes themselves; and should th disproportion in the length and breadth of the conducting body be very great, as in the case of a long wire or other pointed body, a very great concentration will take place upon the points. On this principle we explain the effect of pointed bodies on conductors. If the prime conductor of the machine have a needle or pi fixed upon it, the electricity escapes away into the a visibly in a dark room; and in the same way, if pointe bodies surround the electrical machine, it can not b highly excited, as they rapidly take the charge from its conductor.

These principles may be very well illustrated by taking a long strip of tin-foil, so arranged as to be rolled and unrolled upon a glass axis, and connected with a pair of cork balls, the divergence of which shows its electrical condition. If now to this, when coiled up, a sufficient amount of electricity is communicated to make the balls diverge, on pulling out the tin-foil so as to have a larger surface they will collapse, but on winding the foil up again they will again diverge, showing that the distribution of electricity is wholly superficial, and that when a given quantity is spread over a large surface, it necessarily becomes weaker in effect.

At a very early period electricians had observed the close similarity between the phenomena of the electric

What is the distribution on an ellipsoid? What is the distribution on a wire? What is the effect of pointed bodies on conductors? How may the distribution of electricity be shown by a coil of tin-foil?

spark and those of lightning, but in the year 1752 Dr. Franklin proved that they were identical. He was waiting for the erection of the spire of a church in Philadelphia, on the extremity of which he intended to raise a pointed metal rod, with a view of withdrawing the electricity from the clouds, when the accidental sight of a boy's kite suggested to him that ready means of obtaining access to the more elevated regions of the air. Accordingly, having stretched a silk handkerchief over a light wooden cross, and arranged it as a kite, he attached to it a hempen string, terminating in a silk cord, and, taking advantage of a thunder-storm, raised it in the air. For a time no result was obtained; but the string becoming wet with the rain, and thereby rendered a better conductor, he perceived that the filaments which hung upon it repelled one another, and on presenting his knuckle to a key which had been tied to the end of the hempen string, received an electric spark. The identity of electricity and lightning was proved.

Franklin soon made a useful application of his discovery. He proposed to protect buildings from the effects of lightning by furnishing them with a metallic rod, pointed at its upper extremity and projecting some feet above the highest part of the building, and continuously extending downward until it was deeply buried in the ground. This contrivance, the lightning-rod, is now, as is well known, extensively applied.

There are two theories respecting the nature of electricity: 1st. Franklin's theory, which assumes that there is but one fluid; and, 2d. Dufay's theory, which assumes that there are two fluids.

Franklin's theory is that there exists throughout all space a subtle and exceedingly elastic fluid, called the electric fluid, the peculiarity of which is that it is repulsive of its own particles, but attractive of the particles of other matter; that there is a specific quantity of this fluid, which bodies are disposed to assume when in a natural condition or state of equilibrium, and that if we communicate to them more than their natural quantity

---

Describe Franklin's discovery of the identity of lightning and electricity. What use did he make of the discovery? What theories are there of the nature of electricity? What is Franklin's theory?

they become positively electrified, or if we take away a portion of that which is natural to them they become negatively electrified.

Dufay's theory is that there exists throughout all space a universal medium, the immediate properties of which are not known. It is composed of two species of electricity, the positive and the negative, each of these being self-repellant, but attractive of the other kind. Bodies are in a *neutral* or *natural state* or *condition of equilibrium* when they contain equal quantities of the two electricities; they are *positively electrified* when the positive is in excess, and *negatively* when the negative is in excess.

Electricity is, however, now coming to be regarded as a compound force, remarkable for the peculiar form of action and reaction that it exhibits. This kind of action and reaction follows the same law of equality and opposition in its manifestations as that which is shown more obviously in the phenomena of mechanics. Whenever vitreous electricity appears at one point, a corresponding amount of resinous electricity is invariably developed in its vicinity, reacting against it, and thus enabling its presence to be recognized.

As one kind of electricity can never make its appearance alone, but is always accompanied by an equal quantity of the other, we uniformly find that the rubber and the surface rubbed are in opposite states; if the one is positive, the other is negative. It is on this principle that many machines are furnished with means of collecting from the prime conductor or the rubber and therefore of obtaining the positive or negative electricity at pleasure.

Fig. 101.

In 1745 the Leyden jar was discovered. This consists of a glass jar, *Fig.* 101, coated on its inside with a piece of tin-foil within an inch or two of its upper edge, and also on its outside to the same point. Through the cork which closes the mouth of the jar a brass rod, terminated by a ball, passes. The rod reaches down to the inside coating

What is Dufay's theory? What is the present view of the nature of electricity? Can one kind of electricity alone be produced? Describe the Leyden jar.

and touches it. On holding this instrument by the ex- 'erior coating and presenting its ball to the prime con- 'uctor, a torrent of sparks passes into the jar; and when ıt is fully charged, if, still retaining one hand in contact with the outside, we touch the ball, a bright spark pass- 's with a loud snapping noise, and the operator receives hrough his arm and breast what is called the electric hock.

If we take the discharging rod, *Fig.* 102, consisting two brass arms, *a a*, terminated by balls work- on a joint, *b*, and supported by an insulating ıdle, *c*, by bringing one of its balls in contact h the outside coating of a Leyden jar, and its her ball with the ball of the jar, the discharge vill take place as before, but the operator, protected by the glass handle, receives no shock.

Fig. 102.

If between the outside coating of the jar and one of the balls of the discharging rod a piece of ardboard be made to intervene and the spark assed, the card will be found to be perforated, a rr being raised on both sides of it as though two reads had been drawn through the hole in opposite rections at the same time; ıd from this an argument in ıvor of the theory of two fluids as been drawn.

Fig. 103.

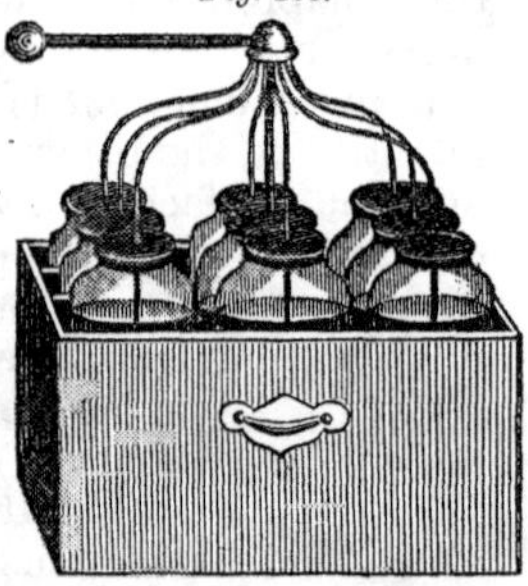

When a great number of jars e connected together, so that ll their inside coatings unite, nd all their outside coatings re also in contact, they constitute what is termed an electric battery, as seen in *Fig.* 103. By this instrument many of the more violent effects of electricity may be illustrated, such as the splitting of pieces of ood, and the ignition and dispersion of metallic wires.

How is it used? Describe the discharging rod? How may a card be perforated by the spark? What is the peculiarity of the perforation? What is an electric battery?

LECTURE XXVIII.

Electrical Instruments and Faraday's Theory oı Electric Polarization.—*Theory of the Leyde Jar.—Electrometers: Quadrant, Gold-leaf, Torsion Peltier's.—Bohnenberger's Electroscope.—Zambon Pile.—Faraday's Theory of Polarization.—Speci Induction.—Methods of Discharge by Conducti Disruption, Convection.—The Brush.—Hydro-e tric Machine.*

The office which is discharged by the metallic co ings of a Leyden jar is illustrated by the apparatı *Fig.* 104. It consists of a conical glass jar, to the interior and exterior of which movable coatings of thick tin plate are adapted, the interior one having a rod and ball projecting from it. Thi may be charged like any other Leyden via' but, on taking off its outside coating and moving its interior, they may be handled a brought into contact with each other, and spark passes; but, on restoring them to the former position, and applying the discharging rod, t. jar is discharged. They therefore only serve to mak a complete conducting communication between all part on the interior and all on the exterior of the jar.

Fig. 104.

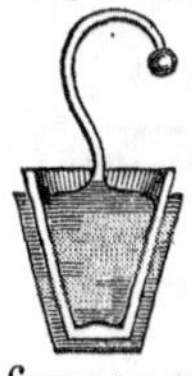

The condensing action of the Leyden vial, which en ables it to hold so large a quantity of electricity, is du to induction. When the inner coating is brought int contact with the prime conductor, it participates in it electrical condition. We may therefore suppose it tc be positively electrified. The positive electricity of the interior repels that of the exterior into the earth, the outside of the vial being in communication with the ground. It therefore appears that the inner coating positive, the outer negative, and the whole jar, taken tc gether, is in the neutral condition. The inner coating, continuing to receive a farther charge from the prime conductor by induction through the glass, continually

What is *Fig.* 104 intended to demonstrate? How is it used? What is the condensing action of the Leyden jar due to?

pels more of the same kind, the positive into the ·ound and the negative accumulates on the outside. this manner an indefinite quantity might be accumu- ·ted, were it not for the fact that, owing to the distance ·hich intervenes between the two coatings by reason the thickness of the glass, the quantity of positive ;ctricity in the interior is never precisely neutralized the quantity of negative on the exterior, for all in- ɔtive actions enfeeble as the distance increases. The eventually refuses to receive any more sparks from machine.

'though, in charging a jar, the interior coating is .monly brought into contact with the prime conduct- yet the charging may be equally well accomplished :he external coating receive the sparks, provided only nat the interior communicates with the ground.

The action of the Leyden vial may be illustrated by ·e following experiments. With- an inch of the ball, *a*, of the me conductor, *Fig.* 105, bring econdary conductor, *b*, sup- ced on an insulated stem, *c*, on putting the electrical ma- ne in activity, two or three ırks will pass from *a* to *b*, but .er that no more. The cause of the refusal on the ırt of the secondary conductor to receive any farther arge is obviously due to the fact that the electricity ıich is already communicated to it repels that upon e ball *a*, and prevents the passage of y more.

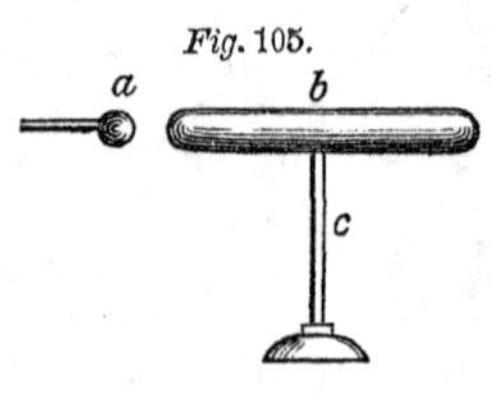

Fig. 105.

Fig. 106.

If now we take a Leyden jar, *b*, *Fig.* 06, and, having insulated it on a stand, oring it within a short distance of the ball, *a*, of the prime conductor, it, in he same manner, will only receive a v sparks. But if we place a conduct- , *c*, which is connected with the round, near the outside coating, it will

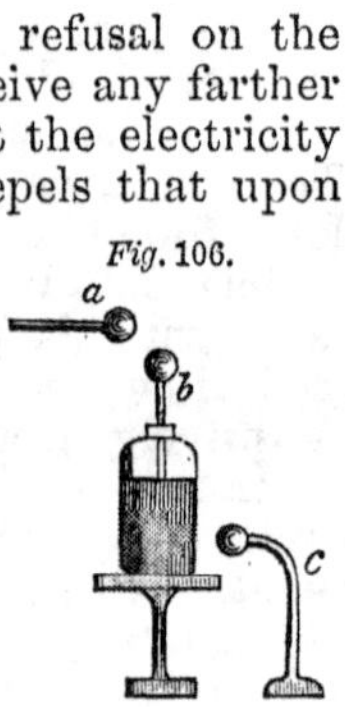

Describe the charging of a Leyden jar. Why is there a limit to the charge? What other method of charging may be used? How does *Fig.* 105 illustrate the action of the Leyden jar? Describe the experiment illustrated by *Fig.* 106.

be found that for every spark that passes between and *b*, one passes between the outside coating and and the sparks follow each other in rapid successi until the jar becomes fully charged. From this, ther fore, we gather that, while positive electricity is passi into the interior of the jar, it is escaping from the ex rior, and that the reason the jar condenses is because sides are in opposite conditions, the positive electric of the interior being nearly neutralized by the negat electricity of the exterior.

Electrometers, or electroscopes, are instruments measuring the intensity of electrical excitement. cork balls which are represented in *Fig.* 92 are one the most simple of these contrivances. The distance which they will diverge is a rough measure the intensity of the electrical force. The quad rant electrometer depends essentially on th same principles. It consists of an upright ste of wood, *Fig.* 107, to which is affixed a semic cular piece of ivory, from the centre of wh there hangs a light ball of cork playing pivot. When this instrument is placed on prime conductor, or other electrified body, stem participates in the electricity, and re the cork ball which hangs in contact with The amount of repulsion may be read off on the gra ated semicircle. But no quantity of electricity can ev drive it beyond 90°; and, indeed, its degrees are n proportional to the quantities of electricity.

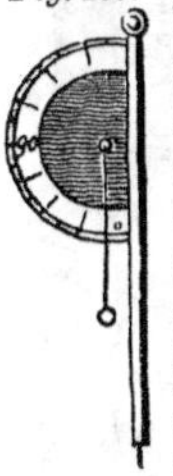

Fig. 107.

Fig. 108.

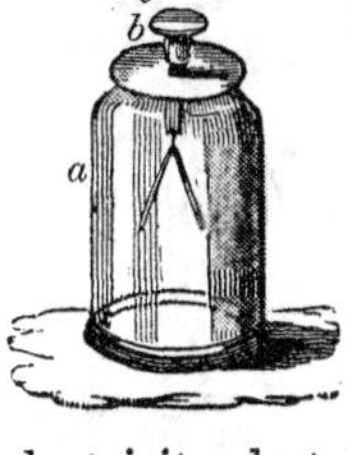

The gold-leaf electrometer, *Fig.* 10 consists of a glass cylinder, *a*, in wh two gold leaves are suspended fron conducting rod, terminated by a ba or plate, *b*. On the glass opposite the leaves pieces of tin-foil are pasted, so that when the leaves diverge fully th may discharge their electricity into t ground. This is a very delicate inst ment for discovering the presence electricity, but the torsion electrometer of Coulomb i

What is an electrometer? What is the most simple of these contrivances? Describe the quadrant electrometer. Describe the gold-leaf electrometer.

› be preferred when it is required to have exact meas-.es of that quantity.

Coulomb's electrometer consists of a glass cylinder, A, *Fig.* 109, upon the top of which ere is fixed a tube, B, in the axis which hangs a glass thread, *b*, the lower end of which a small of shellac, *b d*, with a gilt pith l at one extremity, is fastened. rough an aperture in the top of glass cylinder another shellac *e*, with a gilt ball, may be in-duced. This goes under the ne of the *carrier*.

Fig. 109.

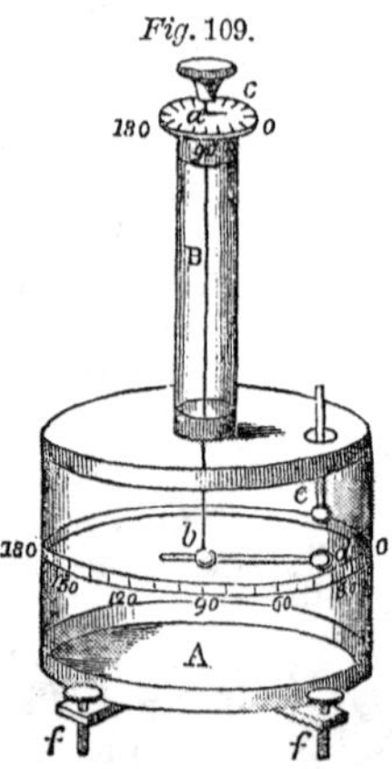

If now the lower ball of the car-ier be charged with the electricity be measured, and introduced to the interior of the cylinder, as en in the figure, it will repel the able ball. By taking hold of button, *a*, to which the upper end of the glass ead, *b*, is attached, we may, by twisting the glass ead, forcibly bring the carrier ball and movable ball o contact. The number of degrees through which thread requires to be twisted represents the amount electricity. To the button, *a*, an index and scale, *c*, e attached. By this we can tell the number of de-ees of twist or torsion which have been given to the read. These angles of torsion are exactly proportion-to the quantities of electricity.

Another very convenient electrometer as invented by Peltier, in which the di-ective force of a small magnet is substi-tuted for the torsion of a glass thread.

One of the most delicate electroscopes that of Bohnenberger. It consists of small Zamboni's pile, *a b*, *Fig.* 110, sup-orted horizontally beneath a glass shade, nd from its extremities, *a b*, curved vires pass, which terminate in parallel

Fig. 110.

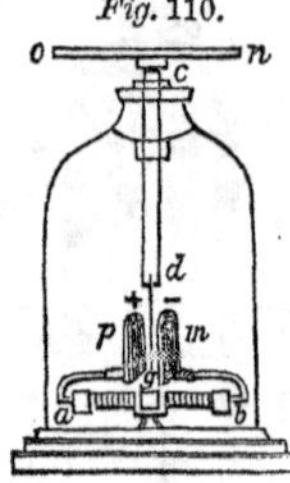

Describe Coulomb's electrometer. How is Coulomb's instrument used? Describe Peltier's electrometer. How is Bohnenberger's electroscope arranged?

plates, *p m*. One of these is therefore the positive, ar the other the negative pole of the pile. Between the there hangs a gold leaf, *d g*, which is in metallic con munication with the plate *o n* by means of the rod *c* If the leaf hangs equally between the two plates, it equally attracted by each, and remains motionless; b on communicating a trace of electricity to the plate *o* the gold leaf instantly moves toward the plate wh has the opposite polarity.

Zamboni's electrical piles are made by pasting zi foil on one side of a sheet of paper and coating the o with finely-powdered peroxide of manganese and ho and then punching out a number of circular pieces h an inch in diameter. If several thousands of these packed together in a glass tube so that the zinc faces look in one direction and the manganese in the other and be pressed tightly together by metallic plates the ends, it will be found that one extremity of the pi is positive and the other negative. With a dry pile 20,000 pairs sparks can be obtai and a Leyden vial charged sufficie to give shocks. If the pile be drie loses activity, but otherwise it will c tinue to work for years. *Fig.* 111 r resents a pair of these piles arranged as to produce what was at one time r garded as a perpetual motion. Tv piles, *a b*, are placed in such a positi that their poles are reversed, and b tween them a ring or light ball, *c*, brates like a pendulum on an axis, It is alternately attracted to the one and then to t other, and will continue its movements for years. It covered by a glass shade.

Fig. 111.

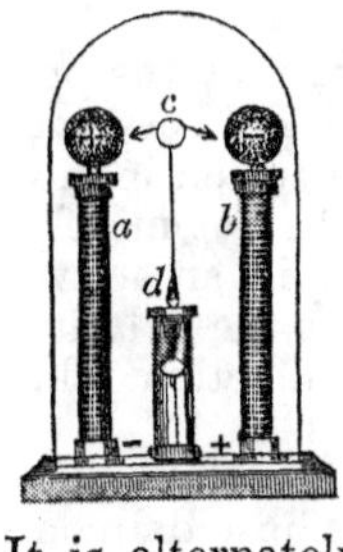

Many of the fundamental phenomena of electric ity have been explained by Faraday upon the hypothe that induction is an action of polarization taking pla in the contiguous molecules of non-conducting medi and propagated in curved lines. Bodies susceptible of this polarization are termed *dielectrics*, for they allow

---

How are Zamboni's piles made? Describe *Fig.* 111. What is Faraday's explanation of induction? What are dielectrics?

ectric power to traverse them, but by a process differ-
'g from conduction. As examples, air, resin, glass, and
ılphur may be mentioned.

Whatever may be the form or constitution of bodies, n electric charge can not be given to them without at ːe same time giving a charge of the opposite kind, but f the same amount, to them or other bodies in their vi-ᵔity. This charge is not confined upon their surfaces the pressure of the atmosphere, but through the po-ation of the aerial or solid particles of the surround- dielectrics, producing in them a charge of the same ..unt, but of an opposite kind. Thus, if a positively ectrified ball be placed in the centre of a hollow me-.lic sphere, the intervening space being filled with at-.ospheric air, the charge is not retained upon the ball by the pressure of the air, but because each aerial parti-le assumes by induction a polarity of the opposite kind ı the side nearest the ball, and of the same kind on the le farthest off. This state of force is therefore com-ıicated to the interior of the hollow sphere, which is ɔtrified to the same amount, but of an opposite kind the ball.

That this polarization takes place is shown by the po-ion which small silk fibres or spangles of gold assume nen placed in oil of turpentine, through which induc-ıon is established. Each particle disturbs not merely ıat which is before or behind it, but it is in an active lation with all surrounding it; and hence the polarity ın be propagated in curved lines, and induction take 'ace around corners and behind obstacles.

On these principles we can easily account for the dis-.bution of electricity on spherical or ellipsoidal con-ctors, the repulsion of bodies similarly electrified, the ıdensing action of the Leyden vial, and many other ıilar phenomena.

By a variety of experiments, Faraday has proved that .ductive action takes place in curved lines, the direc-.ons of which can be varied by the approach of bodies. He has also shown that the particles of solids, as shellac,

---

Mention some examples. What occurs when a body receives an electric charge? Describe the action of an electrical ball in the interior of a sphere. How do we know that polarization takes place? Does induction take place only in straight lines?

glass, etc., assume this character of polarity. Non-co ducting bodies, dielectrics, through which the action o induction takes place, have each a *specific inductive ca pacity*. Thus, if three metallic plates, *a b c*, *Fig.* 112, be insulated parallel to each other, atmos pheric air intervening between *a* and and a plate of shellac between *b* and the shellac will be found to allow ind tion to take place across it twice as rea ly as air. The following table exhi the specific inductive capacity of var bodies:

Fig. 112.

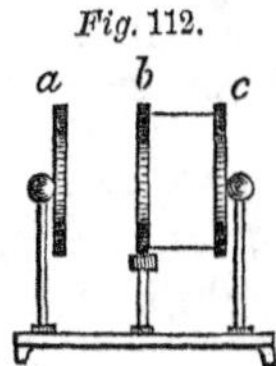

Specific Induction.

Air	1.00	Glass	1.90
Resin	1.77	Sulphur	1.93
Pitch	1.80	Shellac	1.95
Beeswax	1.86		

Fig. 113.

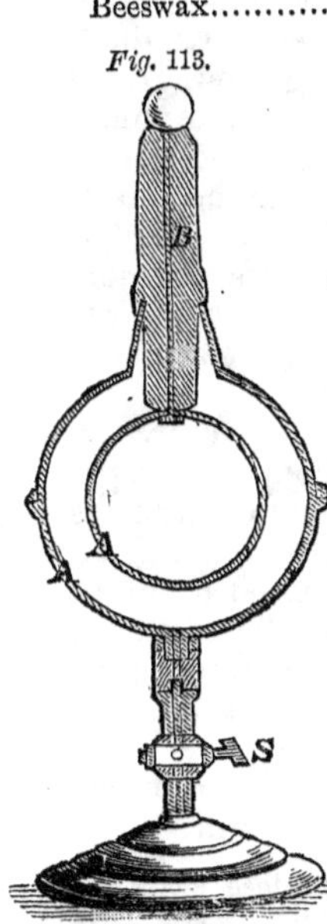

All gases have the same indu tive capacity, whatever their d sity, elasticity, temperature, or grometric condition may be. Fa day's apparatus for this investi tion was a kind of Leyden v *Fig.* 113, consisting of two met lic spheres, *A A*, insulated fro each other by a stem of shellac, *B* The interval between the two coul be filled with any gaseous mediu through the stopcock *S*. Two o the jars were used, one containi air as a standard. When the ot jar was charged, the charge was vided with that containing air, a the relative intensity measured an electrometer.

After the electric equilibrium a body has been disturbed, it ma be restored by conduction, disruption, or convection. An example

What is meant by specific inductive capacity? Give the inductive capacities of air, resin, etc. What is the case with gaseous bodies? Describe Faraday's apparatus.

the first and second methods of discharge is seen in .e use of a discharging-rod applied to a Leyden vial, .e electricity passing quietly along the metallic con- uctor a part of the way, but bursting through the in- erval of air by disruption. Heat is evolved by the ssage of electricity along a conductor, the amount in- easing as the resistance from imperfect conduction in- eases.

Development of Heat by Electricity.

Copper	6	Iron	30
Silver	6	Tin	36
Gold	9	Lead	72
Zinc	18	Brass	18
Platinum	30	Tin 1, Lead 1	54

The amount of heat increases as the square of the quantity transmitted in equal times. If the size of the onductor be sufficiently reduced, it may be deflagrated. a disruptive discharge particles of the solid conduct- are torn off by the spark, and, being ignited, give a to the light. In this way the spectra of many met- have been observed. Not only light, but also heat is oduced in this method of discharge, as may be shown pasting a slip of tin-foil on glass, and, having cut it oss in two or three places, laying wafers thereon. n passing a discharge, the wafers will be thrown off y the expansion produced in the air by heat at those oints.

The velocity of movement of the spark greatly ex- eds that of light, passing through a copper wire at e rate of 288,000 miles in a second. Its duration is mated at less than the millionth of a second. Trees tated violently by the wind, if illuminated at night lightning, seem to be perfectly at rest. The distance ough which disruptive discharge will take place va- with different media, and with their rarefaction. A rk will pass through several inches of flame.

A dilute spark, or *brush*, occurs when a discharge kes place between a good conductor of limited sur-

How is electric equilibrium restored? When does electricity roduce heat? Is the production the same for all metals? Give examples. What is the law of increase of the heat? What occurs during disruptive discharge? State the velocity of transmission of electricity. What is the *brush*?

face and a bad one of larger surface; as, for exampl when a blunt rod discharges into the air. The brush larger from a surface charged vitreously than resinousl

Discharge by convection is where the charge is fe bler, and the brush is replaced by a tranquil glow.

Electricity may be produced in large amount by tl hydro-electric machine, which consists of an insulate boiler, from which steam can escape through long tube The boiler becomes negative, the escaping steam be positive. The smallest quantity of oil or turpen however, reverses these electrical states.

The electrophorus is an instrument which depend its action on induction, and is of frequent use in che istry. It consists of a cake of shellac or sealing-wax *Fig.* 114, on which is placed a flat m tallic plate, *a*, with an insulating han dle, *c*. On exciting *b* with a piece of warm flannel, it becomes negative electric; and *a* being placed on it a a finger brought near, a negative sp driven from *a* by the repulsive influe of *b*, is received. On lifting *a* by its insulating hand a positive spark is obtained; on putting it down on a negative one; and in this manner we may obtain unlimited number of sparks—positive ones when *a* lifted, and negative ones when it is down. A little re flection will show that none of this electricity com from the excited cake *b*, but is merely the effect of i inductive influence on the electric condition of the m tallic plate *a*. The electrophorus may be used wh the weather is too damp for the common machine work.

Fig. 114.

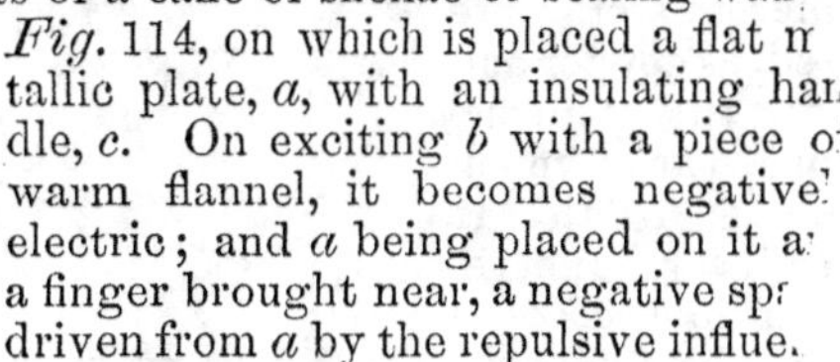

What is the hydro-electric machine, and how does it act? scribe the electrophorus, and its method of action.

LECTURE XXIX.

VOLTAIC ELECTRICITY.— *Galvani's Discovery.— The Simple Voltaic Circle and its Properties.—Direction of the Current.—Different Kinds of Combinations. — Use of the Sulphuric Acid.— Cause of the Voltaic Current.*

N 1790 Galvani observed that if metallic communi-cion is made between the muscles and nerves of a re-ntly killed frog, convulsive movements occur. If two ifferent metals are ised, as copper and zinc, the contractions are much nore energetic. If ie crural nerve, , *Fig.* 115, of a og be exposed nd connected with piece of zinc, *Z*, hile the muscles of the thigh, *m*, are touched with a opper wire, *C*, nothing occurs as long as the metals are kept apart; but, as soon as they are brought into con-tact, a convulsive movement ensues, and the same is re-eated as often as the contact is made. These phenom-na at first went under the name of animal electricity.

Fig. 115.

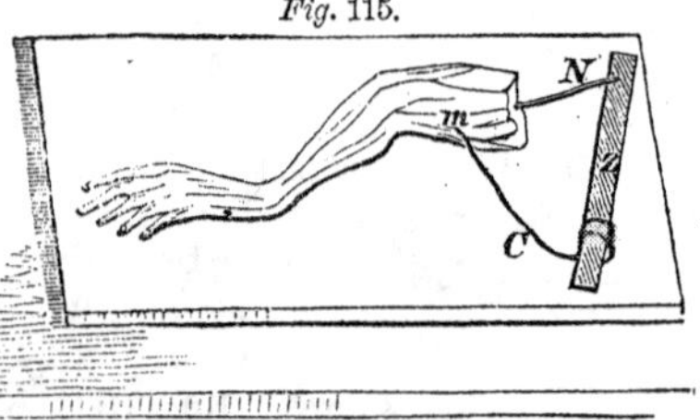

If a piece of zinc and one of silver be placed on op-site sides of the tongue and the overhanging edges ought into contact, a metallic taste is perceived in e mouth. If the silver be between the upper lip and th while the zinc is on the tongue, when the metals uch a bright flash is seen.

The branch of electrical science that arose from such bservations is called Galvanism or Voltaic electricity, and its phenomena are those of electricity in motion. Static electricity, which we have previously considered, exhibits that force in a motionless state.

What fact did Galvani discover? Describe the experiment *Fig.* 115. What electrical experiments may be performed in the mouth? What is the difference between static and voltaic electricity?

It is to be admitted, though of that abundant proof will soon be given, that water is not a simple but a compound body; that it consists of two elements, oxygen and hydrogen. It is also to be understood that metallic zinc may be amalgamated or united with quicksilver, by putting it in contact with that fluid metal under the surface of dilute sulphuric acid. Strips of zinc thus amalgamated exhibit a pure metallic brilliancy.

A simple voltaic circle may consist of a strip of ama gamated zinc, *z*, *Fig.* 116, an inch wide a three or four inches long, and a similar c per strip, *c*, immersed in a vessel of water, slightly acidulated with sulphuric acid. Whi the copper and zinc are kept separated ther is no action; but if we take a metallic rod, *d*, and connect the two together, a series of phenomena arise.

Fig. 116.

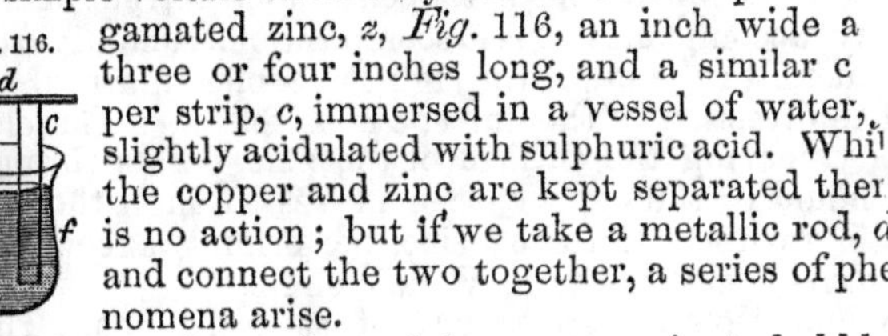

First, from the surface of the copper minute bubbles of hydrogen gas are evolved.

Secondly, the plate of zinc rapidly wastes away, a on examining the liquid in the cup we discover th cause of this waste, for it contains oxide of zinc. Coupling this fact with the former, we infer that, so long a the metallic rod *d* is in place, water is decomposed, it oxygen uniting with the zinc, its hydrogen escaping from the copper. On removing the rod *d*, all these phenomena at once cease.

Thirdly, if, instead of a metallic rod, *d*, a rod of glass or other non-conductor of electricity be employed, no decomposition occurs. This, therefore, indicates th the agent in operation is electricity.

Fourthly, if for the line of communication a piece metal be employed, and we cautiously lift it from th zinc or copper plate, the moment the contact is brok in a dark room we see a minute spark. It has alread been observed that the electric spark can not be co founded with any other natural phenomenon.

Fifthly, if the line of communication be a very slen-

Is water a simple or a compound body? What is amalgamated zinc? Describe a simple Voltaic circle. What occurs on connecting the plates? What gas rises from the copper? What happens to the zinc? What occurs on using a glass rod? How may a spark be produced?

der platinum wire, as long as it remains in position its temperature rises so high that it becomes red-hot, and may remain so for hours. Now, recollecting that the ignition and fusion of metals take place when they are made to intervene between the coatings of a Leyden vial, and considering all the facts which have just been set forth, we see that the following conclusion may be rawn: that in an active simple Voltaic circle water is ecomposed, its oxygen going to the zinc and its hydro- to the copper, and that a continuous current of elec- ity accompanies this decomposition, running from metal to the other through the connecting rod.

The direction of the current is as follows: The elec- icity, leaving the surface of the zinc, passes through ne liquid to the copper, then moves through the con- necting wire back to the zinc, performing a complete circuit. Hence the term Voltaic circle.

Simple Voltaic circles are of several kinds. That which we have been considering consists of two differ- t metals, with one intervening liquid, but similar re- lts can be obtained with one piece of metal and two fferent liquids.

In the foregoing experiment we have used dilute sul- huric acid; this acid discharges a subsidiary duty. inc, when it oxidizes, is covered with a coating imper- meable to water and air. It is this grayish oxide which protects the common sheet zinc of commerce from far- her change. When, therefore, a Voltaic pair gives ise to a current by the oxidation of its zinc, that cur- nt would speedily stop were not the oxide removed fast as it forms. This is done by the sulphuric acid, ich forms with it a sulphate of zinc, a substance very luble in water, and the metal thus continually presents lear surface to the water.

As to the immediate cause which gives rise to the oltaic current, there has been a difference of opinion nong chemical authors. Volta believed that the mere ontact of the metals was the electromotive source, and

---

How may ignition be produced? What conclusion may be drawn from these observations? What is the course of the current? What other kinds of Voltaic circles are there? What is the use of sulphuric acid in these combinations? What did Volta think was the cause of the current?

endeavored to prove by direct experiment that if a piece of copper and one of zinc are brought in contact and then separated, they become excited, the one positively and the other negatively. Upon these principles he was led to the discovery of the Voltaic battery, an instrument which has revolutionized chemistry. But many facts have now indisputably shown that the origin of the current is to be sought in the chemical changes going on, and that the energy of the current is proportionate to the chemical activity. In the instance we have had under consideration it is due to the decomposition of water. The direction of the current is, that the positive electricity, starting from the more oxidizable metal, traverses the liquid toward the less oxidizable metal, and returns to its point of origin along the wire joining the two.

LECTURE XXX.

FORMS OF THE VOLTAIC BATTERY.—*The Voltaic Pile. —Crown of Cups.—Cruickshank's Battery.—Object of Amalgamation.—Daniell's Battery.—Smee's Battery.—Grove's Battery.—Bunsen's Battery.—Voltaic Effects: the Spark, Deflagration of Metals, Ignition of Wires, Arc of Flame.—The Electric Light.—Fusion of Metals.—Decomposition of Water.—Oxygen and Hydrogen evolved.*

THE Voltaic pile used by Volta consisted of a number of pairs of dissimilar metals, as zinc, *Z*, and copper, *C*, *Fig.* 117, separated from one another by pieces of cardboard or flannel, *F*, moistened with acidulated water. In arranging such a pile a regular order must be observed. If a piece of zinc is at the bottom, it must be succeeded by a piece of flannel, then a piece of copper, then zinc again, and so on, the pile terminating on top by a piece of copper.

If the ends of the pile, or wires connected with them, be touched with the moistened hands, a shock is at once

---

What is the view now taken of the electromotive source? What is the direction of the current? Describe the Voltaic pile. What are its effects?

Fig. 117.

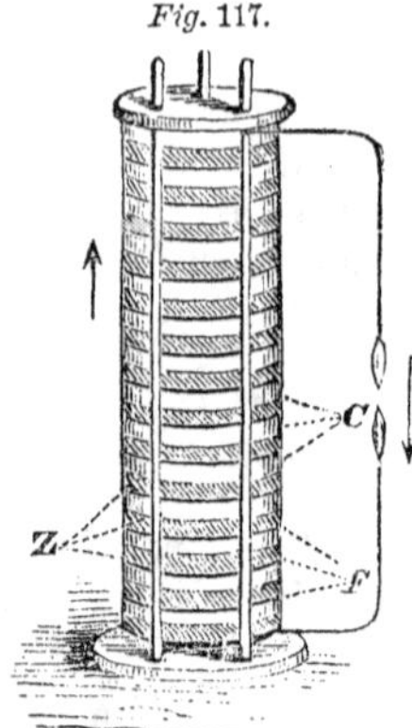

eceived, and, on bringing the vires close together, a spark asses.

There are several inconvenien- ces attending the original con- truction: it is liable to be over- set, is troublesome to put in ac- on, and requires to be taken to eces and thoroughly cleaned ev- time that it is used. Its max- m effect lasts but a short time, ng to the weight of the su- incumbent column pressing out e moisture from the lower pieces cloth, and as soon as they be- ome dry all action ceases.

Volta used another form, which he called a crown of ups, *Fig.* 118, to avoid these difficulties. In it the

Fig. 118.

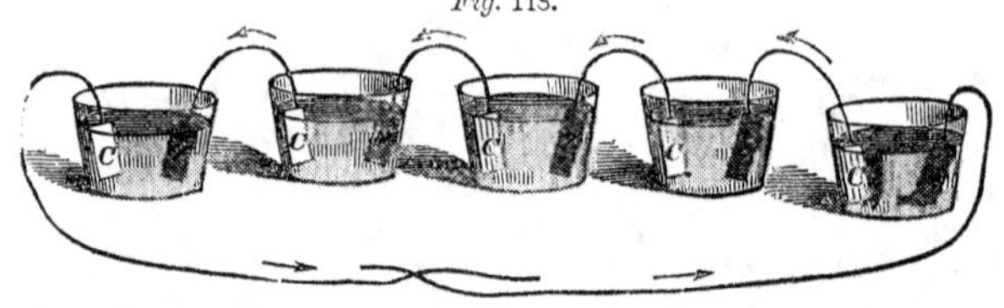

noist flannel is replaced by liquid contained in glass essels, the copper, C, and zinc, Z, being connected by vires soldered to them.

Cruickshank's battery consists of a box or trough, g. 119, three or r inches square the ends, and a t or more long. ooves are made the sides and bot- m of this box, and

Fig. 119.

nto them pieces of zinc and copper, soldered face to face, are fastened water-tight by cement. These grooves are about half an inch apart, and into their interstices acidulated water is poured, care being taken

What are its inconveniences? Describe the crown of cups. Describe Cruickshank's battery.

that the metals are arranged in the same direction, s that if the series begins with a copper plate it ends wit a zinc. The apparatus is obviously equivalent to Vo ta's pile laid on its side, and the facility for charging it and removing the acid when the experiments are over. is very great. From the two extremities flexible co per wires pass; they are called the polar wires, or elec trodes of the battery.

The object of amalgamating the zinc in Voltaic b teries is to prevent what is termed local action, a w in which much metal is consumed, without adding the power of the current, and which likewise dete rates the acid liquid by the accumulation of sulphate zinc. When amalgamated, all the zinc consumed ai in the current.

When it is required to have a current, the intensity of which remains constant for a length of time, Daniell's battery is to be preferred. I consists of a copper cylinder, C, *Fig.* 120, which a solution of acid sulphate of copp is poured. This solution is kept saturat with the salt by means of crystals of s phate of copper resting on the shelf abo P. The interior cylinder P A is filled wit dilute sulphuric acid, and an amalgamate rod of zinc, Z, dips into it. From the cop per and zinc rods project, terminated by binding screws, with which the polar wire may be connected. Twenty or thirty cell of this description furnish a combination of great powe

Fig. 120.

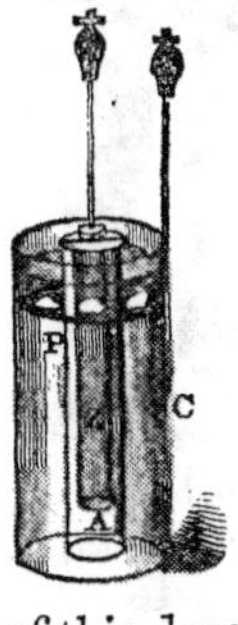

A convenient and inexpensive form of Daniell's b tery is shown in *Fig.* 121. The sulphate of copper contained in glass jars, and the copper cup *C*, *Fig.* 12 is replaced by a sheet of copper, *A*, bent into a cyl drical form. The shelf for the crystals is supported little pieces of copper, turned in as at *d d*. The slip serves to connect the copper with the zinc of the ne cell. At *B C* the method of arranging the cells to gether is seen.

Smee's battery is also a very valuable combination. It consists of a plate of platinized silver, or platinized

What is the object of amalgamating the zinc? Describe Daniell's battery. Describe *Fig.* 121.

Fig. 121.

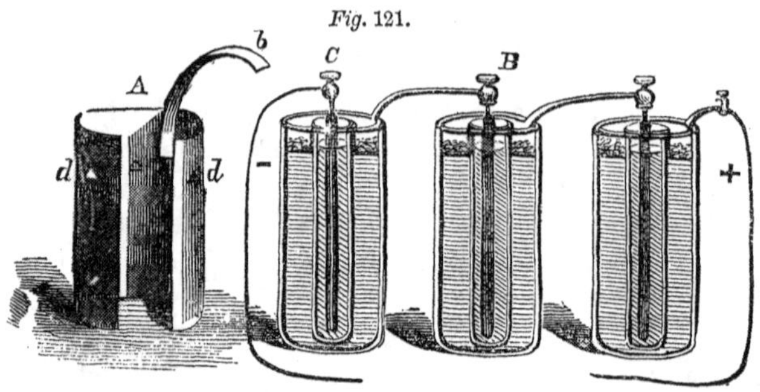

Fig. 122.

;inum, S, *Fig.* 122, on each side of :ich are placed parallel plates of ..nalgamated zinc, Z Z. These plates re held tightly against a piece of wood, V, by means of a clamp, *b*, to which, nd also to the silver plate, binding 'ews, for the purpose of fastening 'ar wires, are affixed. The whole is pended by means of a cross-piece of od in a jar containing dilute sulphu-. acid.

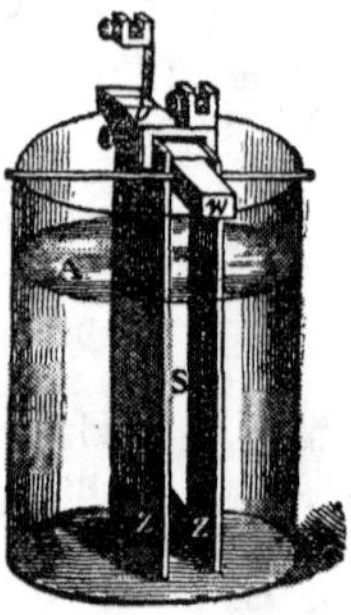

The object of platinizing the silver ate is to facilitate the extrication of ydrogen from it by furnishing a rough instead of a nooth surface. The adhesion of hydrogen to the sil-: plate enfeebles the Voltaic action, owing to the tend-cy of the separated components of the liquid to re-nite.

Smee's compound battery, represented in *Fig.* 123,

Fig. 123.

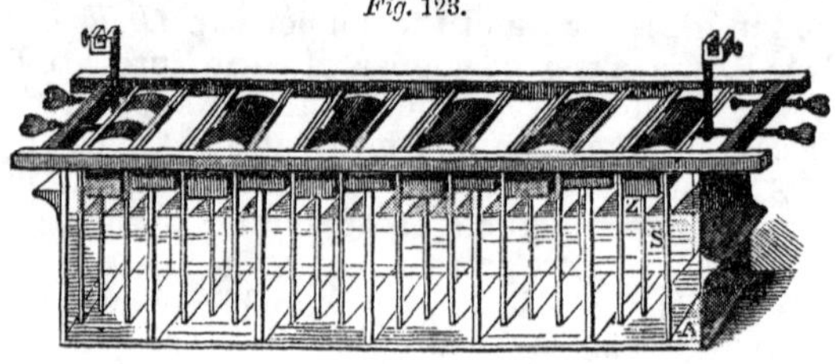

How is Smee's battery constructed? Why is the silver plate platinized?

is a series of the foregoing simple circles. The figure shows one containing six cells. The position of the platinized silver and zinc plates is seen at S and Z.

Fig. 124.

Probably the most powerful of all Voltaic combination's is Grove's nitric acid battery. It consists of two metals and two liquids, amalgamated zinc and platinum, dilute sulphuric acid and strong nitric acid. A jar, P, *Fig.* 12 three quarters of an inch in diameter, ar made of porous or unglazed earthenware, is be filled with strong nitric acid, N, and in slip of platinum is placed. This porous ea enware cup is then set in a glass cup, A, th or four inches in diameter, and is surround by a cylinder of zinc, Z, one quarter of an inc in thickness, and of such a size that it will readi ly pass between the porous cup P and the glass In the glass is placed dilute sulphuric acid.

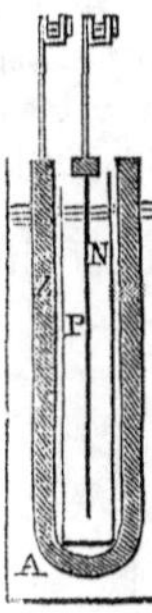

In this manner several cups are to be provide the arrangement being zinc in contact with dilute s phuric acid and pl num, in contact wi strong nitric acid, wi a porous cup interve ing between. The zin cylinder of one ce is connected with th platinum of the ne by soldering. *Fi* 125 represents a bat tery of six cups arranged for action. P is the posi and N the negative pole.

Fig. 125.

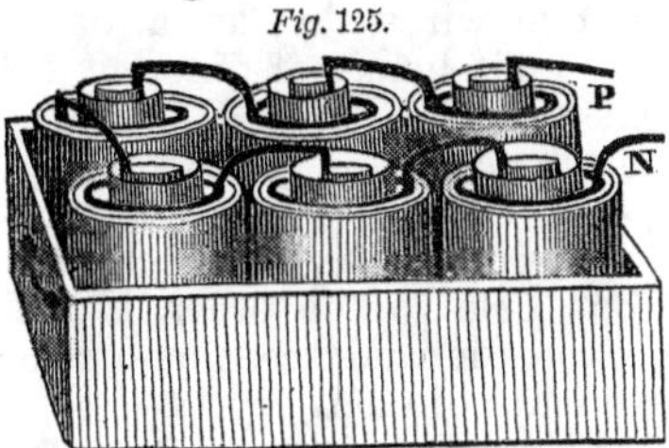

Grove's battery owes its force to the decompositi of water by zinc. But the hydrogen is not evol from the surface of the platinum as it would be in a s ple circle; it is here taken up by the nitric acid, wh undergoes rapid deoxidation, and therefore, during use of this battery, volumes of deutoxide of nitrog are evolved. A battery of fifty cups gives rise to ver striking effects, but five or ten are quite sufficient to re peat most of the following experiments.

What is the most powerful of Voltaic combinations? Describe Grove's battery. What is the source of power in Grove's battery?

In Bunsen's battery cylinders of carbon are used in-tead of the platinum, the best variety being that ob-ained from gas retorts. Another modification of this battery is that in which bichromate of potassa is substi-uted for the nitric acid, and the noxious fumes avoided.

On separating the polar wires of such batteries from ach other a brilliant spark passes, and if the separation gradual, a flame constantly proceeds from one to the her, the light of which, when the wires are of copper, of a beautiful green color.

f on the surface of some quicksilver contained in a ss, *Fig.* 126, we lower a thin piece steel or iron wire, connected with e of the poles of the battery, the ercury being kept in contact with the other, the steel takes fire and de-lagrates beautifully, emitting bright sparks, and the mercury is rapidly vol-ilized.

Fig. 126.

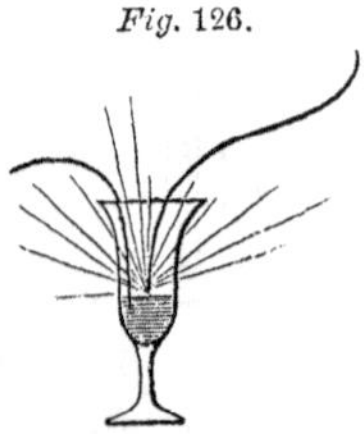

When thin metal leaves are made to tervene between the polar wires, they are at once dis-pated, the flames they emit being of different colors the case of different metals. With a battery of a rge number of cups a file may be in the same way de-agrated.

If a piece of platinum wire be made the channel of ommunication from one pole to the other, if it does ot fuse at once, it becomes incandescent, and remains o as long as the instrument is in activity.

When the polar wires are terminated by pieces of ve ll-burned charcoal or gas carbon, the light which sses between them when they are removed from con-ct is one of the most brilliant that can be obtained by artificial means. With powerful batteries the pieces charcoal may be separated several inches apart with-t the light ceasing, and then it moves from one pole the other in an arched form, *Fig.* 127. Six hundred cells of Bunsen's construction will give an arc nearly

What is the peculiarity of Bunsen's battery? What occurs on separating the polar wires? Describe the experiment *Fig.* 126. How may metal leaves be deflagrated? What is seen on making a platinum wire intervene in the circuit? How is the Voltaic arc formed, and what are its properties?

Fig. 127.

eight inches long, wher the points are in a verti cal position with the neg ative pole below. Witl one hundred pairs th arc is only one inch long. The most intense light i obtained when there but a slight separatic because, the resistance being less, more electricity pa: s in a given time, and the temperature is higher. The s particles of which the poles consist are continually be transported across the interval separating them, a cav being produced in the positive pole, which is in conn tion with the platinum, and a deposit upon the negativ which is in connection with the zinc. The flame ma be blown out by the breath as we blow out a candle.

This light has been utilized for illuminating purposes being employed in some lighthouses, and more exte sively in lecture-table experiments. But the great c and practical difficulties of its application have hithe prevented its general use.

The most refractory metals, which resist the acti of furnaces altogether, may be fused with ease if plac in an excavation in the carbon connected with the pl tinum, or positive pole of the battery. The negativ pole does not attain nearly as great a heat, except whe the secondary current of a Ruhmkorff coil is used.

But, in a scientific point of view, by far the most i teresting experiment to be made with the Voltaic bat tery is the decomposition of water. Through the bot tom of a glass vase or dish, at the point *a b*, *Fig.* 1 two platinum wires are introduc water-tight: they pass into the v as at *a c*, *b d*, parallel to each ot but not touching. Over each of th wires a tube is to be inserted, tube *e* over *c*, and *f* over *d;* the vas and tubes being previously filled with water, acidulated slightly to improv its conducting power. Now let the

Fig. 128.

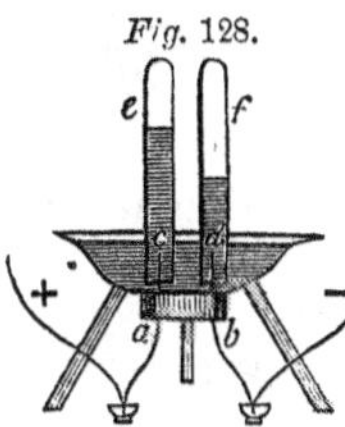

What use is made of this light? How may metals be melted by the Voltaic pile? Describe the decomposition of water.

wire *a c* be connected with the positive pole of the Voltaic battery and *b d* with the negative, bubbles of gas arise in a torrent from their extremities and pass upward in the tubes, displacing the water. The quantity of gas thus collecting in the two tubes is unequal, and whenever we stop the decomposition, there will be found in *f* double the quantity which is in *e*. When a sufficient amount is collected, let the tube containing the smaller portion of gas be cautiously removed, preventing any atmospheric air from getting into its interior by closing it with the finger, and then, turning the tube upside down, let a stick of wood, with a spark of fire on its extremity, be immersed in the gas. In a moment the wood bursts into a flame, proving that this is oxygen gas. Then take the other tube, and allow to pass into it a quantity of atmospheric air equal to the volume of gas it already holds; remove the finger and apply a light, there is an explosion. This shows the gas to be hydrogen. We therefore conclude that in this experiment water has been decomposed and resolved into its constituent ingredients, oxygen and hydrogen; and farther, that in water there is by volume twice as much hydrogen as there is oxygen gas.

LECTURE XXXI.

THE ELECTRO-CHEMICAL THEORY.—*Theory of the Decomposition of Water.—Decomposition of Salts.—Davy's Discovery of Potassium, Sodium, etc.—The Electro-Chemical Theory.—Electrolytes.—Faraday's Law.—Specific Electricity.—Electrotyping and Galvanizing.—Comparison of Frictional and Voltaic Electricity.—The Voltameter.—Different Species of Battery.*

THE prominent fact connected with the decomposition of water is the total separation of the constituent elements on the opposite polar wires or electrodes. From the positive wire oxygen alone escapes, and from

How is it proved that it is formed of oxygen and hydrogen? What are the proportions of oxygen and hydrogen in water? What is the prominent fact in the decomposition of water?

the negative, hydrogen. There is no partial admixture, but the separation is perfect and complete.

Though the polar wires be separated from one another by a considerable distance, the same result is uniformly obtained; and it is to be remarked that the evolution of gas takes place on the wires alone, no bubbles making their appearance in the intervening space. The principle on which this is effected may be easily unde stood by supposing H H and O O, *Fig.* 129, to repr sent atoms of hydrogen and oxyg respectively; each pair of them, the fore, represents a particle of wat Now, if we slide the upper row o atoms upon the lower, as shown *h h*, *o o*, it is obvious that a hydrogen atom will be set free at one extremity of the line, and an oxygen atom at the other; and that, as respects all the intermediate pairs of atoms, though they have changed their places, yet every atom of hydrogen is still assoc ated with an atom of oxygen, constituting, therefore, particle of water, and it is at the extremities of the lin alone that the gases are set free. So in the polar de composition by the pile, all the liquid intervening be tween the poles is affected, decompositions and recom binations successively taking place, the hydrogen atoms moving in one direction, the oxygen in the other, finally to be set free on the surface of the polar wires.

Fig. 129.

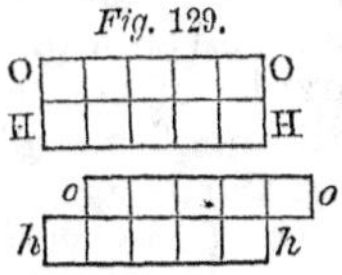

This capital discovery of the decomposition of water by Voltaic electricity was originally made by Nicholson and Carlyle. It is by far the most satisfactory method of demonstrating the constitution of that liquid. After it was made known, any lingering doubts which sti remained on the minds of some chemists, in relation to the composite nature of water, were speedily remove The apparatus that they used was intended originall to determine the conducting power of liquids, and cor sisted of a tube, into which wires were fastened water tight at the two ends. On passing the current, they found that there was an evolution of gas at one wire, and that the copper of which the other consisted wasted

Do any gas bubbles appear in the intervening space? Explain the non-appearance of gas except at the poles. Who first decomposed water by electricity? Describe the apparatus.

away. If the wires are of platinum, hydrogen and oxygen collect in the tube, and may subsequently be fired by a spark and the decomposed water recombined.

In the same manner that water is decomposed by the Voltaic battery, so also many metallic and other salts yield to its influence. Thus, if into a jar containing a solution of sulphate of copper two metallic plates be introduced parallel to each other, and one of them brought in connection with the negative and the other with the positive pole of the battery, decomposition of the salt takes place. The sulphate of copper is resolved into its constituents, sulphuric acid and the oxide of copper; and the latter is reduced to the condition of metallic copper by hydrogen simultaneously evolved with it, arising from the decomposition of a part of the water. In this manner the copper may be deposited, with a little care, under the form of a tough metallic mass.

Becquerel obtained some very beautiful results by the aid of weak but long-continued electrical currents, illustrating the probable mode of formation of mineral substances by such currents traversing the crust of the earth. If we take a glass tube bent in the form of a U, *Fig.* 130, and close the bended part with a plug of plaster of Paris, putting into one of the branches a solution of carbonate of soda and in the other of sulphate of copper, immersing in one of the solutions a zinc plate and in the other a copper, connected together by a piece of bent wire, the liquids communicate together through the porous plug, and crystals of the double carbonate of copper and soda form on the plate immersed in the copper solution. In the same manner, other compound salts and mineral bodies may be produced.

Fig. 130.

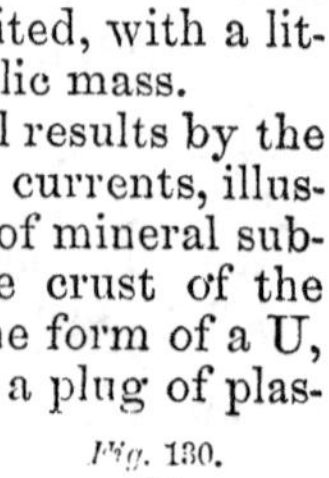

Or, if we take a jar, A, *Fig.* 131, and fill it with a solution of nitrate of copper to *a*, and then with dilute nitric acid to B, and immerse in it a slip of copper, C D, presenting equal surfaces to the two liquids, an electric current is generated; the copper is dissolved in the up-

Is water the only substance that is decomposed by a Voltaic current? Give an example. What were Becquerel's experiments? Describe *Fig.* 130. Describe *Fig.* 131.

Fig. 131.

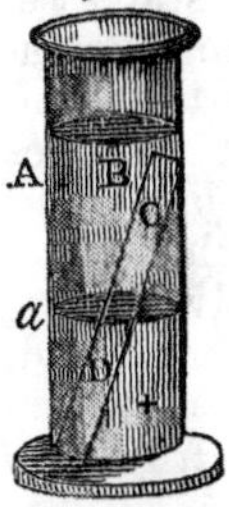

per solution, and is deposited in crystals at D in the lower.

Becquerel has shown that in the strata of the earth similar actions are going on; and others, by connecting the surfaces of twc contiguous lodes of metallic ore by means of wires attached to a galvanometer, have suc ceeded in demonstrating to the eye the e istence of these feeble but continuous cu rents, which are probably the cause of th accumulation of different metals in regu beds, and of their beautiful crystalline arrangement.

As in this manner water and various saline bodies u dergo decomposition by the action of the pile, it occu red to Sir H. Davy that probably other substances, a that time supposed to be simple, might also be decomposed. He accordingly subjected the alkaline and earthy bodies, then reputed to be elementary, to the influence of a powerful battery, and found that his suppo sition was verified. On placing a fragment of causti potassa between the poles it immediately melted; fror the positive pole oxygen gas escaped in bubbles, an from the negative small metallic globules, having th appearance of quicksilver, emerged. These were char acterized, however, by the singular quality of an intense affinity for oxygen, so that they would take fire on being touched by water or even ice, and were so light as to swim upon the surface of that liquid.

The results of Davy's experiments proved that the alkaline substances and all the earths are oxidized bodies, and in most instances oxides of metals. The convulsive spasms of a frog's legs ended in showing that the crus of the earth is made up of metallic oxides, besides revealing the mystery why the magnetic needle points t the north.

On these principles Davy established a division of el ementary bodies into electro-positive and electro-nega tive substances. The former are those which, during a polar decomposition, go to the negative pole, and the latter those that go to the positive. The electro-chemical

What is the electric state of contiguous metallic lodes? What was Davy's great discovery? What did his experiments prove? What division did he make of the elements?

theory assumes that all bodies have a natural appetency for the assumption of the positive or negative states respectively, and that all the phenomena of chemical combination are merely cases of the operation of the common law of electrical attraction, for between particles in opposite states attraction ought to take place; and when in a compound body, such as water, which consists of particles of negative oxygen and positive hydrogen, the poles of an active Voltaic battery are immersed, they will effect its decomposition, the negative oxygen going to the positive pole, and the positive hydrogen to the negative pole.

Davy's theory thus not only accounts for the decomposing agencies of the battery, but also for all common cases of chemical combination, referring both to the fundamental law of chemical attraction. With all its simplicity, it would be easy to show that it is founded on a groundless assumption, and can not account for a great number of well-known facts. The Voltaic pile can not decompose all bodies indiscriminately. An electrolyte, or so a decomposable body is termed, must always be fluid.

It also appears that all electrolytes must have a binary constitution, or contain one atom of each of their constituent ingredients. No elementary body can be an electrolyte.

Faraday's law states that the same current of electricity, when transmitted successively through various electrolytes, decomposes each in the proportion of their respective chemical equivalents. For example, if water, iodide of potassium, and melted chloride of lead be used, and if of the water there be decomposed 9 parts by weight, there will be 165 of iodide of potassium and 139 of chloride of lead. These numbers represent the atomic weights of the bodies in question. The following table, by Daniell, of the *specific electricity* of various bodies is based on the statement that the proportion of electricity which is associated with a given weight of any substance is inversely as its combining proportion:

What does the electro-chemical theory assume? Can the Voltaic pile decompose all substances? What is an electrolyte, and what must be its composition? What is Faraday's law?

Specific Electricity of Bodies.

Electro-Positives.	Equivalent.	Specific Electr'y.	Electro-Negatives.	Equivalent.	Specific Electr'y.
Hydrogen......	1.0	1000	Oxygen...........	8.0	125
Potassium......	39.2	25	Chlorine.........	35.5	27
Sodium.........	23.3	43	Iodine............	126.0	8
Zinc.............	32.5	31	Bromine..........	78.3	12
Copper	31.6	31	Fluorine..........	18.7	55
Ammonia.......	17.0	58	Cyanogen........	26.0	38
Potassa.........	47.2	21	Sulphuric Acid..	40.0	25
Soda.............	31.3	32	Nitric Acid......	54.0	18
Lime............	28.5	35	Chloric Acid.....	75.5	13

Electrolytic action occupies a very important posit in the arts. It is extensively employed as a means precipitating copper, silver, gold, lead, zinc, tin, nick platinum, and other metals from solutions of their salt These processes are called electrotyping or galvanizing. If, for example, it were required to obtain a perfect copy in copper of one of the faces of a medal, let a glass trough, N C, *Fig.* 132, be filled with a solution of th sulphate of copper, and to th negative wire, Z, of a Smee Voltaic battery, let the med N be attached, all those po tions except the face designe to be copied being varnishe over or covered with wax, to protect them from contact with the liquid. To the positive wire S, let there be attached a mas of copper, C. As soon as th battery is in action, decomposition of the sulphate takes place; metallic copper is precipitated on the face of th medal, copying it with surprising accuracy. This cop per is of course withdrawn from the sulphate in the so lution, but, while this is going on, sulphuric acid an oxygen are being evolved on the mass of copper, They therefore unite with it; and thus, as fast as co per is precipitated on N, by oxidation new quantities are obtained from C, and the liquid keeps up its strength unimpaired. In the course of a day the medal may be removed. It will be found incrusted with a tough red

Fig. 132.

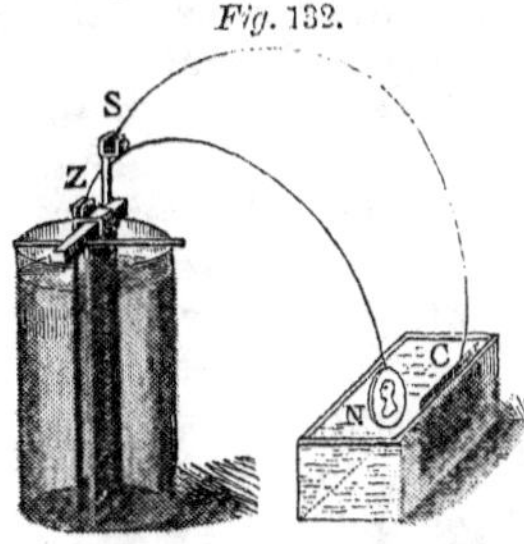

What is meant by the specific electricity of bodies? Describe the process of electrotyping.

oat of copper, which may be readily split off from it. t is a perfect copy of the surface on which the deposition ook place, and in turn it may be used as a mould for btaining a great number of casts. Often it is undesira-le to use the original coin or medal, and in such cases asts must be employed. These may be made in fusible netal, sealing-wax, plaster, gutta-percha, and a variety other substances. The surface may be caused to be-me a conductor by brushing it over with fine plum-go or black-lead, or else by dipping it in a weak solu-ı of phosphorus in ether, and, when the ether has porated, immersing it in a solution of nitrate of sil-. This leaves a thin film of reduced silver on the rface.

Silver-plating is accomplished from solutions of chlo-ide of silver in cyanide of potassium; gilding, from chloride of gold in cyanide of potassium. The metals, s deposited, have a dead or frosted appearance, and re-ire to be polished or burnished. A small quantity of sulphide of carbon in the bath, however, causes the posit to assume the lustre of the polished metal.

The extent to which electrotyping has come to be ap-ed in the arts may be appreciated from the fact that the publishing establishment of Harper and Brothers ooks are now altogether printed from copper electro-ype casts of the type and wood-cuts, the type them-elves being only used in printing the proof-sheets. As consequence, typography is greatly improved.

The electricity developed by frictional means, such as ne plate machine, can produce effects similar to those of Voltaic combinations; the quantity that is required however, very great. Faraday estimates that 800,000 scharges from a Leyden battery of 3500 square inches surface would be required to decompose a single grain water, yet every discharge would be competent to kill small animal. The difference between the two is, that ile, in the Voltaic action, the quantity is great, the in-ensity is but small, while in the electricity of the ma-chine the reverse is the case.

How may casts of medals be employed? From what solution is silver deposited? From what solution gold? What is the appearance of the metal at first? Of what use is electrotyping in printing? What is the difference between frictional and Voltaic electricity?

An instrument, the Voltameter, was invented by Faraday for measuring quantities of Voltaic electricity. It is represented in *Fig.* 133. It consists of a glass jar

Fig. 133.

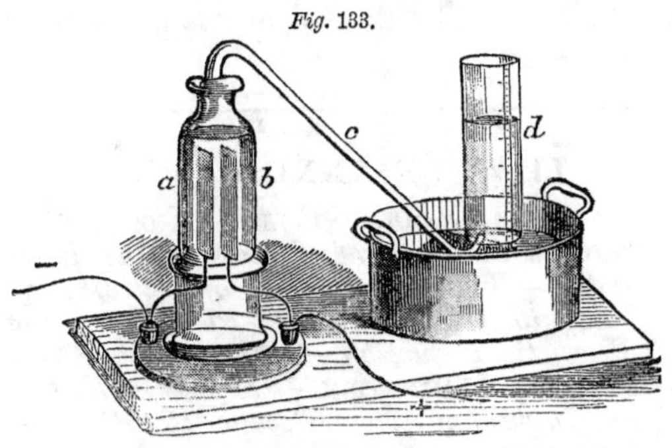

a b, filled with dilute sulphuric acid. From its neck a tube, *c*, conveys the disengaged gases to a graduated jar, *d*. Through the bottom of *a b* two wires, connected with platinum plates, pass. By means of mercury cups connection can be made with the battery to be measured, and the amount of the evolved gases registers the quantity of electricity produced.

With a given amount of metallic surface, we can produce Voltaic batteries having different qualities. Thus, if we take a square foot of copper and a square foot of zinc, and place between them a piece of wet cloth, we shall have a battery which can not give shocks nor effect the decomposition of water, but which will cause a fine metallic wire to become white-hot or even to fuse. If, again, we take a square foot of copper and a square foot of zinc, and cut each into 144 plates an inch square, and arrange them with similar pieces of cloth as a Voltaic pile, the instrument will give shocks and decompose water rapidly. From the same quantity of metal two different species of battery may be made, one consisting of a few plates of large surface, or one of a great number of alternations of small plates.

Of these varieties of battery, the calorimotor of Dr. Hare is an example of the first. It consists of a series

Describe the Voltameter. How may a given surface of metal produce batteries of different qualities? What is Hare's calorimotor?

zinc plates all connected together, and a series of cop- ɔr also similarly connected, constituting therefore, in ·ality, a single pair of very large surface. The great ɹount of heat evolved by this apparatus is its pecu- ·rity.

LECTURE XXXII.

'S THEORY OF THE VOLTAIC PILE.—*Electromotive ɔrce and Resistance.—General Law of the Force the Current.—The Rheostat.—Conductivity of ɔlids and Liquids.*—MAGNETISM.—*Phenomena and ɹaws of Magnetic Induction.—Imparting of Mag- ɹetism.*—ELECTRO-MAGNETISM.—*Oersted's Discov- ɛries.—The Galvanometer.—Electric Rotations.—Electro-Magnets.—Morse's Telegraph.—Magnetism and Diamagnetism.*

'HE phenomena of the electric current are accounted ɔy the aid of an hypothesis of which the following n exposition:

OHM'S THEORY OF THE VOLTAIC PILE.

.st. By ELECTROMOTIVE FORCE, we understand the ːses which give rise to the electric current; this, as ɜ have explained in the simple circle, is the oxidation the zinc.

'd. By RESISTANCE, we mean the obstacles which current has to encounter in the bodies through ich it passes.

hen we affect the electric current in any portion ːs path, either by varying the electromotive force ɔhanging the resistances, we simultaneously affect it oughout the whole circuit, so that in a given space ime the same quantity of electricity passes through : transverse section of the circuit.

ɹ any Voltaic circle, simple or compound, the force the current is directly proportional to the sum of all ɹe electromotive forces which are in activity, and in- ersely proportional to the sum of all the resistances; hat is to say, the force of any Voltaic current is equal

What is meant by electromotive force? What is meant by resistance? In affecting one part of a current, do we affect the rest?

to the sum of all the electromotive forces divided b all the resistances. The resistance to conduction of metal wire is directly as its length and inversely as i section; that is to say, the longer the wire is, the grea er its resistance; and the thicker it is, the less its r sistance.

If we augment or diminish in the same proporti the electromotive forces and the resistances of a V taic circuit, the force of the current will remain same; if we increase the electromotive force, the of the current increases; if we increase the resista the force of the current diminishes.

If in two Voltaic circles of equal force the same sistance be introduced, the forces of the currents may enfeebled in very different proportions; for the new introduced resistance may in one of the circles bear great proportion to the resistance already existing, an in the other a very insignificant proportion.

The following, therefore, is the general law which termines the force of a Voltaic circuit:

1st. The electromotive force varies with the num of the elements, the nature of the metals, and of the uids which constitute each element, but it does not any manner depend on the dimensions of their parts.

2d. The resistance of each element of a Voltaic c cuit is directly proportional to the distance between tl plates as occupied by the liquid, the resistance of t liquid itself, and the length of the polar wire connecti the ends of the circuit; and inversely proportional the surface of the plates in contact with the liquid, a to the section of the connecting wire.

3d. The force of the current is equal to the elec motive force divided by the resistance.

The instrument, *Fig.* 134, serves to illustrate the sistance imposed by different metals in a Voltaic rent, and to demonstrate that heat lessens the cond ing power of a wire. It consists of a Voltaic batte the terminal wires of which go to a brass stand, whe

---

What is the law of resistance in a wire? How does the force o a current change with changes in the electromotive force and th resistance? When a new resistance is introduced into two circles, must they be affected alike? Give the general law of the force of a circuit. What does the apparatus *Fig.* 134 illustrate?

:hey are connected by a thin spiral of platinum wire. As the current passes, the platinum spiral is caused to

Fig. 134.

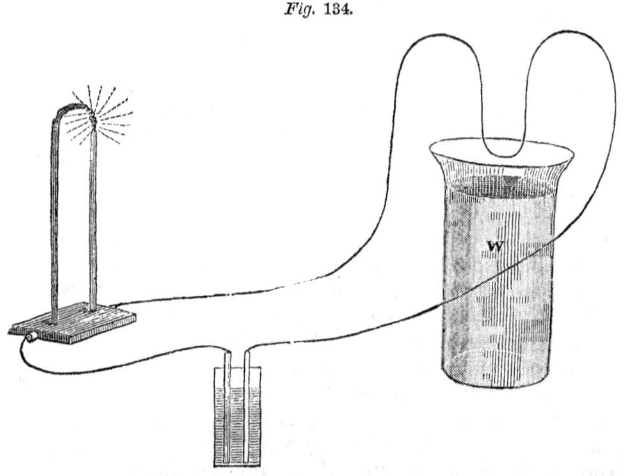

ow. One of the terminal wires is of platinum, and .is, too, is raised to redness. If now that wire be grad-.ally immersed in a vessel of water, W, so as to be re-'uced in temperature, more electricity will pass through, s is shown by the fact that the platinum spiral becomes ore and more intensely ignited, and may even be .sed.

Wheatstone's Rheostat is a contrivance for increasing resistance in a Voltaic circuit. It consists of coils fine wire arranged in lengths of hundreds of feet, and which one or more can be introduced into the circuit. me one combination is used as a standard, and others compared with it by the aid of the deflections of galvanometer, the number of feet of wire necessary reduce the deflections to the same amount being noted.

The following tables exhibit the conducting power of solids and liquids:

What effect has heat on the passage of a current? Describe Wheatstone's Rheostat.

Silver	100	Lithium	19
Copper	77	Iron	14
Sodium	37	Palladium	12
Aluminum	33	Platinum	10
Magnesium	25	Strontium	7
Calcium	22	Mercury	1.6
Potassium	21	Tellurium	.0008

Saturated Solution of Sulphate of Copper	1
(diluted with an equal bulk of Water)	.64
(" " twice its bulk of Water)	.44
(" " four times its bulk of Water)	.31
Distilled Water	.00
Platinum Wire	2,500,000

Magnetism.

Many centuries ago it was discovered that a certa ore of iron, which now passes under the name of th magnet or loadstone, possesses the remarkable quality of attracting pieces of iron. Subsequently it was found that the same power could be communicated to bars of steel by methods to be described hereafter.

Fig. 135.

Bars of steel so prepared pass under the na of artificial magnets; when they are of sm size they are commonly called needles. A ma netic bar bent into the shape represented *Fig.* 135 is called a horse-shoe magnet; and se eral magnets applied together take the name compound magnets, or a bundle of magnets.

The Chinese discovered that when a magneti needle is poised on a pivot, as at C, *Fig.* 136, floated on water by a piece of cork, that it spontaneou ly takes a direction north and south, and if purpo se ly disturbed from that p sition, it returns to it aga after a few oscillation To that end, N, wh points toward the nor the name of north pole given; the other, S, is th south pole.

Fig. 136.

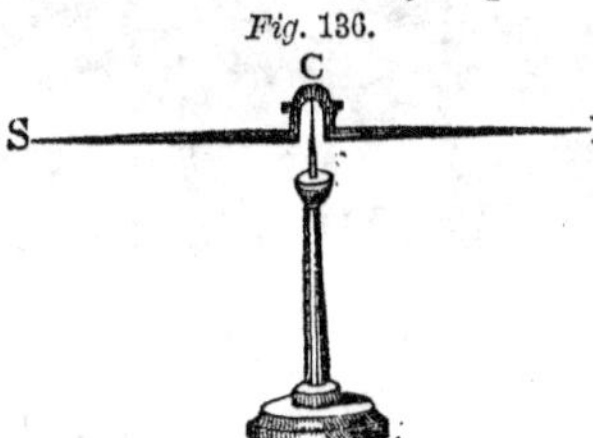

By a needle so suspended the fundamental fact of the

What is the relative conducting power of the metals? What are the properties of a magnet? What is its effect on steel? Describe the properties of the magnetic needle.

ıttraction of a magnet for iron is verified. Present a :nass of iron to either extremity of the needle, and the needle instantly moves to meet it. If a bar magnet be brought near a nail or a mass of iron filings, the iron will be suspended.

That these effects take place through glass, paper, and solid and liquid substances generally, may be thus ,tablished: A quantity of iron filings being laid on a ane of glass, if a magnet be approached beneath, the 'ngs follow its motions; but if a plate of iron inter- ıe, the magnetic influence is almost wholly cut off.

The power of a magnetic bar is not equal in all parts. ere is a point situated near each end which seems to e the focus of action. To these points the names of poles are given, and the line joining them is called the axis.

Fig. 137.

If a bar magnet be rolled in iron filings, they attach themelves, for the most part, at 'e two poles *d d*, *Fig.* 137; if such a bar be placed under a sheet of pasteboard c the surface of which iron filings are dusted, they ar- ,nge themselves in curved lines, as shown in *Fig.* 138, 'hich are symmet- cally situated as espects the poles P P.

Fig. 138.

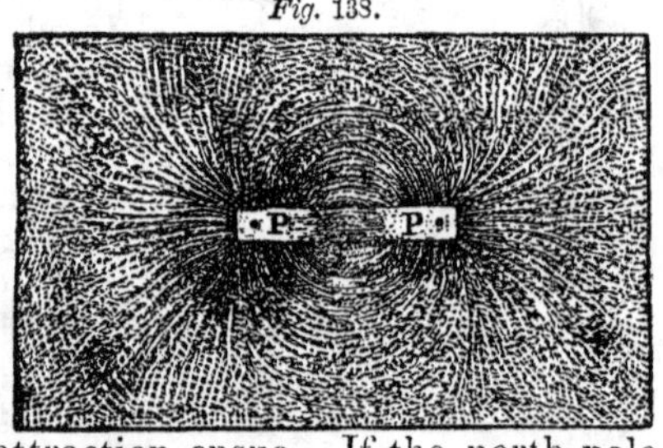

When, instead of resenting to a sus- nded needle a ece of iron, we esent to it an- her magnet, phe- omena of repul- n as well as of attraction ensue. If the north pole one be presented to the north pole of the other, re- ision takes place, and the same occurs if two south poles are presented; but if it be a north and south pole, then attraction takes place.

What occurs on presenting a piece of iron to a needle? Do these effects take place through non-conducting substances? Where are the foci of action in a magnet? What occurs on rolling a bar magnet in iron filings?

These results may be grouped together under the simple law, "*Like poles repel, and unlike ones attract.*"

There is therefore an antagonization of effect between opposite magnetic poles. If a key be suspended to a magnet by its north pole, on the approach of the south pole of one of equal force it drops off.

If we examine the force of a magnet, commencing at either of its poles and going toward its centre, the i tensity gradually declines: it ceases altogether abou midway between the poles. This point is termed t point or line of magnetic indifference.

Magnetism may be excited in both iron and steel; the former with greater rapidity, in the latter m slowly. The magnetism which soft iron has received instantly loses on being removed from the source whic has given it magnetism, but steel retains its virtue per manently. Soft iron is therefore transiently, hard steel permanently magnetic.

When a mass of iron is in contact with the pole of magnet it obtains magnetism throughout its whole ma and can in the same manner communicate a similar qu ity to a second mass brought in contact with it, and th to a third, and so on. Thus, if from the pole of a ma net a key be suspended, this will suspend a second, an that a third, etc., until the weight becomes too great f the magnet to hold. If, having two or three keys thus sus pended, we take hold of the uppermost and gently slid away the magnet, the moment it is removed the keys a fall apart, showing the sudden loss of power in soft iron

Fig. 139.

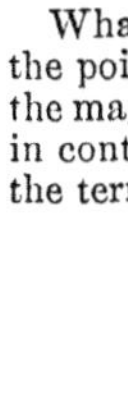

A mass of iron can receive magnetism at a d tance from the magnet itself. To this phenon non the name of *induction* is given; and this d tance through which this effect can take place called the *magnetic atmosphere.* The general e fect of induction may be exhibited by bringing powerful magnet near a large key, as in *Fig.* 1 when it will be found that the large key will su port smaller ones; but as soon as it is remove

What is the law of magnetic attraction and repulsion? What is the point of magnetic indifference? What is the difference between the magnetism of soft iron and steel? What results on bringing iron in contact with a magnet? What is induced magnetism? Explain the term magnetic atmosphere.

from the influence of the magnet these all drop off.

When magnetism is thus induced by the action of a given pole, that end of the disturbed body which is nearest to the pole has an opposite polarity, but the farthest end has the same polarity as the disturbing pole.

The force of magnetic action varies with the distance. It has been proved by Coulomb and others that the intensity of magnetic action is inversely proportional to the squares of the distance. At twice a given distance it is, therefore, one fourth, at three times a ninth, etc.

Both magnetic polarities must always simultaneously occur. We can never have north magnetism or south magnetism alone. Thus, if we take a long magnet, N S, *Fig.* 140, and break it in two, we shall not insulate the north polarity in one half and the south in the other, but each of the broken magnets will be perfect in itself, having two poles, one fragment being N′ S′, and the other N″ S″.

Fig. 140.

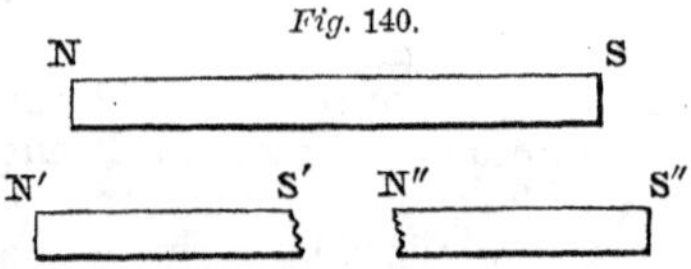

When the poles of a magnet are polished and covered with smooth plates of iron, the magnet is said to be armed. The piece of soft iron which passes from pole to pole of a horse-shoe magnet is called a keeper. The power of a magnet is measured by the weight its poles are able to carry.

There are many ways in which magnetism can be imparted to needles or steel bars, as, for example, by contact, by induction, by certain movements. By the aid of Voltaic currents, hereafter to be described, the most intense magnetic power can be communicated.

The process of magnetization by the single touch is that in which we place one pole of a magnet in the middle of a steel bar and draw it toward the end; then, lifting it up in the air, return it to its former position, and repeat the movement several times. The magnet is now

What is the law of decrease of magnetic action? Can we isolate north or south magnetism? How is a magnet armed? How is magnetism communicated to a steel bar? Describe the process by single touch.

to be reversed, and in that position moved to the oppo site end of the bar, lifted up in the air, replaced, and the movement many times repeated. The bar thus become a magnet, each end having a pole opposite to that by which it was touched. Or we may place two magnet with their opposite ends in the middle of the bar, an then, drawing them apart in opposite directions, the same result arises. A still more powerful magneti may be given, if the bar to be magnetized is laid on t poles of two magnets, so that the contrary poles of magnets and bar coincide.

In the *double touch*, two bar magnets are so tied gether that their opposite poles may be maintaine short distance from one another. This combination then placed on the middle of the bar to be magnetize and drawn toward its end; but as soon as it reache that without passing over it, it is returned to the other end with a reverse motion and then back again; an after this has been done several times, the process ended by drawing the combination off sideways wh it is at the middle of the bar.

The magnetism of a bar is reduced by percussi scratching, or filing, and by heat a steel bar may be tally demagnetized. So, too, a mass of iron, if rais to redness, becomes indifferent to the presence of a ma netic needle, though on cooling it is as active as ever.

Of Electro-Magnetism.

In 1819 it was discovered by Oersted that if a ma netic needle be brought into the neighborhood of a w along which an electric current is passing, the needl at once disturbed from its position, and tends to se self at right angles to the wire. If there be an elect current moving in the direction A B, *Fig*. 141, i wire, and directly over the wire and parallel to it th be a suspended needle, as soon as the current passes needle is deflected from its position, and, if the curr be sufficiently powerful, comes at right angles to th wire. The direction in which the transverse motio takes place depends on the relative position of the nee

Describe magnetizing by double touch. How may magnetism b reduced? Describe what occurs on bringing a needle near a co ducting wire. How does the position of the needle affect the result?

dle and the wire. Thus, 1st, if the wire be above the needle and parallel to it, the pole next the negative end of the battery moves westward; 2d, if the wire be beneath the needle, it will move eastward; 3d, if the wire be on the east side of the needle, the pole is elevated; 4th, if on the west, it is depressed. In all these positions the tendency is to bring the needle at right angles or transverse to the wire.

Fig. 141.

A

S N

B

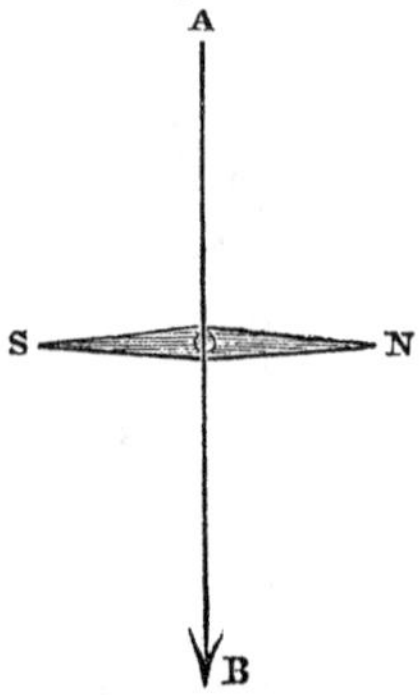

It follows, from these facts, that if a magnetic needle be placed in the interior of a rectangle of wire, *Fig.* 142, through which a current is made to flow, all the portions of the wire conspire to move the needle in the same direction. The effect, therefore, becomes much greater than in the case of a single continuous wire.

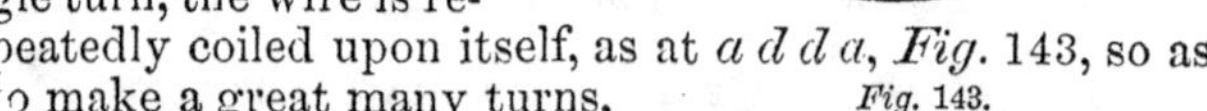

Fig. 142.

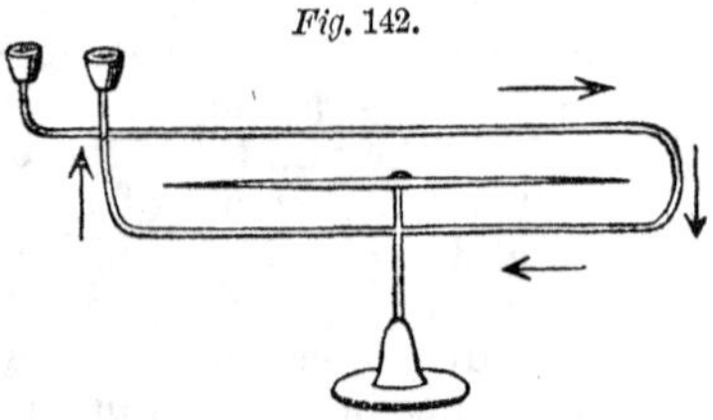

On the same principle, if, instead of a single turn, the wire is repeatedly coiled upon itself, as at *a d d a*, *Fig.* 143, so as to make a great many turns, the effect upon the included needle, *n s*, is greatly increased; and when the needle is made nearly astatic, that is to say, its tendency to point north nearly destroyed, by arranging it upon an axis with another needle similar to it in all respects, but with its poles reversed, as N S, *s n*, *Fig.* 144, the directive tendency of the one needle neutralizing the other, but both tending to turn in the same direction by the current in the coil of wire, inasmuch as one is within the coil and the other above it, the arrangement forms a most delicate means of discovering and measuring an electric current. It is called a galvanometer.

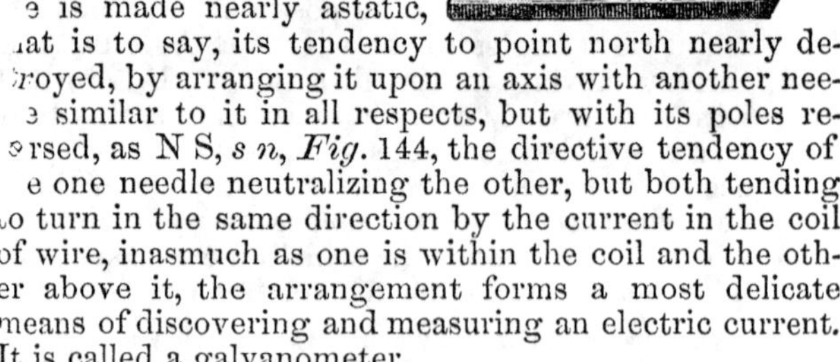

Fig. 143.

What is the effect on a needle in the interior of a rectangle? What is the principle of the galvanometer?

As action and reaction are always equal and contrary, it is obvious that if a conducting wire be movable and the magnet stationary, the latter can be made to impress motions on the former.

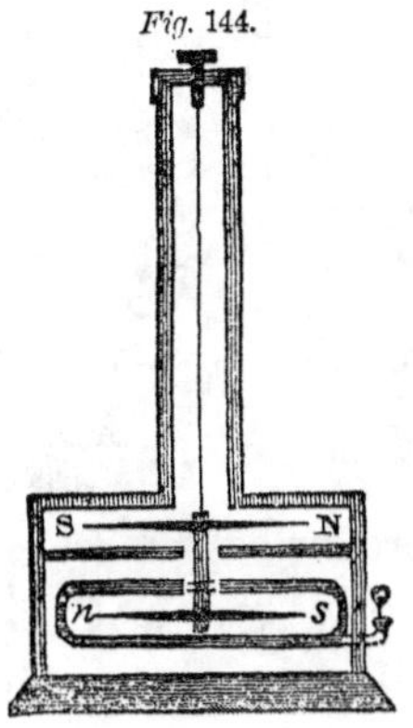

Fig. 144.

Conducting wires may be made to revolve around the poles of a magnet, or the pole of a magnet around a conducting wire. Thus in a glass cup, *Fig.* 145, let a ma net, *n*, be fixed vertically, and t cup filled with mercury. By mea of a loop, *a*, let a conducting wire, be suspended, having perfect free dom of motion. If an electric current is made to pass down this wire through the mercury and escape by the path *d*, the wire rotates round the pole *n* as long as the current passes. From th and similar experiments it therefore a pears that the force exerted between conducting wire and a magnet is not direct attractive or repulsive power, bu one continually tending to turn th movable body round the stationary one, deflecting it continually and acting in a tangential direction. Hence it is sometimes spoken of as a tangential force.

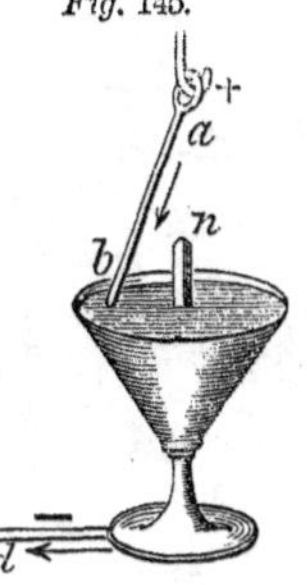

Fig. 145.

If round a bar of soft iron a conducting wire, covered over with silk, be spirally twisted, as in *Fig.* 146, whenever an electric current is passed the ir becomes intensely magnetic, and lose its magnetism as soon as the curren stops. A bar an inch in diamete bent so as to represent a horse-sho *Fig.* 147, with a wire covered wi silk for the purpose of separating its successive strands from each other,

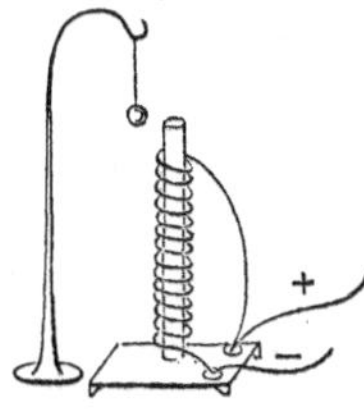

Fig. 146.

On the same principles can the wire be made to move? Describe a method of showing rotation of a wire round the pole of a magnet. What is the nature of the force exerted between a conducting wire and a magnet? Describe the construction and properties of a straight electro-magnet. Describe the horse-shoe electro-magnet.

may be made to give rise to very striking results. Professor Henry, by a modification of the conducting wire, succeeded in imparting so intense a degree of magnetism to a piece of soft iron that it could support more than a ton weight. If under one of these ELECTRO-MAGNETS a dishful of small iron nails be held, the moment the current passes the nails are attracted, and, while they are held by poles, may be moulded by the hand various shapes, but as soon as the current stops they all fall off.

Fig. 147.

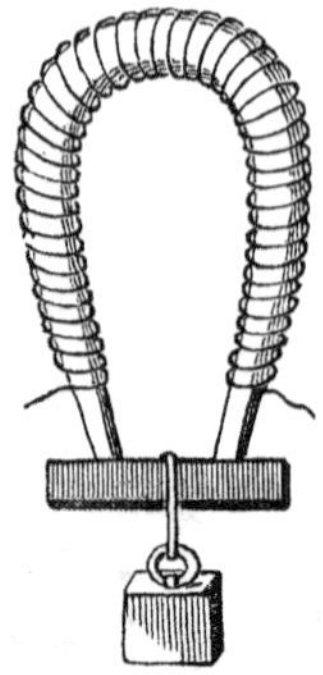

It is upon this principle of producing temporary magnetism by an electric current that Morse's electric telegraph depends. In *Fig.* 148, m m is an electro-magnet, with its

Fig. 148.

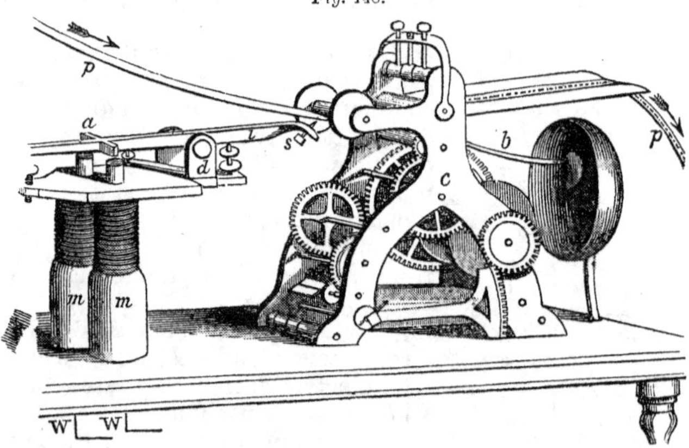

poles upward. The wires, W W, from the distant station induce magnetism in it, and draw down the soft iron keeper, a. At the same time the clock-work, c, is set going, and a strip of paper, p p, is drawn steadily forward in the direction of the arrows. Whenever a current passes, and the keeper a is depressed, the point s is forced upward against the moving paper. If the cur-

What is the construction of Morse's telegraph?

rent is but momentary, a dot only is made on the paper, but if it continue, a line is formed. Morse's tele graphic alphabet consists of a set of such dots and lines which by varied grouping represent the various letters and figures. Experienced operators do not, however require the paper strip, but learn to distinguish, by th clicking sound of the keep er falling on the magne what the transmitted m sage is.

Fig. 149.

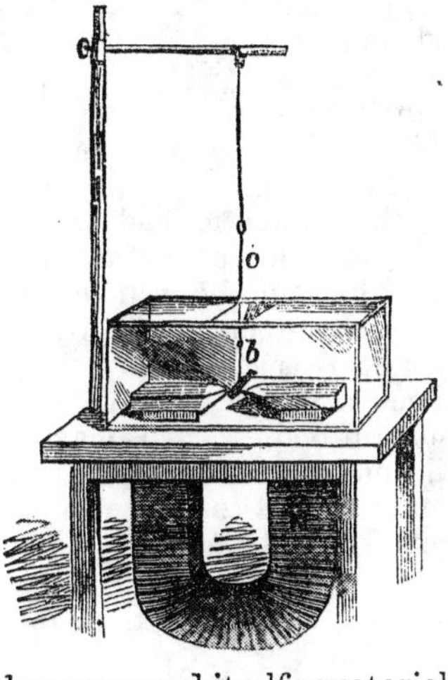

When different stances are suspended tween the polar termi tions of a horse-shoe ele tro-magnet, in the magnet ic field, as it is termed, i is found that some arrang themselves from pole pole—that is, *axially*, a others transversely to t position, *equatorially*. Th former are called magne ic, and the latter diama netic bodies. In *Fig.* 14 *b* is a bar of bismuth which, being diamagnetic has arranged itself equatorially between the poles of th electro-magnet, N S. It is suspended by a fibre of u spun silk, *c*, so as to turn freely. The following table a list of magnetic and diamagnetic bodies, those at the beginning having the properties most strongly marke

Magnetic Bodies.		*Diamagnetic Bodies.*	
Iron.	Palladium.	Bismuth.	Flint Glass.
Nickel.	Crown Glass.	Phosphorus.	Mercury.
Cobalt.	Platinum.	Antimony.	Lead.
Manganese.	Osmium.	Zinc.	Wood.
Chromium.	Oxygen.	Tin.	Beef.
Cerium.	Etc.	Cadmium.	Bread.
Titanium.		Sodium.	Etc.

The diamagnetism of gases was shown by Faraday by

Is the paper strip necessary? What are magnetic and diamagnetic bodies? Describe *Fig.* 149. How did Faraday show the diamagnetism of gases?

the aid of the apparatus, *Fig.* 150. A tube conveyed the gas to be examined into the magnetic field. Above it three other tubes were arranged so that, under ordinary circumstances, the gas passed into the middle one; but, when the iron was magnetized, if the gas were diamagnetic, it passed into the side tubes. The passage was shown by placing a little ammonia in the lower tube, and strips of paper moistened with hydrochloric acid in the others.

Fig. 150.

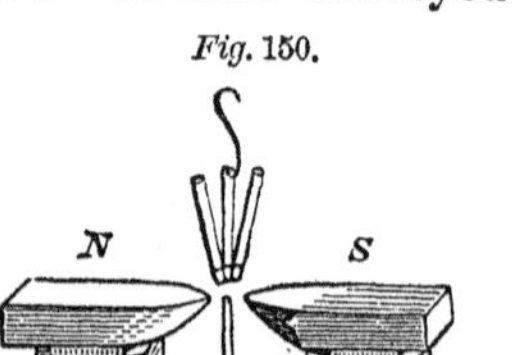

A taper burning in the magnetic field has its flame spread out equatorially, and the flame may even be divided into two parts. The same body may be caused to become either magnetic or diamagnetic, by changing the medium by which it is surrounded. Magnetism also exerts an influence on polarized light transmitted through certain transparent bodies. This may be shown by placing them, while under examination, in the magnetic field. The beam is caused to rotate to an extent which depends on the nature of the body and the intensity of the magnetism.

If a copper cylinder, *Fig.* 151, filled with fusible metal, be so arranged, by the aid of the band S S, as to rotate rapidly in the magnetic field, the resistance to its motion is so great that it will seem to be grasped by an invisible hand. On continuing the rotation for a few minutes, the heat developed will be sufficient to melt the contained alloy, which may be poured out upon the table.

Fig. 151.

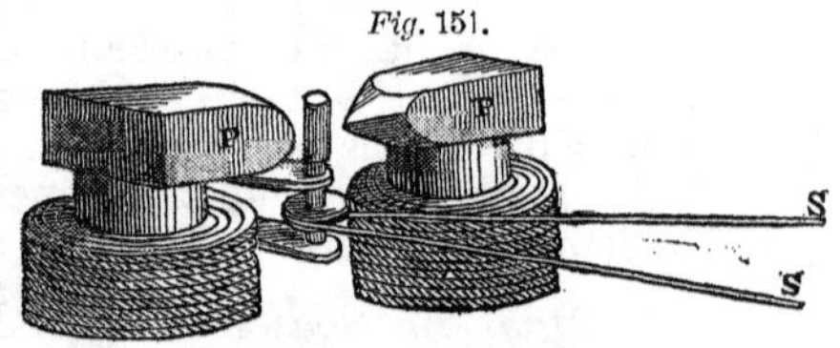

What is the effect on a taper in the magnetic field? How may the effect of magnetism be demonstrated by polarized light? Describe the experiment, *Fig.* 151.

LECTURE XXXIII.

ELECTRO-DYNAMICS.—*Ampère's Discovery.—Properties of a Helix.—Cause of Magnetism.—Faraday's Discovery of Magnetic Electricity.—Magnetic Machines.—Ruhmkorff's Coil.— Geissler's Tubes.*—THERMO-ELECTRICITY.—*Production of Electricity by Heat.—Thermo-Electric Piles.—Melloni's Pile and Thermometer.—Draper's Improvements in Thermo-Electric Pairs.*—ANIMAL ELECTRICITY.—*The Torpedo.*

SOON after the relation between electricity and magnetism was established by Oersted, Ampère discovered that there are reactions between the currents themselves. Two electric currents flowing in the same direction attract each other, but two electric currents flowing in opposite directions repel. "Like currents attract, unlike ones repel."

If a conducting wire, *a b*, *Fig.* 152, be bent in the form of a helix, *n s*, its ends returning toward its middle, it exhibits all the properties of an ordinary magnetized bar. As soon as the current passes, being freely suspended, it points north and south, and is attracted and repelled by the pole of a magnet, just as though it were a magnet itself. Another arrangement, called De la Rive's ring, for illustrating the same effect, is seen in *Fig.* 153. A small simple circle, consisting of a zinc, Z, and a copper plate, C, connected by a coil, is suspended in a vessel of acidulated water, which is floated by a cork, D, on water.

Fig. 152.

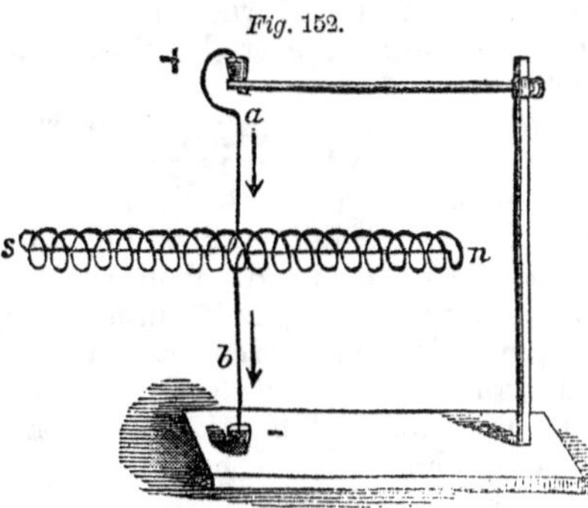

What is the law of reaction between electric currents? Describe the electro-dynamic helix. Describe De la Rive's ring.

The current runs round the coil in the direction of the arrows, and the arrangement, obeying the magnetic influence of the earth, turns with its plane pointing north and south, just as a magnetic needle would do. If the north end of a magnet, *n s*, be presented toward the loop, the wire will be attracted, and will place itself midway between the ends of the magnet; but if the south end be presented, the wire will be repelled, the floating combination will turn half way round so as to reverse its direction, and will then be attracted.

Fig. 153.

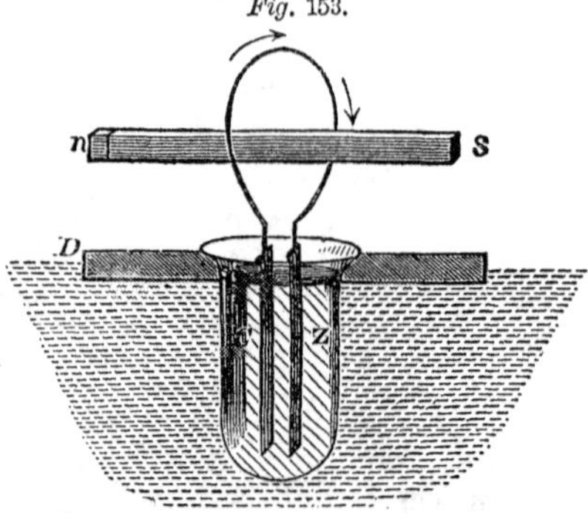

Ampère inferred from the analogy of these instruments that the magnet owes its qualities to electric currents circulating in it in a transverse direction. The directive action of the magnetic needle or the electric helix depends on the action of electric currents circulating in the earth, due to the unequal heating of its surface by the rays of the sun, the earth being regarded as an electro-magnet, the poles of which are *nearly* in the line of the axis of rotation. The angle between the two is not constant, the *variation* of the needle from the true north point exhibiting slow increases and diminutions.

We have seen that an electric current can develop magnetism in a bar of iron or steel—in the former transiently, in the latter permanently. Thus, if the iron bar, *n s*, *Fig.* 154, be placed in the axis of a helix of copper wire along which a current is flowing, the current develops magnetism in the bar. It was discovered by Faraday that the converse also holds good, and that a

Fig. 154.

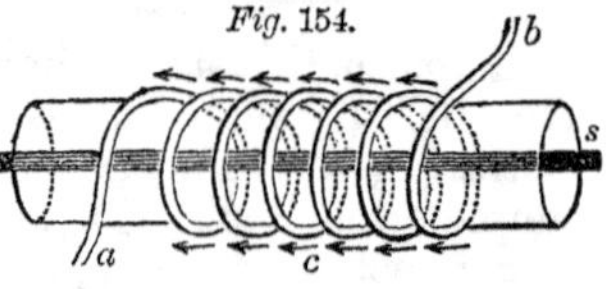

How does it act toward a magnet? What is Ampère's theory of the nature of the magnet? What is the effect of a current running round a bar of soft iron?

magnet can give rise to an electric current. Thus, in *Fig.* 154, let the terminations, *a b*, of the helix, *c*, be brought into contact, and, having placed a soft iron bar, *n s*, within it, let the bar be made magnetic by the approach of a strong magnet. It at once generates a current which runs through the helix, *c*; and if at this moment the wires *a b* be drawn apart, a bright spark, sometimes called the magnetic spark, passes. It doe not come, however, from the magnet itself, but is due t the electric current *induced* in the helix by the distur ing action of the magnet. If between the terminatio *a b* a slender wire be placed, it may be made red-hot, water may be decomposed, or any of the phenomena o a Voltaic battery may be exhibited by the aid of this *magneto-electric* current.

Fig. 155.

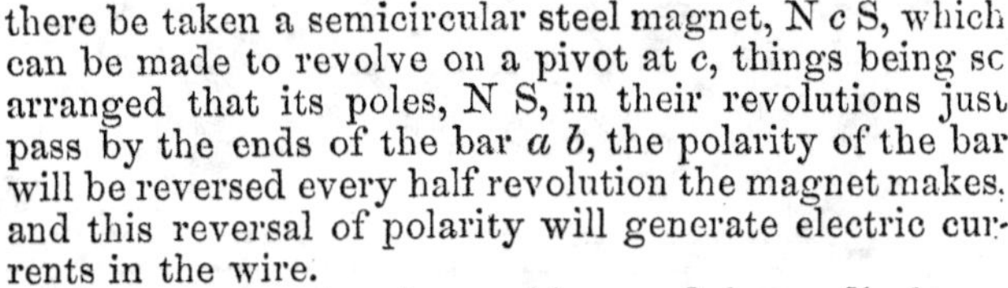

The same results would occur if, instead of introducing and removing a permanent steel magnet we continually change the polarity of a stationary soft iron bar. Thus. if *a b*, *Fig.* 155, be a rod of sof iron surrounded by a helix, and there be taken a semicircular steel magnet, N *c* S, which can be made to revolve on a pivot at *c*, things being sc arranged that its poles, N S, in their revolutions just pass by the ends of the bar *a b*, the polarity of the bar will be reversed every half revolution the magnet makes, and this reversal of polarity will generate electric currents in the wire.

The magneto-electric machine used for medical pu poses, *Fig.* 156, is constructed on the principle that we coil round a piece of soft iron a conducting wire, often as the iron is magnetized by the permanent ma net, S, a wave of electricity flows through the wire The soft iron being rapidly rotated by the multiplyin wheel, and the patient being brought in relation with the machine by the binding-screws and handles, A B, he experiences a succession of shocks.

If two conducting wires be placed parallel and near each other, when an electrical current is passed through

How may a magnet produce an electric current? Describe *Fig.* 155. What is the principle of the magneto-electric machine?

one of them a wave of electricity flows in the opposite direction through the other. On the first current stop-

Fig. 156.

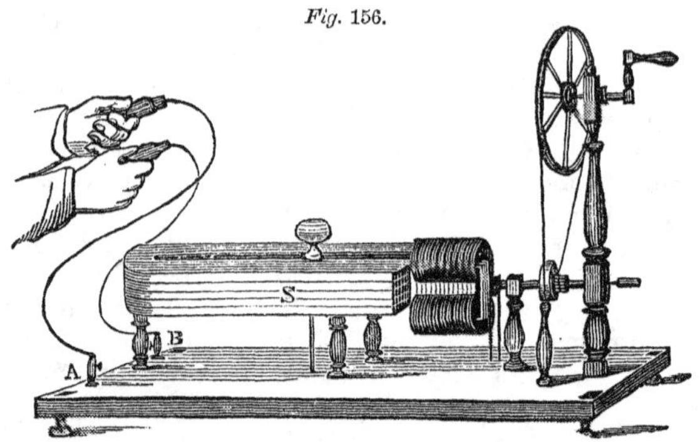

ping, another wave, called the secondary current, passes through the second wire. These momentary currents are called, from the name of their discoverer, Faradian currents. The effects are much increased by using

Fig. 157.

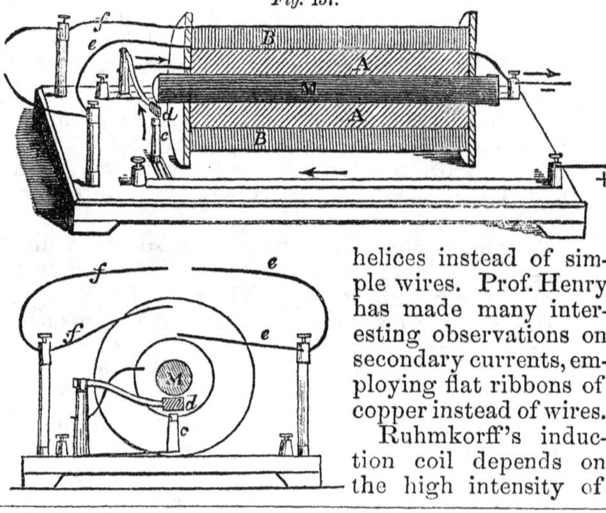

helices instead of simple wires. Prof. Henry has made many interesting observations on secondary currents, employing flat ribbons of copper instead of wires.

Ruhmkorff's induction coil depends on the high intensity of

What are Faradian currents? What is the fact on which Ruhmkorff's coil depends?

the secondary currents produced by magnetic induction. It consists of two concentric helices of copper wire, the inner or primary coil, A A, *Fig.* 157, being of thick wire and only two or three hundred feet long, while the outer one, B B, is several thousand feet long, and of ver thin silk-covered wire. In the axis is a bundle of sof iron wires, M. The primary coil is not continuous, bu may be broken at *c d*, the keeper, *d*, being raised whe M becomes magnetic and the circuit interrupted. Tl making and breaking occurs several hundred times a minute, a powerful secondary current being induc each time in B B. A continuous stream of sparks pa. es between the ends of the secondary wire, *e f*. No. is an end view, No. 1 being a longitudinal section.

In addition, there is attached to the primary wire a species of Leyden jar termed a condenser, which increases the power of the instrument. It consists of a band of oil-silk or paper coated on each side with tin foil, the two coatings being connected with *c* and *d*.

By this instrument sparks twenty inches long ma be procured, and by causing these sparks to traverse an air-pump vacuum an auroral light will be seen. By placing in the air-pump receiver a tumbler of uranium glass, lined partly inside with tin-foil, a cascade of electric light will flow over on the air-pump plate.

If the discharge take place into a space more rarefied than the ordinary receiver, the luminous portion is observed to be stratified, or crossed by dark bands.

Fig. 158 represents hermetically-sealed glass tubes into which platinum wires have been fused, and in which the rarefaction is progressively more perfec The rarefaction is produced by filling the tubes wi dry carbonic acid gas, and at the same time puttir pieces of caustic potassa, *P P P P*, into them. Th bands become wider and change their shape, until, whe a perfect void is obtained, they disappear. Materi particles seem to be necessary to the transfer of the current, as in the case of the Voltaic arc. By inclosing a variety of substances in such tubes beautiful effects

---

Describe Ruhmkorff's induction coil. What length of spark may be obtained? How may auroral light be produced? What are Geissler's tubes?

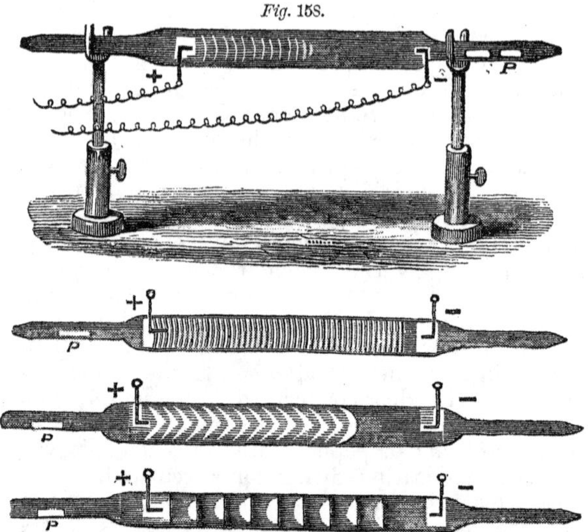

Fig. 158.

color may be obtained. These are called Geissler's bes.

THERMO-ELECTRICITY.

If we take a bar of antimony, *a*, *Fig.* 159, and one of smuth, *b*, and, having soldered them end end at *c*, pass a feeble current through em in a direction from the antimony to ne bismuth, the temperature of the com- on ınd bar rises, but if the current pass in opposite direction, cold is produced. fixing thermometers into the substance he bars these facts may be verified; and the latter case, when water is placed in lepression made for it in the bar, and the reduction temperature slightly aided, it may be frozen by the ırrent.

Fig. 159.

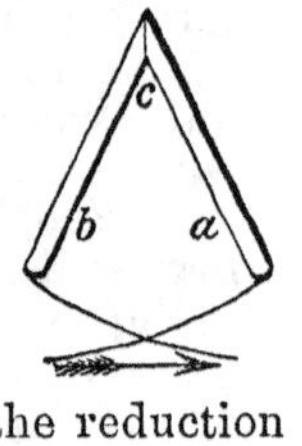

The same compound bar of bismuth and antimony, having its extremities connected together by a wire, whenever heat is applied to the junction an electric cur-

How may an electric current produce heat and cold? How is an lectric current produced by the apparatus *Fig.* 159?

rent sets from the bismuth to the antimony, and whe. cold is applied, from the antimony to the bismuth These important facts were discovered by Seebeck i 1822, and the current designated thermo-electric cu rents.

If a rectangle, *Fig.* 160, be composed of a bar of a timony, A A, and one of bismuth, *B B*, on app ing heat to one of junctions a current run around the comb tion, and a magnetic dle suspended within deflected.

Fig. 160.

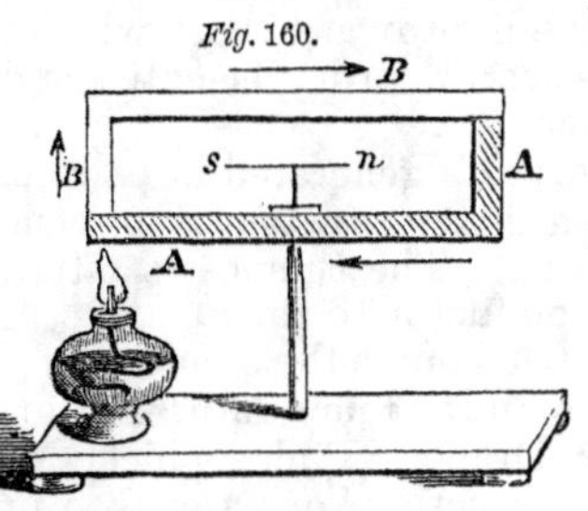

For the production these thermo-electric fects two metals are n. necessarily required. One end of a thick metallic w being made red-hot, and brought in contact with other, a current instantly passes from the hot to cold portion, and continues to flow in diminishing qua tities until the two ends have reached the same temp ature. Or if a metallic ring be made red-hot in a. limited portion of its circumference, as long as the h passes with freedom to the right hand and the left el tric development does not appear, but if we touch w a cold rod the hot portion, abstracting thereby a pa of its heat, a current in an instant runs round it.

It is not alone in metals that these thermo-elect currents can be induced ; other solids, and even liquids may originate them. Among metals associated tog er, the relation often exhibits singular changes. C per and iron form a very active couple until their t perature approaches 800° ; the current then stops, on continuing the heat another current is develo passing in the opposite direction. The same takes pl with a pair of silver and zinc at 248°.

Describe the thermo-electric rectangle. How can thermo-ele tric effects be produced with one metal? Describe the peculiaritie of a copper and iron couple.

Thermo-Electric Order of Metals.

Bismuth.	Copper and Silver.
Platinum.	Zinc.
Lead.	Iron.
Tin.	Antimony.

The current proceeds, when heated together, from ose at the end of the list toward those which precede em. The thermo-electric order is entirely different om the Voltaic order.

Thermo-electric currents generated in pairs of metal- bars, experiencing little resistance to conduction, ve very little tension. The thinnest stratum of wa- r is a perfect non-conductor to them.

In any thermo-electric couple the quantity of electrici- y evolved depends upon the temperature. But, as was hown by Professor Draper (Philosophical Magazine, une, 1840), it is not directly proportional to it, except rough limited ranges of temperature; and we can not, erefore, make use of these currents for the determina- on of temperatures with accuracy, on the hypothesis f the proportionality of the quantities of electricity to e quantities of heat.

By joining a system of bars alternately together, we ay reduplicate the effects of a single pair. As might ave been predicted on the theory of Ohm, and as has een shown in the memoir just quoted experimentally, where the conducting resistances remain the same, the uantity that passes the circuit is directly proportional the number of pairs. By thermo-electric batteries of ufficient number of pairs of German silver and iron, ated intensely at one set of junctions by a coal fire, d kept cool at the other by water, water may be de- mposed with rapidity, and all the effects of Voltaic mbinations produced. There seems to be no reason hy such combinations should not eventually displace l other sources of electricity, and be used for chemical, galvanic, and magnetic purposes. Electricity could then be utilized in the construction of prime movers, in warm- ing apparatus, and for a thousand other applications.

What is the tension of thermo-electric currents? Why are they unreliable for thermometric purposes? How is a thermo-electric pile or battery made? What are the properties of such piles?

Melloni's thermo-electric pile, which is by far th most sensitive of thermometers, consists of a number of bars of bismuth, *B*, and antimony, *A*, with the alternate ends soldered together, as in *Fig.* 161. If both enc are heated no effect is producec but if only one, then a current sent along the wire which connec the last bismuth at one end w' the first antimony at the otL Melloni used 30 or 40 such alter tions, arranged to present a squa or circular face. The position of ti galvanometer is indicated at *G*.

Fig. 161.

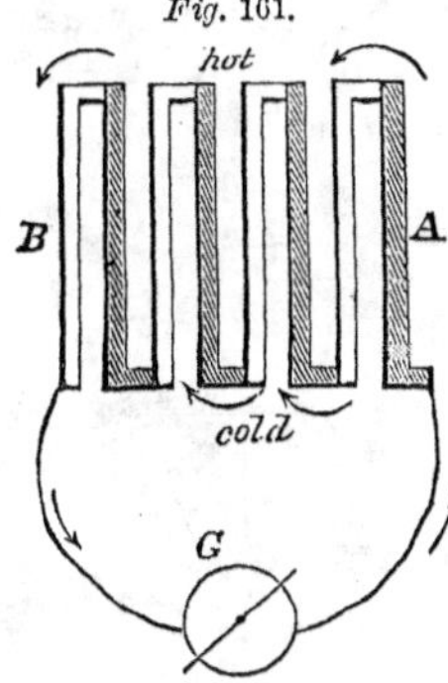

The galvanometer is seen in *Fig.* 162, in section an perspective. A B C is the coil of the multiplier, its ter minal wires resting in the connecting cups, F F′. Th coil rests on a plate, D E, which can be made to revol by means of a wheel and screw connected with the bu ton G. An astatic combination of needles is supporte by the frame Q M N by a silk thread, V L. To pro tect the instrument from currents of air, it is covered b a glass cylinder, F L, strengthened by brass rings, P S Y Y. K T is the basis on which the cylinder rest. The angle of deflection of the needle is taken as the measure of the temperature.

Professor Draper introduced certain improvements in the construction of the thermo-electric element. Let *a*, *Fig.* 163, be a bar of antimony, and *b* a bar of bismuth. Let them be soldered along *c* and at *d* let the temperature be raised. A current is immediately excited; but this does not pass around the bars *a b*, inasmuch as it finds a shorter and readier channel through the metals between *c* and *d*, as in-

Fig. 163.

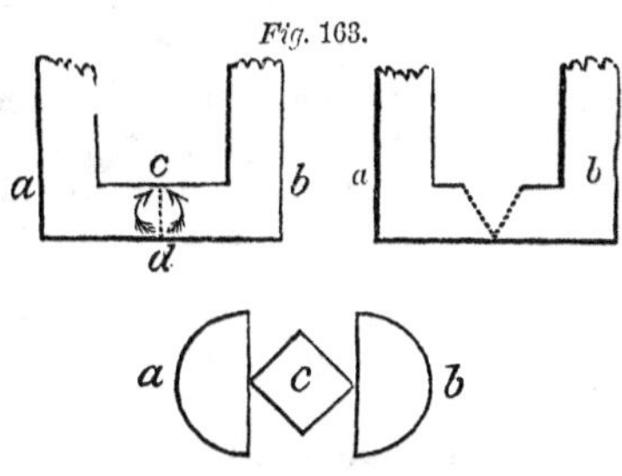

Describe Melloni's pile. Describe the galvanometer. What are Draper's improvements in thermo-electric pairs?

.cated by the arrows; nor will the whole current pass ound the bars until the temperature of the soldered sur-

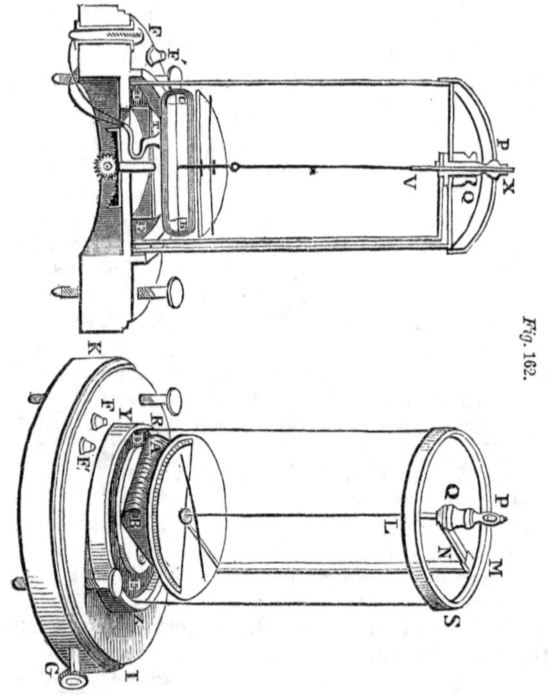

Fig. 162.

.e has become uniform. An improvement on this con- ıction is therefore such as is represented at a' b', ich consists of the former arrangement cut along the .ted lines. Here the whole current, as soon as it ex- ;, is forced to pass along the bars. One of the best "ms of a thermo-electric pair is seen in the lower fig- ɾe, where the antimony, *a*, and bismuth, *b*, are united ,y a lozenge-shaped piece of copper, *c*. The heat is made ɔo fall on *c*, which becomes hot and cold with prompti- ıude, and determines a current.

Why should the junction be small?

Animal Electricity.

Besides the various sources of electricity to which have referred, there are certain animals which posses the power of controlling the equilibrium of the electric fluid in their neighborhood at will, being accommodate for this purpose with a special ner ous apparatus. The torpedo, a fis living in the Mediterranean and our coast, and the gymnotus ei tricus, which is found in some the fresh-water streams of So America, have this property. electric organs of the torpedo shown in *Fig.* 164, at *a a*, the perficial tissues having been r moved. They are composed o prisms, *Fig.* 165, presenting divid ing diaphragms on which nerves ramify, and are fille with an albuminous sal liquid. About 470 of the are found in each of the t organs. The dorsal surfa of the animal is positive a the ventral negative. T nervous supply of these organs comes from a speci lobe, the electrical lobe, IV, *Fig.* 166. I is the cere brum; II, optic lobes; III, cerebellum. U T S a branches of the pneumogastric, and L a branch of th trifacial, distributed in the electric organ.

Fig. 164.

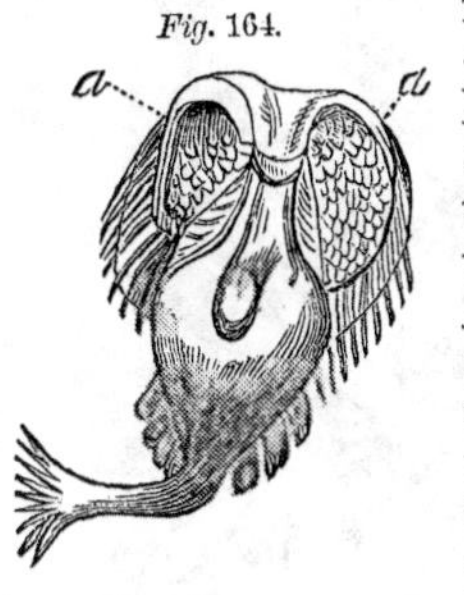

Fig. 165.

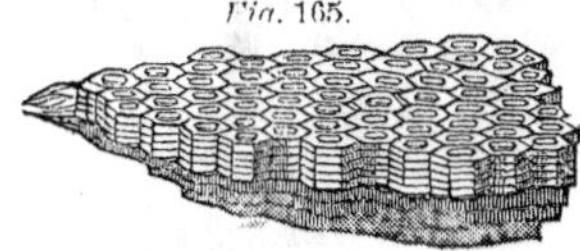

The shock of the torpedo passes through conduct bodies, but not through non-conductors. A gymnc which was exhibited in London was found to deflec magnetic needle powerfully by its discharge. A st wire was magnetized by it, and iodide of potassium composed. In an interrupted metallic circle a sp was seen, and the induced spark was also obtained by coil. The current passed from the anterior to the po terior parts of the animal. Faraday, who performe these experiments, calculates that the quantity of elec tricity passing at each discharge of the fish was equa

What is the torpedo, and what are its powers? How are the electric organs composed? Whence do they derive their nervous supply? Describe the phenomena shown by the gymnotus.

, that of a Leyden battery of 3500 inches surface charged to the utmost, and this could be repeated two or three times with scarcely a sensible interval of time.

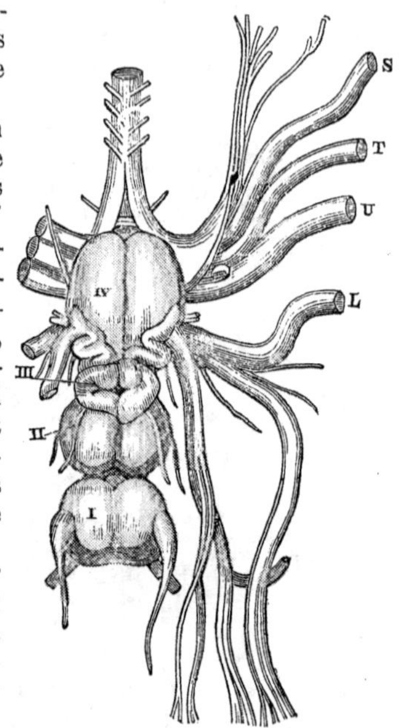

Fig 166.

As the electricity which these animals discharge pends on their nervous ion, the production of attended with a corponding nervous exustion. Matteucci rerds the prisms as piles, which a secondary elecric current is generated y the nervous current ent to them. An intite connection is thus ablished between the o forces.

The same philosopher s also shown that in all ing animals there is a ırrent of positive elecricity from the interior the exterior of every muscle; and by arranging a series of half muscles so that the interior of one ouched the exterior of the next, as in *Fig.* 167, exhibited magnetic efts and chemical decomitions.

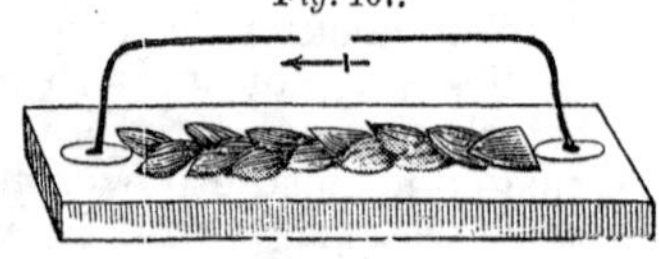
Fig. 167.

Fig. 168 enables us to demonstrate the effect of the oltaic current on the muscles of a frog. The animal prepared by dissection, so that the legs are only conected by the nerve and a part of the spinal cord. On plunging the terminal wires of a battery into the cups of water in which the feet are, the frog will leap entirely out.

What is Matteucci's explanation of the action of the prisms? Describe the muscle battery.

Fig. 168.

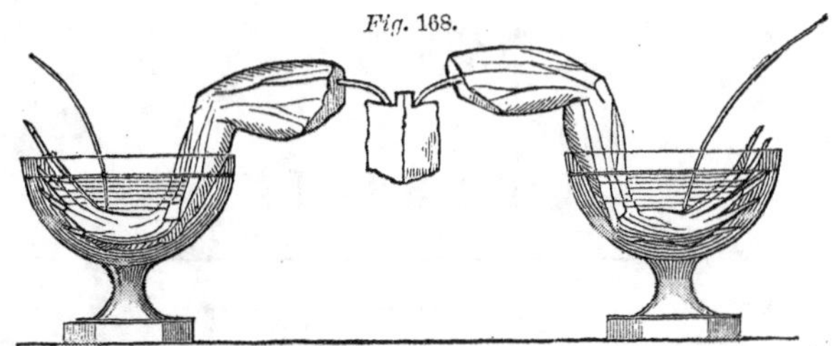

In *Fig.* 159 the muscular contractions that are

Fig. 169.

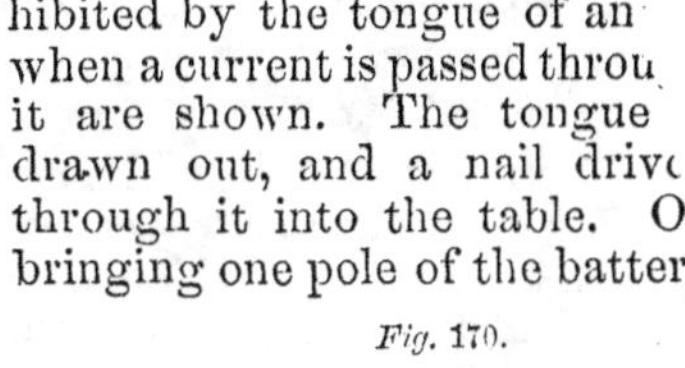

hibited by the tongue of an when a current is passed throu it are shown. The tongue drawn out, and a nail drivc through it into the table. O bringing one pole of the batter

Fig. 170.

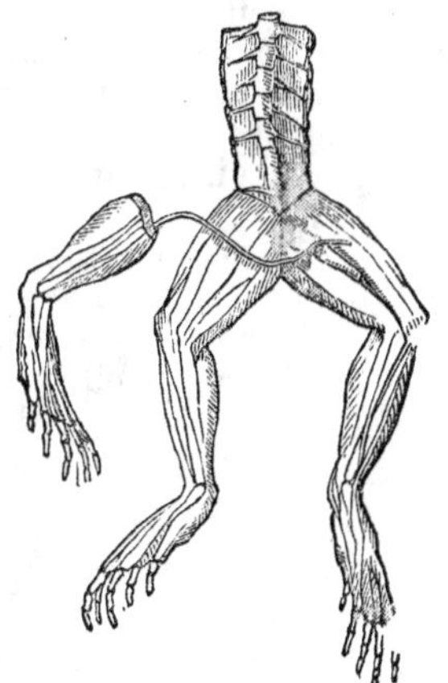

in connection with the tip of the tongue, and the other pole in connection with the spinal cord, the tongue contracts, and either the nail is drawn out or the head pulled forward.

Fig. 170 shows the manner in which, when the thighs of a frog have been excited, they may be caused to produce induced contractions in the leg of another frog. When the two poles of a battery a applied to the spine joining the thighs, we observ whenever the thighs contract, convulsions simultaneously occur in the foot whose nerve lies across them.

What does *Fig.* 168 demonstrate? How may the contraction of an ox's tongue be shown? How may induced contractions be exhibited?

PART II.

LECTURE XXXIV.

NOMENCLATURE.—*The French Nomenclature.—able of Elementary Bodies.—Nomenclature for 'ompound Bodies, Acids, Bases, and Salts.—The Binary Hypothesis.*

UNTIL after the discovery of oxygen gas, the nomenlature of chemistry was very loose and complicated. 'he trivial names which were bestowed on various bod·s had frequently little connection with their proper-; sometimes they were derived from the name of discoverer, or sometimes from the place of his resiice. Glauber salt takes its designation from the emist who first brought it into notice, and Epsom salt m a village in England, in which it was at one time de.

It is obvious that such a system of nomenclature, as oon as the number of compound bodies increased, ould not only become unmanageable, but, by reason f the impossibility of carrying in the memory such a lass of unconnected terms, offer a very serious impedint to the progress of the science. Lavoisier and his ciates, about the close of the last century, concted a new nomenclature, with a view of avoiding se difficulties. Its principles, with some modifica-s, are now universally received. The following is a f exposition of it:

Natural bodies may be divided into two classes, sim- and compound; the former are also called element-y. By simple or elementary bodies we mean those which have not as yet been decomposed.

Among simple substances, those which have been

What was the nature of the nomenclature used by the older hemists? When was the system now in use invented? What is meant by simple or elementary bodies?

known for a long time retain the names by which th: are popularly distinguished; thus, gold, iron, coppe etc.; and when new bodies belonging to this class ar discovered, they are to receive a name descriptive o one of their leading properties; thus, chlorine takes i: name from its greenish color, and iodine from its purpl vapor. It is to be regretted that this rule has ofte been overlooked.

Some doubt exists as to the exact number of the mentary bodies. It may be estimated at 68, inclu metals recently discovered, the titles of which have yet been completely established.

Of the list of elementary bodies, the metals form far the larger portion, there being 55 of them; the maining 13 are commonly spoken of as non-meta substances. By some authors these are called meta loids, in contradistinction to the metals, an epith which, however, is very objectionable. (*See* opposite

Compound bodies may, for the most part, be divi into three groups: acids, bases, and salts. By an we mean a body having a sour taste, reddening ve table blue colors, and neutralizing alkalies; by a bas body which restores to blue the color reddened by acid, and possessing the quality of neutralizing the pı erties of an acid; by a salt, the body arising from union of an acid and a base. These definitions, howeve are to be received with considerable limitation.

The nomenclature for acid substances is best se from an example. Thus, sulphur and oxygen unite form an acid: it is called sulphuric acid, the termina in *ic* being expressive of that fact. But very freq ly two substances will form more than one acid uniting in different proportions; in this case the ter ation in *ous* is used; thus we have sulphurous aci called because it contains less oxygen than sulph The prefix "hypo" is also used, as in hyposulphu and hyposulphuric acids: it indicates acids contain

---

What is the rule for naming the elements? What is the numb of the elementary bodies? Of these, to what class do the greate number belong? Into what groups may compound bodies be vided? What is the definition of an acid? What is a base? Wh is a salt? What do the terminations *ic* and *ous* mean? What the meaning of the prefixes *hypo* and *hyper?*

Table of Elementary or Simple Substances, with their Symbols and Atomic or Equivalent Weights.

Non-metallic Elements.	Symbols.	At. Wts.	Metallic Elements.	Symbols.	At. Wts.
Oxygen	O.	8	Indium	In.	36
Hydrogen	H.	1	Manganese	Mn.	28
Nitrogen	N.	14	Iron	Fe.	28
Sulphur	S.	16	Nickel	Ni.	30
Phosphorus	P.	32	Cobalt	Co.	30
Carbon	C.	6	Zinc	Zn.	32
Chlorine	Cl.	35.5	Cadmium	Cd.	56
Bromine	Br.	78	Tin	Sn.	59
Iodine	I.	126	Chromium	Cr.	26
Fluorine	F.	19	Vanadium	V.	68
Boron	B.	11	Tungsten	W.	92
Silicon	Si.	22	Columbium	Ta.	184
Selenium	Se.	40	Niobium (?)	Nb.	—
			Ilmenium	Il.	—
Metallic Elements.			Norium	No.	—
Potassium	K.	39	Pelopium (?)	Pe.	—
Sodium	Na.	23	Dianium (?)	—	—
Lithium	Li.	7	Molybdenum	Mo.	48
Cæsium	Cæ.	133	Uranium	U.	60
Rubidium	Rb.	85	Titanium	Ti.	24
Barium	Ba.	69	Arsenic	As.	75
Strontium	Sr.	44	Antimony	Sb.	129
Calcium	Ca.	20	Tellurium	Te.	64
Magnesium	Mg.	12	Copper	Cu.	32
Aluminum	Al.	14	Lead	Pb.	104
Glucinum	G.	7	Bismuth	Bi.	213
Zirconium	Zr.	34	Silver	Ag.	108
Thorium	Th.	60	Mercury	Hg.	100
Yttrium	Y.	32	Gold	Au.	197
Erbium	Er.	?	Palladium	Pd.	54
Terbium	Tb.	?	Platinum	Pt.	99
Cerium	Ce.	46	Iridium	Ir.	99
Lanthanum	La.	44	Rhodium	Ro.	52
Didymium	Di.	48	Osmium	Os.	100
Thallium	Tl.	203	Ruthenium	Ru.	52

oxygen than sulphurous and sulphuric acids. The prefix "hyper" is used in the same way; thus, hyperchloric acid, an acid containing *more* oxygen than chloric acid.

With respect to bases, the general termination is in *ide*. If oxygen and lead unite, we have oxide of lead; and in the same manner we have chlorides, bromides,

What are the symbols for the elementary bodies? What are their atomic weights? What does the termination *ide* signify?

I

iodides, and fluorides. And if these elements forn compounds in more proportions than one, we indicate their proportion by the Greek numerals, protos, deuteros, tritos; thus we have protoxides, deutoxides, tritoxides; the protoxide of lead contains one atom of oxygen and one of lead, the deutoxide of nitrogen two atoms of oxygen and one of nitrogen, etc. In the same manner the prefixes sub, sesqui, and per are used; thus a suboxide contains the lowest proportion of oxygen, peroxide the highest proportion, and a sesquioxide intervenes between a protoxide and a deutoxide, its oxygen being in the proportion of one atom and a half.

By an alloy we mean the substance arising from t union of two metals; thus copper and zinc unite t form brass, which is an alloy. If one of the metals mercury, the compound is called an amalgam. And when sulphur, phosphorus, carbon, and selenium unite with metals, or with each other, the termination *ide* is used; thus we have sulphides, phosphides, carbides, et

With respect to the nomenclature for salts, the terminations *ate* and *ite* are used to indicate acids in *ic* an *ous* respectively. The sulphate of potash contains su phuric acid, and the sulphite of potash sulphurous aci And as we have already seen that different oxides aris by the union of oxygen in different proportions, and these bodies frequently give rise to different series of salts, the operation of the nomenclature may be readily traced; thus the protosulphate of iron is the sulphate of the protoxide of iron, but the persulphate of iron is a sulphate of the peroxide, and the deutosulphate o platinum a sulphate of the deutoxide of platinu When the relative quantity of the acid and base vari Latin numerals are employed; thus the bisulphate potassa contains two atoms of sulphuric acid and one potassa.

Salts are said to be neutral if neither their acid n base be in excess. If the acid predominates, it is a acid or super-salt; if the base, it is a basic, or sub-salt.

What do the prefixes *protos, deuteros, tritos, sub, sesqui,* and *per* signify? What is an alloy? When is the termination *ide* employed? What do the terminations *ate* and *ite* indicate? What is the nomenclature for the salts? What is a neutral, an acid, a basic salt?

When a metalloid is united to another metalloid or ıetal, or when a compound radical is united to a metal ɔr metalloid, the combination is called *binary*, from its onsisting of two elements. The binary compounds ɔrmed by chlorine, bromine, iodine, and fluorine with :he alkaline metals are called *haloid* salts. It has been uggested that the constitution of such oxacid salts as itrate of silver, for example, is similar to the haloid ˈlts, and that they consist of the metal united with a ˋothetical radical, instead of an acid united with an ie. This *binary hypothesis* would, however, involve ɔmplete change in the present nomenclature. Car- :ate of potash, KO, CO_2, for example, called by its vocates the carbonate of potassium, would be, in real- , $K CO_3$, the teroxycarbide of potassium, a radical O_3, which is not carbonic acid, being united to the ıetal. The old nomenclature will be adhered to in this vork.

LECTURE XXXV.

ɪᴇ Sʏᴍʙᴏʟs.—*Failure of the Nomenclature in the Case of Complex Compounds.—Failure in Difference of Grouping.—Symbols for Elementary Bodies.—Expressions for several Atoms.—Use of the Plus Sign.—Expressions for Grouping.*

So long as the constitution of compound bodies is imple there is no difficulty in applying the nomencla- ˑˑe, or in recognizing from the name of the compound nature and proportions of its constituents. Thus, toxide of hydrogen clearly indicates a body in which atom of oxygen is united with one of hydrogen— ılphate of potassa, a body composed of two atoms of ɹhuric acid and one of potassa; and even in more nplicated cases, such as the sulphato-tricarbonate of ıd, etc., the same principles will serve as a guide.

But when compound bodies consist of a great num- ɔer of atoms, the nomenclature ceases to be of any serv- ˈce. Thus starch is composed of twelve atoms of car-

What is a binary compound? What is a haloid salt? What is the binary hypothesis? When does the nomenclature apply? When does it fail?

bon, ten of hydrogen, and ten of oxygen. Fibrin composed of 216 atoms of carbon, 169 of hydrogen, 6 of oxygen, 27 of nitrogen, 2 of sulphur, and a trace of phosphorus. On the principles of the nomenclature, it would be difficult to give to the first a technical name and in the case of the latter impossible.

The peculiarity of organic compounds is, that they contain but few of the elementary bodies, being chiefly made up of carbon, hydrogen, oxygen, and nitrogen; but these, as in the case of fibrin, unite in a very complicated way, very often hundreds of atoms being involved. The nomenclature is therefore inapplicable to organic chemistry. (*See* Lecture LXVIII.)

There is also another very serious difficulty in its w It has been discovered that compounds may consist of the same elements, united in precisely the same proportions, so that when they are analyzed they yield precisely the same results, and yet they may, in reality, be very different substances. Identity in composition is no proof of the sameness of bodies. Thus we may have the same elements uniting together in the same proportion, and yielding a solid, a liquid, or a gas indifferent. This result may depend on several causes, as will be presently explained; but among these causes I may here specify what is termed by chemists "Grouping." Thus, suppose four elementary bodies, A B C D, unite together, there is obviously a series of compounds which may arise by permuting or grouping them differently, as in the following example:

(1) A + B + C + D.
(2) A C + B D.
(3) A D + C B.
etc. etc.

The method of symbols which is designed to remove these difficulties, and is, in reality, an appendix and improvement upon the nomenclature, was originally introduced by Berzelius; but the form which is now most commonly adopted is that of Liebig and Poggendorff. The advantages which have been found to accrue from

What is the peculiarity of organic compounds? Why is the nomenclature inapplicable? Is identity of composition any proof of the identity of bodies? What is meant by grouping? Give an example.

. are so great, that it is now introduced into every part)f chemistry, so that it is impossible to read a modern work on this science without having previously master- ;d the symbols.

The student should not be discouraged at the mathe- matical appearance of chemical formulæ. He will find, by a little attention, that they are founded upon the simplest principles, and involve merely the arithmetical erations of addition and multiplication. The follow- is a brief exposition of their nature:

For the symbol of an elementary substance we take first letter of its Latin name, as is shown in the ta- given in the last lecture. Those symbols should be mmitted to memory. But as it happens that several ubstances sometimes have the same initial letter, to dis- inguish between them we add a second small letter. Thus carbon has for its symbol *C.;* chlorine, *Cl.;* cop- er (cuprum), *Cu.;* cadmium, *Cd.*, etc. It may be ob- rved that in the case of recent Latin names the Ger- n synonym is always used; thus potassium is called lium in Germany, and has for its symbol *K.;* sodium called natrium, and has for its symbol *Na.*, etc.

But a symbolic letter standing alone not merely rep- sents a substance; it farther represents *one* atom of ; thus *C* means one atom of carbon, and *O* one atom of oxygen.

If we wish to indicate that more than one atom is present, we affix an appropriate figure, as in the follow- ng examples: $C_{12}.H_{10}.O_{10}$. Thus nitric acid is com- posed of one atom of nitrogen united to five of oxygen, and we write it NO_5.

When a compound, formed of several compounds, is be represented, we make use of an intervening com- ; thus strong oil of vitriol is composed of one atom sulphur and three of oxygen, united with one atom water, which is composed of one atom of oxygen and ne of hydrogen, and we write it SO_3, HO.

If we desire to indicate that compounds are united with a feeble affinity, we make use of the sign +; thus

What are the symbols for elementary bodies? When is a second letter necessary? What does a single symbol standing alone repre- sent? How are more atoms than one represented? How is the comma employed? What is the use of the *plus* sign?

the composition of sulphuric acid may be written SO_3 or SO_2+O, the latter formula implying that one of th. atoms of oxygen is held by a feebler affinity than the other two.

When a large figure, or coefficient, is placed on the same line as the symbol, and to the left of it, it multiplies that symbol as far as the first comma or + sign or, if the formula be placed in a parenthesis, it multiplie every letter under the parenthesis; thus $2SO_3$, K HO or $2SO_3+KO+HO$ mean two atoms of sulph acid united with one of potassa and one of water, fo ing the bisulphate of potassa; but $2(SO_3, KO, H$ would represent two atoms of a salt composed of c of sulphuric acid, one of potassa, and one of water, t figure here multiplying all under the parenthesis.

The advantages which arise from the use of these simple rules are very great; we can, even with the mos complex bodies, not only express their composition, b also the molecular arrangement or grouping of their oms; we can follow them through the most intric changes, and without difficulty trace out their me morphoses. For example, analysis shows that alcoh is composed of C_4, H_6, O_2;
but many facts in its history lead us to know that i molecular constitution is

$$(C_4H_5)O+HO;$$

that is to say, it contains a compound radical, C_4H_5, t which the name of ethyle has been given, and this fac being understood, we see at once that upon the princ ples of the nomenclature the true name for alcohol is hydrated oxide of ethyle; moreover, alcohol is deri by processes of fermentation from sugar. The const tion of dry fruit sugar is

$$C_{12}, H_{12}, O_{12}.$$

This complex atom, under the influence of active ye is split into

$$2(C_4H_6O_2)\ldots\ldots 4(CO_2),$$

that is to say, into two atoms of alcohol and four c carbonic acid gas; and, accordingly, we find, during fermentation, that the sugar disappears, alcohol forming in the liquid, and carbonic acid gas escapes.

How far does a coefficient multiply? What are the advantages arising from the symbols? Give an example in the case of alcohol.

The student should accustom himself to the translation of the nomenclature into symbols, and symbols into the nomenclature, in cases where it is possible, for it is absolutely essential that he should be perfectly familiar with the process.

LECTURE XXXVI.

(E LAWS OF COMBINATION.—*Law of Fixed Proportions.—Numerical Law.—Multiple Law.—Modes of expressing Composition.—Proportions, Equivalents, and Atomic Weights.—Relation between Combining Volumes and Atomic Weights.—Table of Specific Gravities and Atomic Weights.—Gerhardt's System of Notation.*

IT has been shown in the first lectures that material ubstances possess an atomic constitution, and all the henomena of chemistry bear out this conclusion. It llows, therefore, when substances combine with each ther and give rise to new products, the union takes lace by the atoms of the one associating themselves ith the atoms of the other; and as these atoms possess weight and other properties which are specific, there are certain circumstances, easily foreseen, which must attend such combinations.

1st. The constitution of a compound body must always be fixed and invariable. This arises from the fact of the unchangeability of the properties of atoms; one at, om of water will always be composed of one atom of ygen and one of hydrogen; one atom of carbonate of ne will always consist of one atom of carbonic acid d one of lime. Or, more generally, if a good analysis water has shown that nine grains of that substance ntain eight grains of oxygen and one of hydrogen, every subsequent analysis will correspond therewith.

2d. The proportions in which bodies are disposed to unite with each other can always be represented by certain numbers; these numbers being, in fact, the relative

---

In what manner does combination of bodies take place? What is meant by the law of fixed proportions? What by the numerical law?

weights of their atoms. Thus water is composed of an atom of oxygen and one of hydrogen; and inasmuch as the oxygen atom is eight times heavier than that of hydrogen, it necessarily follows that in every nine parts of water we shall have eight of oxygen and one of hydrogen. These numbers are, therefore, spoken of as the combining proportion or equivalents of the substances to which they are attached. If, farther, we examine, when oxygen and sulphur unite, what are the relative quantities, we shall find that eight parts of oxygen combine with sixteen of sulphur, forming hyposulphurous acid. And if sulphur and hydrogen unite, it will be found that sixteen of sulphur combine with one of hydrogen. In this manner, by examining the various elementary bodies, we find that certain numbers are expressive of the proportions in which they are disposed to unite, and these numbers represent the relative weights of their atoms; thus, if 1 be taken as the atomic weight of hydrogen, that of oxygen is 8, that of sulphur 16, etc. the atomic weights of the elementary bodies have been given in Lecture XXXIV.

3d. If two substances unite with each other in more proportions than one, those proportions bear a very simple arithmetical relation to one another; thus 14 grains of nitrogen will successively unite with 8, 16, 24, 32, 40 grains of oxygen, forming successively the protoxide of nitrogen, the deutoxide, hyponitrous acid, nitrous acid and nitric acid. And when the numbers expressing the amount of oxygen are examined, it is seen that they are in the second twice, in the third thrice, in the fourth four times, and in the fifth five times the amount of the first; they are, therefore, simple multiples of it. The reason of this is plain when we write the constitution of these bodies in symbols; they are successively,

$$NO \,..\, NO_2 \,..\, NO_3 \,..\, NO_4 \,..\, NO_5;$$

and if one atom of oxygen weighs 8, two must weigh 16, three 24, four 32, etc.; the multiple law, therefore is a necessary consequence of the combination of atoms.

Observation has shown that there are two series according to which bodies may unite with each other:

Give an example in each case. What do the numbers represent? Give examples of these numbers. What is meant by the multiple law? Give an example of it in the case of nitrogen and oxygen.

(1.) 1 atom of A may unite with 1, 2, 3, 4, 5, etc., atoms of B.
(2.) 1 atom of A may unite with $\frac{1}{2}$, 1, $1\frac{1}{2}$, 2, $2\frac{1}{2}$, 3, etc., atoms of B.

But as an atom is indivisible, there can be no such thing as a half atom; consequently the second series becomes,

(3.) 2 atoms of A may unite with 1, 2, 3, 4, 5, etc., atoms of B.

The three foregoing laws are known under the name of the laws of combination; they are the law of definite proportions, the law of numbers, and the multiple law.

There are three ways in which the composition of a substance may frequently be expressed: 1, by atom; by weight; 3, by volume. Thus the constitution of water, by atom, is one of oxygen to one of hydrogen; by weight, it is one of hydrogen to eight of oxygen; and by volume, two of hydrogen to one of oxygen. These different modes of expression involve nothing contradictory; they are all reconciled by the statement that the atom of oxygen is eight times as heavy as that of hydrogen, but only half the size.

By some authors the terms combining proportion and equivalent are used; they are to be understood as having the same signification as atomic weight. And as we know nothing of the absolute weight of atoms, but only their relative proportions to each other, we may select any substance with which to compare all the rest, and make it our unit or term of comparison. In this book hydrogen is employed for this purpose, and its atomic weight is marked 1; on the Continent of Europe oxygen is selected, and marked 100. It is obvious that this does not affect the relationship of the numbers, for it is the same thing whether we state the atomic weights of hydrogen and oxygen as 1 to 8, or as $12\frac{1}{2}$ to 100.

Combinations may take place in two different ways: 1st, in definite proportions; 2d, in indefinite proportions. It is to the former that all the foregoing observations and laws apply. One grain of hydrogen will not unite

---

What are the two series in which bodies may unite? In what ways may the composition of a body be expressed? How is the contradiction reconciled? What do proportion and equivalent signify? What substances are used as standards of comparison? What are the two modes of combination?

with nine or seven grains of oxygen, but only wit eight. But one drop of spirits of wine may combin with one of water, or with a pint, or a quart, or te gallons. This is what is understood by union in indefi nite proportions.

When two gaseous bodies unite, their combining proportions bear a simple relation to each other; one volume of hydrogen unites with one of chlorine, and produces two volumes of hydrochloric acid. And in t case of the five compounds of nitrogen just referre two volumes of that gas combine successively with 3, 4, 5 of oxygen.

A relation, therefore, exists between the combir volume and the atomic weight of gaseous bodies. the weight of a given volume of oxygen be called 10 that of an equal volume of hydrogen will be 62.5, thes numbers representing, of course, the specific gravity o the two gases. The proportion in which they unite one volume of oxygen to two of hydrogen to form w ter; the relative weights of these quantities, therefo would be 100.0 to 6.25×2; that is, 100.0 to 12.50; these numbers are the atomic weights of the bodies r spectively. From such considerations, it was at c time supposed that, in the case of all gases, the spec gravities would correspond to the atomic weights. E perience has, however, shown that this is not the case as is seen in the following table:

Gas or Vapor.	Specific Gravities.		Chemical Equivalents	
	Air=1.	Hydrogen=1.	By Volume.	By Weigh
Hydrogen	.0690	1	100	1
Nitrogen	.9727	14.12	100	14.
Carbon (hypothetical)	.4213	6.12	100	6.
Chlorine	2.4700	35.84	100	35.
Iodine	8.7011	126.30	100	126.
Bromine	5.3930	78.40	100	78.
Mercury	6.9690	101	200	202
Oxygen	1.1025	16	50	8
Phosphorus	4.3273	62.8	25	15.7
Arsenic	10.3620	150.8	25	37.7
Sulphur	6.6480	96.48	16.66	16.1

From this it is seen that if the combining volume of hydrogen, nitrogen, or chlorine be taken as unity, tha

What relation is observed when gases combine by volume? Wha is the relation between specific gravities and atomic weights?

of oxygen is one half, of vapor of phosphorus one fourth, and of vapor of sulphur one sixth.

A new system of notation has been proposed by Gerhardt, with a view to establish a constant relation between the atomic weight of bodies, their specific gravities, and vapor volumes. He suggests that hydrogen should be the standard unit for the atomic weight, combining volume and specific gravity, and that the equivalents of certain bodies should be doubled, as follows:

	Symbols.	At. Wts.
Hydrogen....................	H.	1
Oxygen........................	*O.*	16
Sulphur........................	*S.*	32
Selenium......................	*Se.*	79
Tellurium.....................	*Te.*	128
Carbon	*C.*	12

The symbols that are italicized represent the doubled atomic weights.

The specific gravity is referred by Gerhardt to hydrogen instead of air, and this has the advantage of generally representing by one set of figures both specific gravities and atomic weights. Thus oxygen is 16 times heavier than hydrogen; its specific gravity would therefore be 16; and as it combines with hydrogen in the proportion of 8 to 1, that is, as 16 to 2, it will take 2 atoms of hydrogen to form water. Hydrogen is thus supposed to unite in the proportion of two atoms to one of oxygen, and hence water is a suboxide of hydrogen, H_2O. Nitric acid becomes NO_3H, and sulphuric acid SO_4H_2.

The radical change that this system requires in the nomenclature, together with the straining of facts that it demands, have prevented its general adoption.

On the principles which have just been developed, we can often calculate the specific gravity of a compound gas with more accuracy than it can be determined experimentally. Thus hydrochloric acid, which consists of equal volumes of chlorine and hydrogen united, with-

What is the object of Gerhardt's nomenclature? What suggestions does he make? What is the advantage of referring specific gravities to hydrogen? On this theory, what is the composition of water? How may the specific gravity of a compound gas be determined?

out condensation, must have a specific gravity of 1.2695, because the specific gravity of hydrogen being 0.0690, and that of chlorine 2.4700, the sum of which, 2.5390, is the weight of two volumes of hydrochloric acid, and, therefore, if we divide by 2, the quotient, 1.2695, is equal to the weight of one volume; or, in other words, the specific gravity of the compound gas.

Sometimes, also, we can determine the specific gravity of a vapor by calculation when it is impossible to do so experimentally. Assuming that one volume of carbonic acid gas contains one volume of oxygen and one of carbon vapor, we have,

Specific Gravity of Carbonic Acid	1.5238
" " Oxygen	1.1025
" " Carbon Vapor................	.4213

The hypothetical specific gravity of the vapor of carbon is therefore .4213.

The rule for the calculation of specific gravities, on the foregoing principles, is, "Multiply the specific gravities of the simple gases or vapors respectively by the volumes in which they combine; add those products together, and divide the sum by the number of volumes of the compound gas produced."

LECTURE XXXVII.

Constitution of Bodies. — *Crystallization.* — *Systems of Crystals.* — *Dimorphism.* — *Isomorphism.* — *Isomorphous Groups.* — *Isomerism.* — *Metameric and Polymeric Bodies.*—*Allotropic States of Bodies.*

It frequently happens that substances assuming the solid form from the liquid or vaporous states, take on a geometrical figure, being terminated by sharp edges and solid angles; under such circumstances, they are said to crystallize. Thus common salt will crystallize in cubes, and nitrate of potassa in six-sided prisms.

The various geometrical forms which crystals can

How is the hypothetical specific gravity of the vapor of carbon determined? Give the rule for calculating specific gravities of compound gases.

thus assume may be divided into six classes or systems:

(1.) The Regular system.
(2.) The Rhombohedral system.
(3.) The Square Prismatic system.
(4.) The Right Prismatic system.
(5.) The Oblique Prismatic system.
(6.) The Doubly Oblique Prismatic system.

This division is founded on the relations of certain lines or axes which may be supposed to be drawn through the centre of the crystal round which its parts are symmetrically arranged.

THE REGULAR SYSTEM.

This has three equal axes at right angles to each other.

The letters *a a* show the direction of the axes. The figure (*Fig.* 171) represents: 1. *The cube;* 2. *Regular*

Fig. 171.

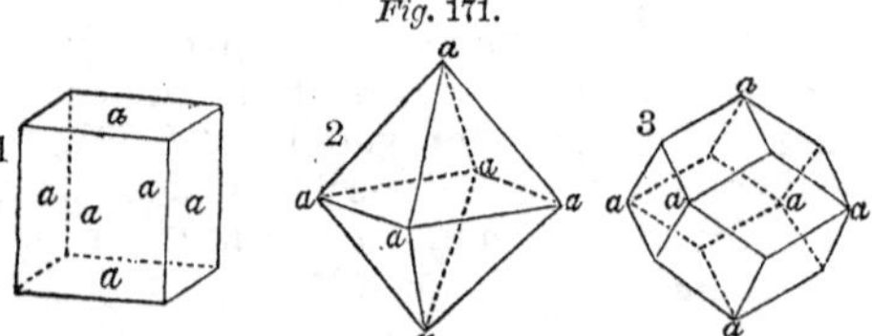

octahedron; and, 3. *Rhombic dodecahedron.*

THE SQUARE PRISMATIC SYSTEM.

This has three axes, two of which are equal and the third of a different length.

Fig. 172.

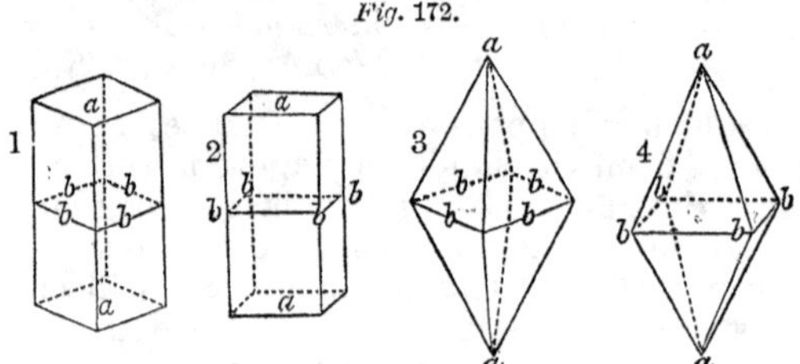

What are the six systems of crystallization? Upon what is this division founded? In the regular system, what is the relation of the axes? In the square prismatic system, what is their relation?

a a is the principal axis, *b b* the secondary one. In the figure (*Fig.* 172), 1 is a *right square prism*, with the axes on the center of the sides, *b b;* 2 is a *right square prism*, with the axes in the edges; 3 and 4, corresponding *right square octahedrons.*

THE RIGHT PRISMATIC SYSTEM

has three axes, *a a*, *b b*, *c c*, of unequal lengths, at right angles to each other.

In the figure (*Fig.* 173), 1 is *a right rectangul*ʋ

Fig. 173.

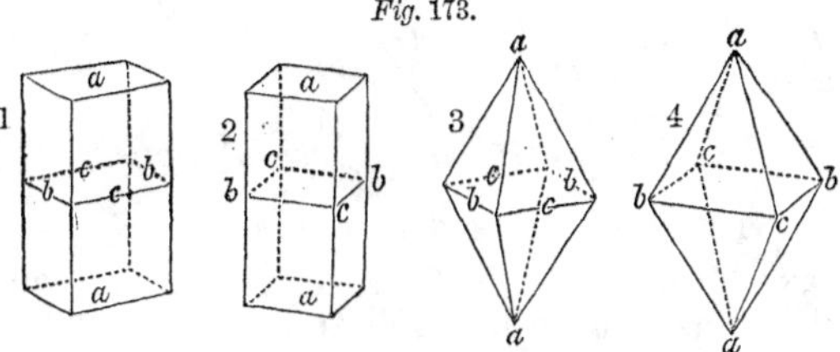

prism; 2. *Right rhombic prism;* 3. *Right rectangula based octahedron;* 4. *Right rhombic-based octahedron.*

THE OBLIQUE PRISMATIC SYSTEM

has three axes, which may be unequal; two are placec at right angles to each other, and the third is oblique to one and perpendicular to the other.

In the figure (*Fig.* 174), 1 is an *oblique rectangular*

Fig. 174.

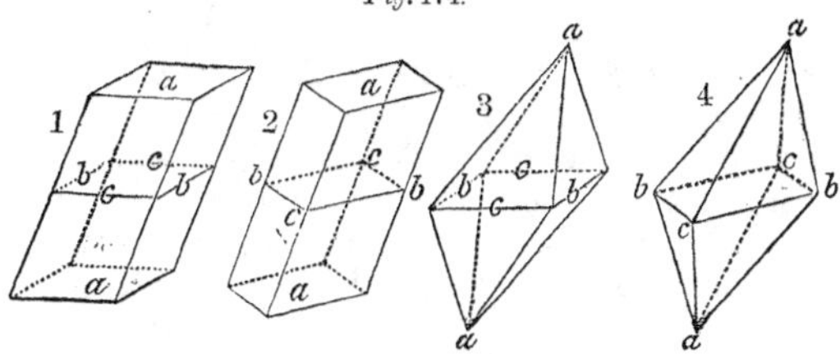

prism; 2. *Oblique rhombic prism;* 3. *Oblique rectangular-based octahedron;* 4. *Oblique rhombic-based octahedron.*

What is the relation of the axes in the right prismatic system? In the oblique prismatic, what is it?

THE DOUBLY OBLIQUE PRISMATIC SYSTEM

has three axes, which may be all unequal and all oblique.

In the figure (*Fig.* 175), 1 and 2 are *doubly oblique*

Fig. 175.

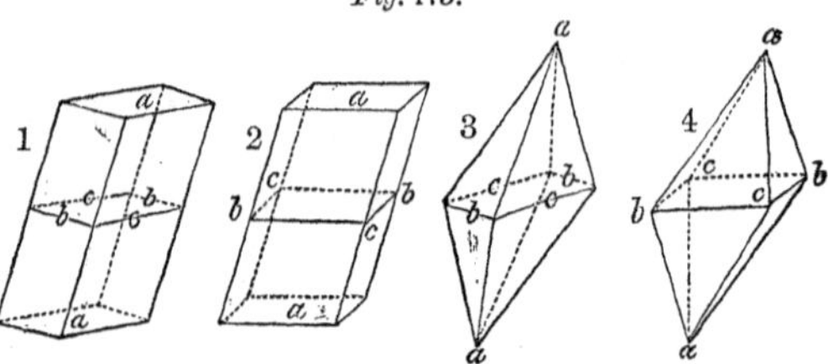

prisms, and 3 and 4 *doubly oblique octahedrons.*

THE RHOMBOHEDRAL SYSTEM

has four axes, three of which are equal in the same plane, and inclined at angles of 60°; the fourth, which is the principal axis, is perpendicular to all.

In the figure (*Fig.* 176), 1 is the *regular six-sided*

Fig. 176.

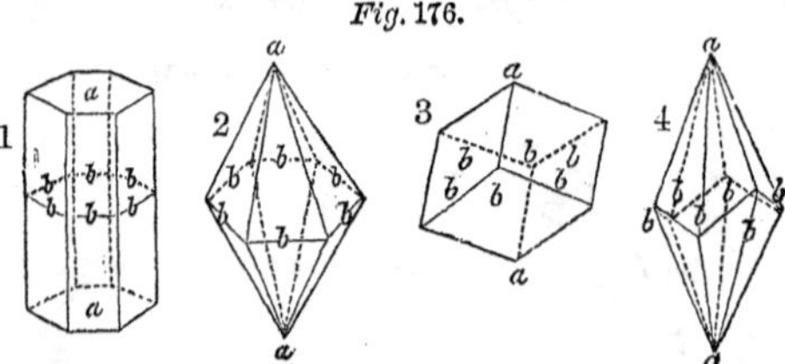

prism; 2. the *dodecahedron*; 3. *Rhombohedron*; 4. another *dodecahedron.*

It often happens, owing to a change in the deposit of new matter on a crystal while forming, that other figures than the proper one is produced; thus the cube may pass into the octahedron, as shown in *Fig.* 177.

The effect may perhaps be better conceived by imagining the solid angle of the cube 1 to be cut off by planes equally inclined to the constituent faces; 2 rep-

---

What is the relation in the doubly oblique prismatic? How many axes are there in the rhombohedral system, and what is their relation? In what manner may crystals of one form pass into another, as the cube into the octahedron?

Fig. 177.

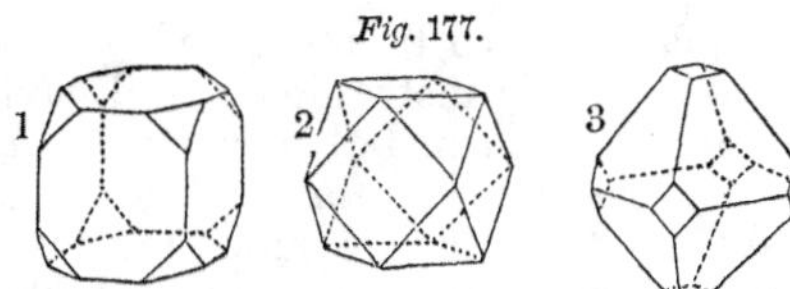

resents an increased removal of the same kind; 3, one still farther advanced.

Sometimes it happens that each alternate plane of crystal grows at the expense of the adjacent one, givin rise to *hemihedral*, or half-sided crystals, as is shown *Fig.* 178, which represents the tetrahedron, arising

Fig. 178.

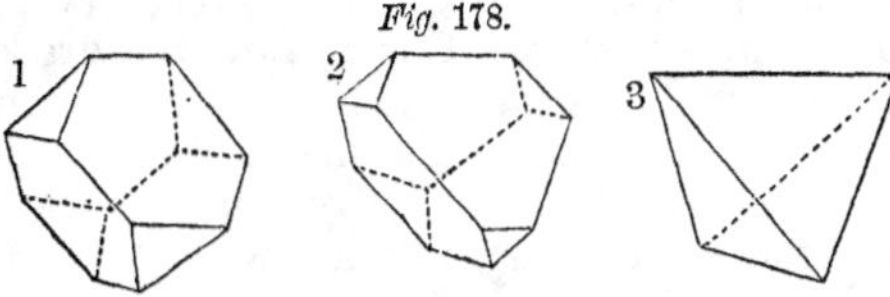

this manner from the octahedron by the growth of eac alternate face. 1. The octahedron partially modifie 2. The change farther advanced; 3. The tetrahedro completed.

The angles of crystals are measured by goniometers of which there are sever al kinds; as the common goniometer, and Wollaston's reflecting goniometer. This instrument is represented in *Fig.* 179 The crystal to be mea ured, *f*, is fixed upon movable support, *d*, whic is in connection with th button-headed axis of th goniometer, *o*, which pas es through a larger axis in the upright, *b*; *a* is a divided circle, and *e* its vernier, which is fixed immovably on the upright, *b*.

Fig. 179.

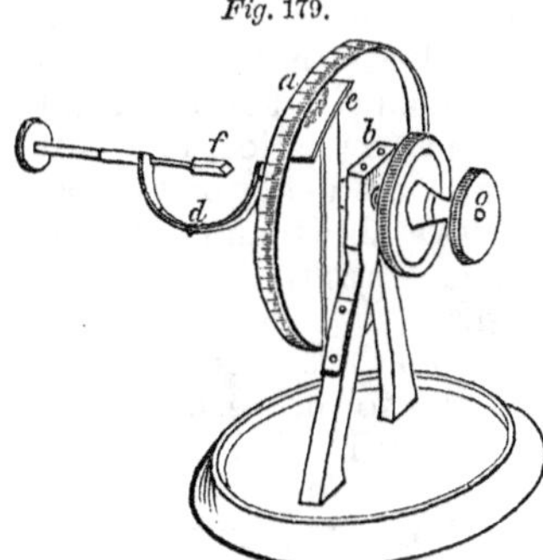

The edge of the crystal, which is formed by the two

What are hemihedral crystals, and how are they produced? Describe the reflecting goniometer.

faces whose inclination is to be measured, is to be set parallel to the axis of the instrument; and having, by means of the button, *o*, turned the crystal until some definite object, such as the bar of a window, is seen distinctly reflected from it, the larger milled head is turned, and with it the divided circle and crystal, until the same object is again seen by reflection from the second face. The angle through which the great circle has moved, subtracted from 180°, gives the angle included between the two crystalline faces, or their inclination to each other.

As a general rule, the same substance, crystallizing under the same circumstances, will produce crystals belonging to the same system. Cases, however, are known in which the same substance belongs to different systems. Thus sulphur will crystallize in rhombic prisms, and also rhombic octahedrons. By dimorphous bodies we therefore mean substances which will afford crystals belonging to two different systems.

Dimorphism is frequently connected with the temperature at which the crystals were produced. Thus carbonate of lime, at ordinary temperatures, yields rhombohedrons, but at the boiling point of water right rhombic prisms; and with this difference of form a difference of chemical qualities may occur; the deutosulphide of iron, for example, crystallizes in cubes which remain unacted upon by water or air; but in the right rhombic form it undergoes rapid oxydation in moist air, producing sulphate of iron. Commonly one of the forms of a dimorphous body is less stable than the other, and if the transition takes place abruptly, it is sometimes attended by a flash of light.

It was discovered by Mitscherlich that when different compound bodies assume the same form, we are often able to trace a remarkable analogy in their chemical composition. Thus the chloride of sodium, the iodide of potassium, the fluoride of calcium, etc., crystallize in the first system. These substances are all constituted

---

How is the goniometer used? What is meant by dimorphous bodies? What effect has temperature on the formation of crystals? What connection is there between chemical qualities and dimorphism? What relation is there in the form and composition of iodide of potassium and chloride of sodium?

upon a common type, in which we have one atom of a metal united to one atom of an electro-negative radical: or, taking M as the general symbol for the metals, and R for the electro-negative radicals, the class is constituted upon the type

$$M, R,$$

and thererefore includes such bodies as

$$KCl..NaCl..KBr..KF..CaF..AmCl..., \text{ etc.}$$

Such substances are called isomorphous bodies, and designations isomorphous elements, isomorphous gro are used, being derived from ισος, equal, μορφη, form.

Let us take a second more complicated case. T formula for common alum, the sulphate of alumina a potash, is,

	$KO, SO_3+Al_2O_3, 3SO_3+24HO$
Ammonia alum is	$AmO, SO_3+Al_2O_3, 3SO_3+24HO$
Chrome alum is	$KO, SO_3+Cr_2O_3, 3SO_3+24HO$
Iron alum is	$KO, SO_3+Fe_2O_3, 3SO_3+24HO$

And in the same way an extensive family of alums m be formed by the substitution of a limited number various other bodies comprised in the general formul

$$mO, SO_3, +M_2O_3, 3SO_3+24HO,$$

in which m represents any metal belonging to the p tassium group, and M any one belonging to the aluminum group.

All these alums crystallize with the same form, and such illustrations afford us reason to believe that tha similarity of form is due, in a great measure, to the *grouping* or *arrangement* of the constituent atoms; *th at in a compound molecule the substances which can place one another without giving rise to a change external form must have certain relationships to ea other. We call them, therefore, isomorphous.*

From the external forms of bodies we may next tu to their internal constitutions, calling to mind what h been already observed in Lecture XXXV., that identity of composition by no means implies identity of character. Two substances may be composed of the same

Why are they called isomorphous bodies? Give an example of isomorphism in the case of the alums. What general conclusion may be drawn from these facts?

elements, united in the same proportions, and yet be to-ally unlike; and it is obvious that this may be due to two different causes: 1st. Difference of grouping; 2d. Difference in the absolute number of atoms.

Difference of grouping I have already explained in the lecture just quoted; and with respect to difference in the absolute number of atoms, the effect is obvious rom an example. Thus we have as the constitution of

Aldehyde	$C_4H_4O_2$
Acetic Ether	$C_8H_8O_4$

d these bodies, if analyzed, would, of course, yield ecisely the same proportions in 100 parts, the true fference being that the atom of acetic ether contains twice as many constituent atoms as that of aldehyde, and is, therefore, exactly twice as heavy, though equal weights of the two will yield equal quantities of their constituents.

To these peculiarities the term isomerism is applied, d by isomeric bodies we mean bodies composed of e same elements in the same proportion, but differing properties. When isomerism arises from difference grouping, the bodies are said to be metameric; and hen it arises from difference in the absolute number of atoms, they are called polymeric.

There is a third cause which gives rise to the phenomena of isomerism: it is the allotropic condition of elementary bodies. Carbon, for example, exists under number of different forms; we find it as charcoal, umbago, and diamond. They differ in specific gravi-, in specific heat, and in their conducting power as spects caloric and electricity. In their relations to ght, the first perfectly absorbs it, the second reflects it e a metal, the third transmits it like glass. In their lation to oxygen they also differ surprisingly; there e varieties of charcoal that spontaneously take fire in he air, but the diamond can only be burned in pure

---

What two causes may give to bodies of the same composition different characters? Give an example of the effect of difference of the absolute number of atoms. What is meant by isomerism? What are metameric and polymeric bodies? What is meant by the allotropic condition of bodies? What allotropic states does carbon present?

oxygen gas. The second and third varieties do not belong to the same crystalline form.

It is now known that a great many elementary substances are affected in this manner. Professor Draper showed that this is the case with chlorine gas, which changes under the influence of the indigo rays (*Philosophical Magazine*, July, 1844). In the same manner it has been long known that iron exists in two states 1st. In its ordinary oxidizable state; 2d. In a conditio in which it simulates the properties of platinum or go

There can be no doubt that these peculiarities a carried by these bodies when they unite to form co pounds; thus, for example, if carbon and hydrog unite, it is possible we may have three different cor pounds; one containing charcoal carbon, a second plumbago carbon, a third diamond carbon; or, if we designate these respectively as $C\alpha$, $C\beta$, $C\gamma$, we may have

$$C\alpha H \ldots C\beta H \ldots C\gamma H;$$

and perhaps, as M. Millon has suggested, carbureted hydrogen gas and otto of roses, which have the same co stitution, differ in the one containing charcoal and th other diamond.

These peculiarities are known under the name of all tropic states, and the phenomenon itself under the des ignation of allotropism.

LECTURE XXXVIII.

CHEMICAL AFFINITY. — *Phenomena accompanyin Chemical Affinity.—Disturbance of Temperature.—Production of Light.—Evolution of Electricity.—Change of Color.—Change of Form.—Change o Chemical Properties.—Change of Volume and De sity.—Tables of Geoffroy.—Measure of Affinity.—Disturbing Causes.—Catalysis.*

BY chemical affinity we mean the attraction of atoms of a dissimilar nature for each other, an attraction which is exhibited upon the apparent contact of bodies.

How may an allotropic change be impressed on chlorine? What are the allotropic states of iron? Are the peculiarities continued in their compounds? What is meant by chemical affinity?

There are certain striking phenomena which very frequently accompany chemical action. They are the evolution of Light, Heat, and Electricity; and, as respects the bodies engaged, they may exhibit changes of color, of form, of volume, of density, or of their chemical properties.

If in a glass vessel, *a*, *Fig.* 180, a mixture of strong sulphuric acid and water be stirred together by means of a tube, *b*, containing some sulphuric ether, so much heat will be evolved by the acid and water as they unite, that the ether will be made to boil rapidly.

Fig. 180.

If upon some water contained in a shallow dish, *Fig.* 181, a piece of potassium be thrown, the potassium decomposes the water with the evolution of a beautiful lilac flame.

Fig. 181.

As respects the evolution of electricity during chemical action, the Voltaic battery, and, indeed, all Voltaic combinations, are examples. In the simple circle we have already, in Lecture XXIX., traced the production of electricity to the decomposition of the water.

We have observed that the evolution of Light, Heat, and Electricity is not the only phenomenon to be remarked during the play of chemical affinity; the ponderable substances themselves undergo changes.

If in a glass containing litmus water a drop of sulphuric acid be poured, the blue color of the litmus is at once changed to a red, and if into the reddened liquid so produced a little ammonia be poured, the blue color is restored. This simple experiment is of considerable interest, for the reddening of litmus is commonly received as one of the attributes of acid bodies, and the restoration of the blue color of those belonging to the alkaline type.

On adding to a solution of sulphate of copper a small quantity of ammonia, a pale green precipitate is thrown down; a greater quantity of ammonia redissolves this precipitate, and gives rise to a splendid purple solution.

What phenomena accompany chemical action? What changes are exhibited by the bodies engaged? Give examples of the evolution of heat, light, and electricity. Give examples of changes of color.

A similar solution of sulphate of copper gives rise, under the action of a solution of ferrocyanide of potassium, to a deep chocolate-colored precipitate.

A solution of the nitrate of lead, which is colorless, acted on by a solution of iodide of potassium, also colorless, gives rise to the production of a beautiful yellow precipitate, the iodide of lead.

And, lastly, if sulphuric acid be placed in a solution of a soluble salt of lead, or of baryta, a white precipitate at once goes down.

These are all instances of changes of color, and such changes are of the utmost importance in practical chemistry, inasmuch as the art of testing depends, for the most part, upon a knowledge of them.

Changes of form in the same manner are exhibited; thus, when gunpowder explodes, a large proportion of the ingredients, from being in the solid, escapes in the gaseous state. If upon fragments of chalk, carbonate of lime, we pour hydrochloric acid, a violent effervescence takes place, due to the escape of carbonic acid, which, from being in the solid, assumes the gaseous form.

The converse of this is sometimes seen, vapors passing into the solid state. In the glass *a*, *Fig.* 182, place some strong hydrochloric acid, and in *b* some strong ammonia; both these bodies yield vapors at ordinary temperatures in abundance, and those vapors, meeting in the air over the glasses, give rise to a dense fume, or smoke, which, if examined, proves to be solid sal ammoniac.

Fig. 182.

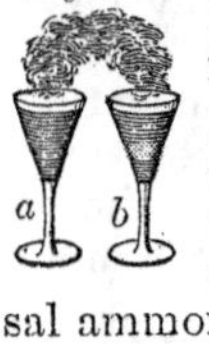

Fig. 183.

Very often change of form is accompanied by change of color; thus, if under a large bell jar, *Fig.* 183, there be placed a wine-glass containing a few copper or iron nails and nitric acid, a gas of a deep orange color makes its appearance, filling the whole bell.

Perhaps no better instance of an entire change of properties could be cited than that of the combustion of phosphorus in atmospheric air.

On what does the art of testing depend? Give an example of the production of a gas from a solid, and of a solid from gases.

This substance phosphorus is a body of a waxy appearance, possessing so great a degree of combustibility that it requires to be kept under the surface of water to prevent the action of the air. If a piece of it be set on fire beneath a clear and dry bell-jar, as shown in *Fig.* 184, it unites with great energy with the oxygen of the included air, producing white flakes, which, as the combustion is ceasing, descend in the jar, giving a miniature representation of a fall of snow. On collecting some of this phosphoric snow, its properties will be found to be in striking contrast with the phosphorus which produced it; for instance, far from being unacted on by water, it has such an intense affinity for that substance that it hisses like a red-hot iron when brought in contact with it. It reddens litmus solution, and possesses the qualities of a powerful acid. Nor is the change confined to the phosphorus; if we examine the air in which it was burned, we find it has lost its quality of supporting combustion.

Fig. 184.

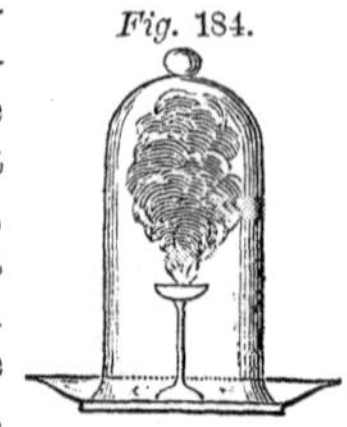

Changes of volume, and, consequently, changes of density, constantly attend chemical action; a pint of water and a pint of sulphuric acid, mixed together, form less than two pints; and the same may be observed of alcohol and water.

When to two substances already in union a third, having a stronger affinity for one of the other two, is presented, decomposition ensues. Thus, if to the carbonate of soda nitric acid be presented, the soda and nitric acid combine, and the carbonic acid is driven off in the form of a gas. And, again, if upon the nitrate of soda so produced sulphuric acid be poured, the nitric acid is driven off, and sulphate of soda results. It was at one time thought that, by examining a number of such cases, we might discover the order of affinity of bodies for one another and arrange them in tables; these are sometimes called the Tables of Geoffroy. Thus the table

---

What are the changes which phosphorus undergoes when burned in the air? Give an example of change of volume and of density. Under what circumstances does decomposition take place? What are the Tables of Geoffroy?

Soda.	
Sulphuric	Acid.
Nitric	"
Hydrochloric	"
Acetic	"
Carbonic	"

presents us with the order in which a number of acids stand in relation to soda, the most powerful being the first on the list, and the salt which results from the union of any one of those acids with the soda can be decomposed by the use of any other acid standing higher on the list.

But it is now known that these tables are far from representing the order of affinities; a weaker affinity often overcomes a stronger by reason of the intervention of disturbing extraneous causes; and tables so constructed lead, therefore, to contradictory conclusions. Some very simple considerations may illustrate this. Potassium can take oxygen from carbon at low temperatures, or, in other words, decompose carbonic acid gas, but it by no means follows that the affinity of potassium for oxygen is greater than that of carbon, and accordingly we find that at higher temperatures carbon can take oxygen from potassium. Indeed, under the influence of heat, light, and electricity, we find all kinds of chemical changes going on, and in the same manner the condition of form exerts a remarkable influence in these respects, so that cohesion and elasticity may be placed among the predisposing causes producing chemical results. If a number of bodies exist in a solution together, they will at once arrange themselves in such a way under the influence of cohesion as to produce *insoluble* precipitates, if that be possible; or, under the influence of elasticity, to determine the evolution of a gas; if the carbonate of soda be decomposed by acetic acid, it by no means follows that the latter has the stronger affinity for soda, the decomposition being probably determined by the fact that the carbonic acid can take on the elastic form and escape away as a gas. The sulphate of

How may it be shown that these are not the tables of affinity? What may be enumerated among these disturbing causes? What is the influence of cohesion? What is the influence of elasticity? Give examples of the action of these disturbing agencies?

oda may be decomposed by baryta, the cause of the ecomposition being probably due to cohesion, for the sulphate of baryta which results is a very insoluble ody. We have, therefore, no true measure of affinity, or the relation of bodies in this respect changes with xternal conditions, and the Tables of Geoffroy are only ables of the order of decompositions, but not of the order of affinity.

Catalysis is a term which was first employed by Berus to explain those cases in which two bodies are sed to unite by a third, which itself suffers no change. is is also called action by presence. In the producn of oxygen from chlorate of potassa and black oxide manganese, the latter substance, while it causes the ction to go on at a lower temperature, itself remains nchanged. In the same manner, platinum black will ause the union of hydrogen and oxygen.

What do the Tables of Geoffroy express? What is meant by catsis? Give examples.

K

PART III.

INORGANIC CHEMISTRY.

LECTURE XXXIX.

PNEUMATIC CHEMISTRY.—*Ancient Opinions on the C stitution of Gases.—Doctrine of the Unity of the A —The Pneumatic Trough.—The Gasometer.—T. Mercurial Trough.*—OXYGEN.—*Modes of Prepar tion.—Properties.—Origin of its Name.—Relatior to Atmospheric Air and Combustion.—Burning o Metals.*

IN the catalogue of the elementary bodies of the a cients four substances were enumerated—earth, air, fi and water. Of these, three are now known to be co pounds, and the fourth a state into which bodies are casionally thrown. Modern researches have shown tl there are about 68 different elements.

For a length of time it was supposed that the vario exhalations and vapors were nothing more than vitiate forms of atmospheric air; and though, from time t time, first one and then another of the gaseous bodie was discovered, chemists were slow to admit that the were any thing more than modifications of one commc principle. Thus Roger Bacon, in the thirteenth century, discovered one of the carburets of hydrogen, a Van Helmont, in the sixteenth, carbonic acid. The visibility of these bodies—their remarkable chemical lations in extinguishing flame and producing death— great mechanical force to which they often gave when generated in pent-up vessels—their occurrence mines, the bottom of wells, in church-yards, and lone places, suggested to a superstitious mind a supernatur origin, and Van Helmont gave them the name of gas, which signifies a ghost or spirit.

What opinions were formerly held respecting the different gases? What was the original meaning of the term gas?

But it is to the researches in the properties of fixed air, or carbonic acid, which Black made in 1757, that pneumatic chemistry owes its origin. These were soon followed by the discoveries of Priestley, Scheele, and others. That of oxygen gas by the former of these philosophers in 1774 forever destroyed the ancient notion of vitiated airs, for this gas can support combustion and respiration far better than the atmosphere. It may be said that modern chemistry dates its origin from the discovery of oxygen gas.

For the purpose of confining gases and manipulating with them, a contrivance of Priestley's, the pneumatic trough, *Fig.* 185, is of continual use. It consists of a

Fig. 185.

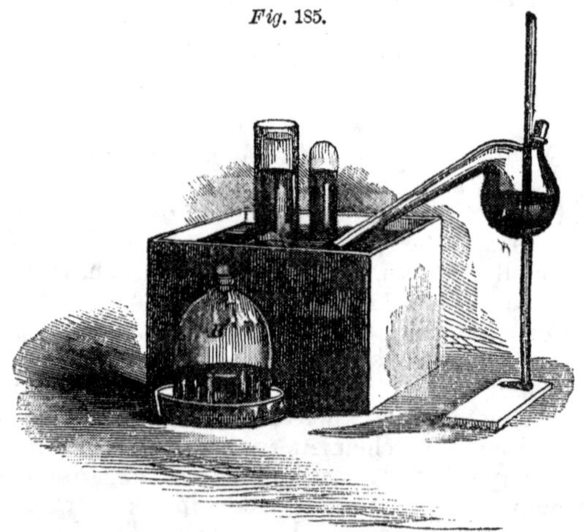

water-tight trough or box, in which a shelf is placed at a distance of an inch or more under the level of the contained water. The jars to hold the gases are filled with water, and placed mouth downward on this shelf. When gas is disengaged from materials contained in a retort, the end of the retort neck is caused to project under the mouth of one of the jars, and the gas rises, bubble by bubble, eventually filling the jar. When that is accom-

By whom was the doctrine of the plurality of gases established? Who invented the pneumatic trough? Describe it. How are the jars filled?

plished the jar may be moved to one side, or, a plate being inserted under its mouth, it may be taken out of the trough altogether, as at *B*.

When large quantities of gas are to be preserved, a gasometer, *Fig.* 186, is employed. It consists of a large

Fig. 186.

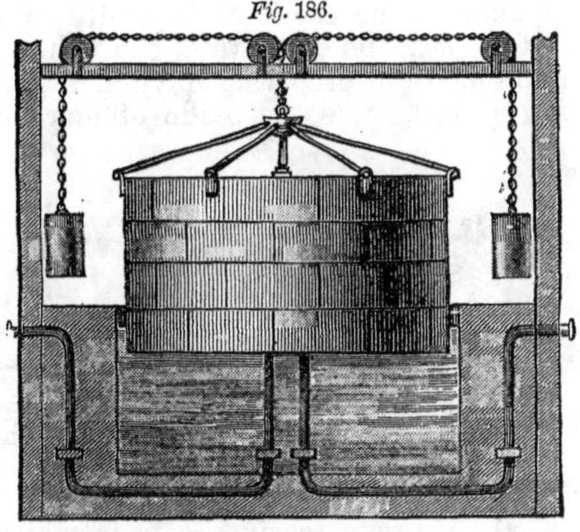

cylindrical vessel, counterpoised in a tank of water by the aid of chains and weights. Two pipes open into it below, one to bring in the gas from the factory, the other to distribute it where wanted. As gas accumulates, the cylindrical vessel rises out of the water more and more, until, when its lower edge is on the surface, it is full. The water is kept from freezing in winter by the aid of jets of steam.

Fig. 187.

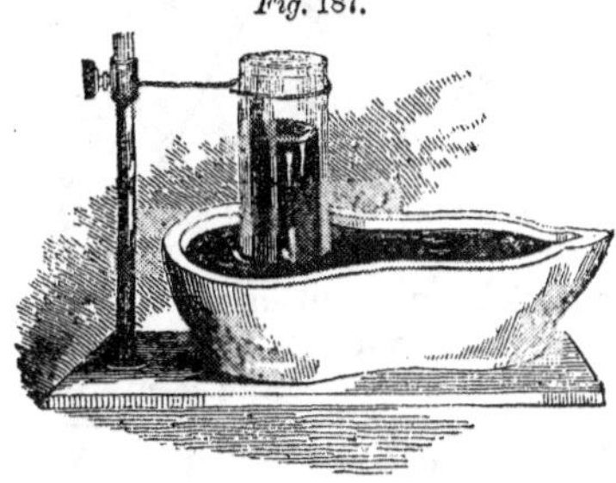

As all gases are more or less soluble in water, and some to such an extent that they can not be collected over that fluid, the mercurial trough, *Fig.* 187, has occasionally to be employed. It is in all respects similar to the pneumatic trough, except that its size is

Describe the gasometer. What is the use of the mercurial trough?

limited by the expense and weight of the quicksilver.

OXYGEN. O=8.

Oxygen gas is probably the most abundant of the elements. It constitutes about one third of the weight of the solid mass of the earth, eight ninths of that of the waters of the sea, and one fifth of the volume of the air. A simple mode of preparing oxygen is to place in a retort, *a*, *Fig.* 188, some red oxide of mercury, con-

Fig. 188.

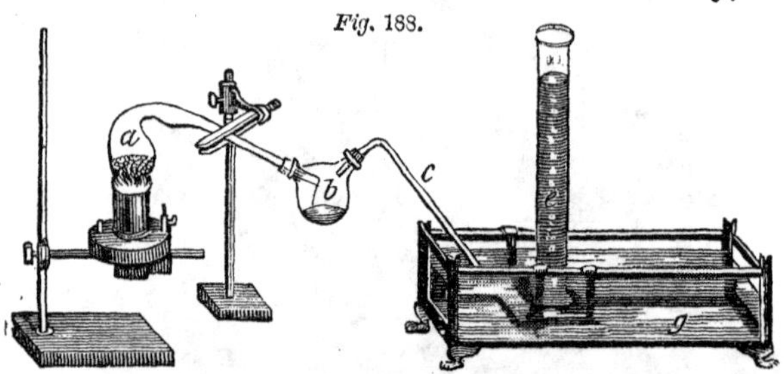

necting with the retort a receiver, *b*, from which there passes a bent tube, *c*, dipping below the water of a pneumatic trough, *g*. On raising the temperature by a lamp, the oxide is decomposed into metallic mercury and oxygen gas; the former distills into the receiver, *b*, and the latter collects in the inverted jar in the trough. This was the process resorted to by Priestley.

It may also be readily procured by heating in a flask, *a*, *Fig.* 189, a mixture of 1 part of peroxide of manganese and 3 parts of powdered chlorate of potassa. To the mouth of the flask a tube, *b*, is adapted by means of a tight cork, the lower end of the tube dipping beneath a jar, *c*, on the pneumatic trough. On raising the temperature of the flask by a spirit-lamp, oxygen is freely evolved. It is derived from the chlorate of potassa altogether, the peroxide of manganese merely

Fig. 189.

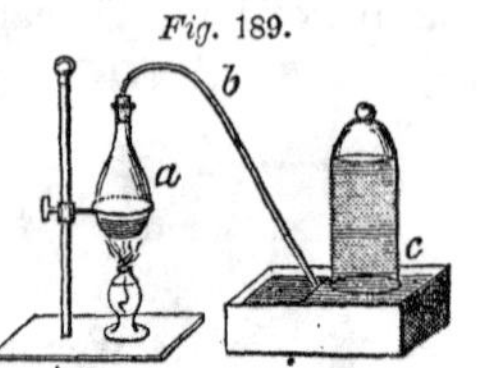

What limits its size? In what bodies does oxygen occur? Describe its preparation from red oxide of mercury; from chlorate of potassa.

acting by presence (catalysis). The change is as follows:

$$KO + ClO_5 = KCl + O_6,$$

that is, chloride of potassium arises, and six atoms of oxygen are given off. One ounce of the chlorate will yield about two gallons of the gas. A mixture of chlorate of potassa with one tenth its weight of sesquioxide of iron also yields oxygen with facility and at a low temperature.

On a large scale oxygen is prepared by heating peroxide of manganese ($Mn\ O_2$) to a red heat in a wrought iron bottle or retort. One pound, if pure, will yield five gallons of the gas, the manganese parting with about one third of its oxygen.

It may also be produced by heating a mixture of bichromate of potassa and sulphuric acid, or peroxide of manganese and sulphuric acid.

Oxygen gas may be procured directly from the atmosphere by the aid of baryta. If a current of air containing a small amount of water, and free from carbonic acid, be passed over baryta heated to a low redness, it absorbs oxygen and becomes peroxidized. After a time the current of air is to be shut off and the temperature raised, when the baryta is reduced again to the condition of protoxide, and oxygen is given off. The operation may then be repeated. A pound of baryta will yield nine gallons of pure oxygen gas.

This gas can also be obtained by passing the vapor of sulphuric acid through a porcelain tube heated to redness. The products are oxygen, water, and sulphurous acid. The sulphurous acid is removed by passing it through a solution of carbonate of soda, sulphite and bisulphite of soda arising. The latter is valuable in the arts for the removal of chlorine. A fluid ounce of sulphuric acid produces 360 cubic inches of oxygen.

Oxygen is a colorless, inodorous, and insipid body. It is a non-conductor of electricity, and has the least refractive power of any of the gases. Its specific gravity is greater than that of air in the proportion of 1.1057 to 1.000. It is 16 times as heavy as hydrogen. One

Describe its preparation from peroxide of manganese. How is oxygen prepared by the aid of baryta; of sulphuric acid? What are its leading physical properties? What is its specific gravity?

hundred cubic inches of it weigh 34.24 grains. It is a powerful electro-negative element, and is the most magnetic of all gases. It has never been condensed into the liquid form.

It is slightly soluble in water, 100 cubic inches of water at 60° dissolving 3 cubic inches of it. Upon this supply all water-breathing animals depend for the aeration of their blood.

Oxygen is a neutral gas, not altering the tint of blue or red litmus water, but it gives rise, on uniting with many other bodies, to powerful acids.

Atmospheric air owes its power of supporting combustion and respiration to the presence of oxygen. If a stick of wood with a spark on the end be plunged into this gas, it bursts out at once into a flame. A piece of tow saturated with ether, if inflamed in oxygen, causes an intense combustion throughout the jar.

Fig. 190.

A lighted taper, *Fig.* 190, immersed in oxygen, consumes away rapidly, the wax itself, in a melted state, burning as well as the wick. The combustion, however, soon comes to an end and the taper dies out, because of the production of carbonic acid from the union of the oxygen with the carbon of the candle.

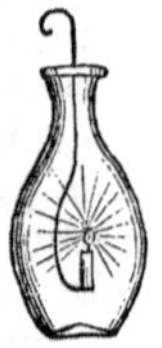

If a piece of charcoal-bark in an ignited state be put in a vessel of oxygen, *Fig.* 191, the combustion is greatly accelerated, and a multitude of scintillating fragments are thrown off. When the charcoal dies out, a little lime-water poured into the vessel and shaken becomes white, from the production of carbonate of lime by the union of the carbonic acid which had arisen during the burning, with the lime.

Fig. 191.

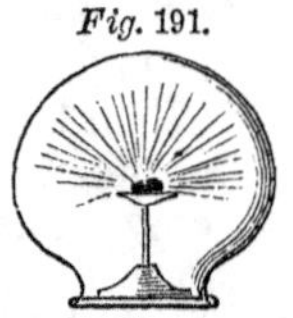

Fig. 192.

Sulphur burns in oxygen with a beautiful purple flame, becoming converted into sulphurous acid (SO_2). If the jar be placed in a dish of water tinged with blue litmus, the liquid will gradually redden as the acid is produced.

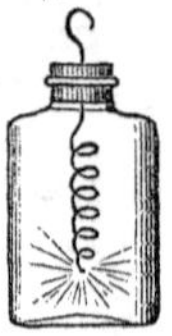

Many substances commonly regarded as incombustible burn with brilliancy in oxygen. If

---

Describe its effect on a lighted stick; on a taper; on charcoal bark. What is its effect on burning sulphur?

a piece of iron wire be coiled into a spiral, *Fig.* 192, and a fragment of ignited wood attached to its lower part, on immersing it in oxygen the iron takes fire, white-hot beads of oxide of iron accumulating and falling off from the end of the wire. It is necessary to put water in the bottom of the jar, to prevent the heated globules melting their way through.

Fig. 193.

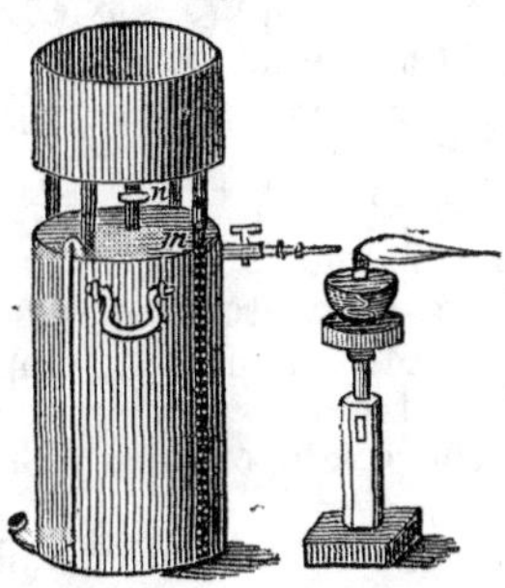

A stream of oxygen from a gas-holder being thrown upon an iron nail made red-hot in the flame of a spirit-lamp, *Fig.* 193, causes the iron to burn with rapidity, emitting a shower of sparks. By the same means a platinum wire may be fused.

LECTURE XL.

OXYGEN CONTINUED.—*Drummond's Light.—Combustion of Phosphorus and Zinc.—Changes during Combustion.—Lavoisier's Doctrine.—Oxides, Basic, Indifferent, and Acid.—Physiological Relations of Oxygen.—Supporter of Combustion and Combustible.—Flame.—Constancy of Heat evolved.—Vegetable Production of Oxygen.*—OZONE.—*Properties.—Production.—Detection.—Antozone.*

IF a piece of lime be placed in the flame of a spirit-lamp through which oxygen is directed by a blowpipe, the lime emits a light of such brilliancy that the eye can scarcely bear it. This is known as the Drummond light. A much more intense light is produced where a stream of ignited oxygen and hydrogen is thrown upon the lime, the calcium lights used in open-air illuminations being produced in this way.

A vivid light also originates when phosphorus is burnt in oxygen, *Fig.* 194. If the spoon in which the

---

What is its effect on an ignited iron wire? What is the Drummond light? What are the phenomena of combustion of phosphorus in oxygen?

melted phosphorus is contained be shallow, the entire interior of the jar becomes filled with flaming phosphorus vapor, and the effect upon the eye is the same as that caused by looking at the sun. The white flakes which fall like a miniature snow-storm in this experiment consist of phosphoric acid (PO_5). If they are scraped from the interior of the jar and dipped into water, a hissing sound is heard, caused by the intense affinity of the acid for water.

Fig. 194.

Pieces of zinc foil or shavings of that metal, if tipped with sulphur and introduced lighted into oxygen, burn with an intense green light, the oxide of zinc (ZnO) being formed.

When any combustible is burned in oxygen two phenomena are witnessed—a change in the oxygen, and a change in the combustible. A piece of charcoal, for example, wastes away, and the gas in which it is burning loses the power of supporting combustion. Until the time of Lavoisier, it was supposed that burning was due to the escape of a substance, phlogiston, from bodies. He proved that there is no loss of weight, but that the products of combustion weigh just the same as the combustible and oxygen taken together. He supposed that no combustion could go on without oxygen, an idea now known to be erroneous, for many substances, as phosphorus, antimony, brass, iron, will burn in chlorine, their respective chlorides being formed. Bromine and vapor of sulphur will also support combustion.

In Lavoisier's system of chemistry oxygen was regarded as being the essential principle of acidity, and hence received its name (from *oxus*, acid, and *gennao*, to form). But many acids do not contain oxygen, as, for instance, hydrochloric acid (HCl). It has since been said that there was no strong acid which did not contain hydrogen, and that therefore that body was more truly the acid former; but this statement is also untrue, fluoboric (BF_3) and fluosilicic acid (SiF_3) being illustrations.

Describe the burning of zinc. What changes take place in the oxygen and the burning body during combustion? What was Lavoisier's theory of combustion?

To the compounds which arise from the union of oxygen with other bodies the designation of oxides is given. There are three classes: 1st. Basic oxides; 2d. Indifferent oxides; 3d. Acids. The following oxides of manganese furnish good examples of these classes:

Protoxide of Manganese...........	MnO	Basic Oxides.
Sesquioxide "	Mn_2O_3	
Deutoxide "	MnO_2	Indifferent Oxides.
Manganic Acid.......................	MnO_3	Acids.
Permanganic Acid....................	Mn_2O_7	

From which it may be inferred that among the oxid of an electro-positive body, the most powerful base that containing one atom of oxygen; that as the quan tity of this element increases, indifferent bodies may be formed—that is to say, those in which neither basic nor acid qualities are well marked; and in a still further increase, acids are produced. In this respect the original idea of Lavoisier is substantiated to some extent.

In its physiological relations, oxygen, which form about one fifth of the atmospheric air, is a most interes ing body. It is for the purpose of introducing this e ement to the interior of the system that the respirator mechanism of animals is necessary. The form of th apparatus may vary, from the gills in fish to the lungs in man, but the object is in each the same. The gas introduced into the body arterializes the blood, and, uniting with combustibles, such as carbon, hydrogen, phos phorus, sulphur, produces the various manifestations of force that are exhibited—muscular motion, intellectu operations, etc., and serves, at the same time, to keep u the temperature to a standard point (98° in man), a con dition essential to the well-being of the individual. Th heat, light, or other force set free in such cases, or in t combustion of fuel, has all originally been derived fro the sun and stored up in the structures of plants. Th desire for eating so continually felt by animals does no arise from a mere wish to gratify the appetite, but from

What is the relation of oxygen to acid and basic bodies? What is the generic designation for its compounds? What are the three classes of compounds it yields? What is the respiratory mechanism for? What effect does oxygen produce in the body? Whence does the desire for eating arise?

the necessity of securing a fresh supply of force-producing material.

The terms supporter of combustion and combustible are expressive of an erroneous idea. No substance is intrinsically either one or the other. Hydrogen will support the combustion of a jet of oxygen just as well as oxygen will that of a jet of hydrogen. In fact, both bodies are equally engaged in producing the result, combustion only taking place on the mutual surface of contact. Oxygen has come to be regarded as the principal supporter of combustion, because most combustions take place in the air, which owes its active qualities to this gas.

Fig. 195.

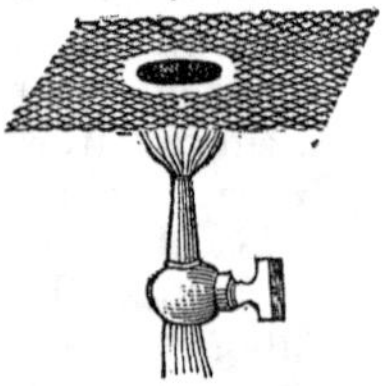

An ordinary flame is not incandescent throughout, but is a mere superficies or luminous shell, as may be seen on lowering a wire gauze down upon it, as in *Fig.* 195. In such a flame several distinct parts may be traced. Around the wick, *a*, *Fig.* 196, at the points *i i*, the light is of a blue color, for here, the air being in excess, the combustion is perfect. From this toward *c* the combustible matter predominates, and the light is most intense. A faint exterior cone, *e e*, surrounds the more luminous portion, but the interior at *b* is totally dark. It is probable that the light arises chiefly from the ignition of solid matter, for incandescent gases are only faintly luminous. The hydrogen of the flame is first burned, and for a moment carbon is set free in the solid form at a very high temperature, its oxidation instantly ensuing. The temperature of the oxy-hydrogen flame is 8061°, but this is surpassed in heat by the Voltaic arc. The colors that substances communicate to flames furnish the means of ascertaining their presence by spectrum analysis (Lecture XVIII.).

Fig. 196.

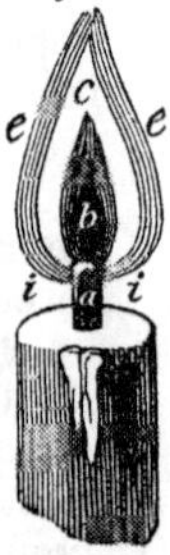

A given weight of a combustible always yields a constant amount of heat. It does not matter whether the

---

Why are the terms supporter of combustion and combustible erroneous? What is the structure of a flame? Why do the different regions of a lamp flame differ in luminous power?

combustion be accomplished in a few minutes or requires a year; an iron wire yields as much heat in rusting slowly as in being burnt in oxygen. A fixed quantity of oxygen can only produce a certain amount of heat, but different combustibles require very different proportions of it to effect their complete burning. One pound of hydrogen requires 8 pounds of oxygen; one of sulphur only 1 of oxygen.

The most frequent products of oxidation, both in animals and flame, are carbonic acid and water, and it might be supposed that these would eventually accumulate in the air to such an excess that living beings could no longer exist. Plants, however, continually correct this tendency. They take those oxidized products, decompose them under the influence of the yellow ray of light (a fact shown by Professor J. W. Draper), appropriate their carbon and hydrogen, and set the oxygen free once more to run the same course. This deoxidizing power of vegetable matter was discovered by Priestley, who found that green leaves, placed in a flask, *Fig.* 197, containing water saturated with carbonic acid, if put in the sunshine, liberated bubbles of a gas rich in oxygen. In the dark no such effect took place.

Fig. 197.

OZONE.—Oxygen gas exists in two different states, a passive and an active. In this latter it is called Ozone, from its peculiar phosphoric odor. Ozone is produced when electric sparks are passed through pure dry oxygen, or when water is decomposed by a Voltaic current, or when phosphorus is allowed slowly to oxidize in damp air. It will also originate from the slow oxidation of ether. A mixture of two parts of permanganate of potassa and three parts of sulphuric acid will give off ozone for months.

Ozone is insoluble in water, alcohol, and ether. It has intense bleaching and oxidizing powers, attacking substances that ordinary or passive oxygen leaves untouched. It decomposes iodide of potassium, the protosalts of manganese, cyanide and ferrocyanide of potas-

---

Is there any difference in the amount of heat evolved in rapid and slow combustion? Why should the air become unfit to support life? How is this tendency compensated? What is Ozone? How is it produced? What are its properties?

sium, and oxidizes silver, metallic arsenic, and antimony. It is reconverted into oxygen by a heat exceeding 450°, but loses its characteristic properties on being subjected to a much lower temperature or to certain hydrocarbons.

The presence of ozone may be detected by slips of paper impregnated with iodide of potassium and starch, which become brown, and, on being wetted, assume shades of color varying from pink to blue.

Schonbein states that ozone itself exists in two different states, positive and negative, and calls the former ANTOZONE, but the difference is not well marked.

LECTURE XLI.

HYDROGEN. — *Preparation.* — *Properties.* — *Effect on Sounds.*— *Combustibility.*— *Lightness.*—*Balloons.*— *Explosive Combustion.*—*Musical Combustion.*— *Water always produced.*— *Oxyhydrogen Blowpipe.*—*Hydrogen a Metal.*

HYDROGEN. $H=1$.

THE name of this gas is derived from *udor*, water, and *gennao*, to produce. It forms one ninth by weight of all the water on the globe, and is a large constituent of animal and vegetable matter. It was at first called inflammable air.

If a piece of potassium be wrapped in paper and rapidly immersed beneath an inverted jar at the water-trough, a violent reaction sets in, a gas collects in the upper part of the jar, and the potassium, oxidizing, dissolves in the water. The gas produced is hydrogen, and the decomposition is shown as follows:

$$HO+K=KO+H;$$

that is, water acted upon by metallic potassium yields oxide of potassium and hydrogen gas.

A more economical process is usually resorted to, depending on the fact that metallic zinc can decompose

---

How may ozone be detected? What is Antozone? What is the origin of the name hydrogen? What was its ancient name? What is the principle of the decomposition of water by potassium? How is hydrogen generally made?

water at ordinary temperatures. As the oxide of zinc produced is insoluble in water and would soon cause the action to cease, it is necessary to add an acid, as, for example, sulphuric acid, to unite with the oxide and form a soluble salt. The surface of the zinc is then kept clear, and the gas is continuously liberated.

Fig. 198.

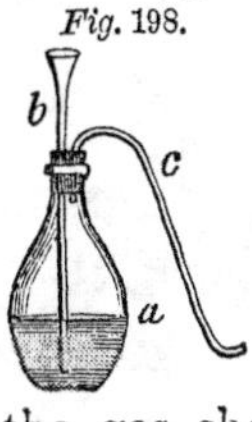

To make hydrogen, a bottle, *a*, *Fig.* 198, is taken partly filled with water and strips of zinc. The mouth of the bottle is closed with a cork perforated by two tubes, the one, *b*, for a funnel, the other, *c*, for the exit of the hydrogen. Through the tube *b* sulphuric acid is gradually poured until a brisk effervescence sets in. The first portions of the gas should not be collected, as they are contaminated with air and are explosive. When the action slackens more acid must be added. Half an ounce of water will produce five gallons of the gas. The chemical change is thus represented:

$$SO_3, HO + Zn = ZnO, SO_3 + H.$$

Hydrogen may be prepared by substituting iron for zinc in the apparatus *Fig.* 198, or by passing the vapor of water over iron turnings heated to redness in a porcelain tube.

As usually prepared it is quite impure, being contaminated with arsenic, sulphur, or phosphorus; or, if made from iron, with an offensive volatile hydrocarbon. When obtained from the Voltaic decomposition of water it is perfectly pure.

Hydrogen is a transparent colorless body, having the highest refracting power of all the gases. Compared with air it is as 6614 to 1000. When pure it has neither taste nor smell. It is the lightest known body, one hundred cubic inches weighing 2.14 grains — 11,000 times less than an equal bulk of water. The weight of its atom is taken as the standard of other atomic weights; it is therefore 1. It exerts no action on vegetable colors, is only soluble to the extent of 1.5 per cent. in water, and has never been liquefied. During experiments for that purpose, it appeared that there is reason to believe that its molecules are smaller than those of

What is the use of the sulphuric acid? Describe the apparatus for its production. What are the properties of hydrogen?

any other body, as it could leak through stopcocks impervious to other gases.

In the animal economy hydrogen does not exercise any deleterious effect. When respired, it causes the voice to assume a feeble, shrill tone like that of a child, and a tendency to sleep. A bell rung in this gas emits a more feeble sound than when ringing in an air-pump vacuum. If a jar, *Fig.* 199, with a stopcock at its upper end, be filled with hydrogen, and, being depressed in the water of the trough, the cock opened and a light brought near the hydrogen, as it escapes it takes fire at once, burning with a pale yellow flame. Or if to the mouth of a bottle containing the materials for generating hydrogen, *Fig.* 200, a cork, through which a glass tube is passed, be adjusted, and, after allowing the air in the bottle to be displaced to avoid an explosion, a light be applied to the issuing gas, it takes fire and burns in the same manner. This is called the philosopher's lamp. Although the light is faint the heat is very great—sufficient, indeed, to melt fine platinum wire.

Fig. 199.

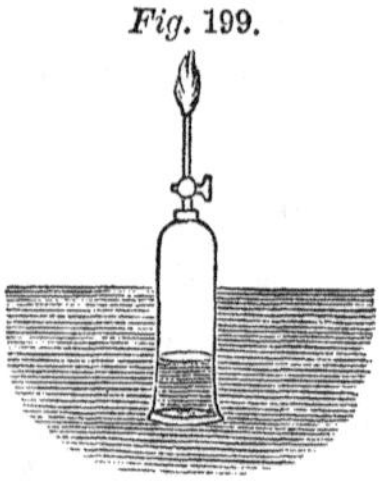

Fig. 200.

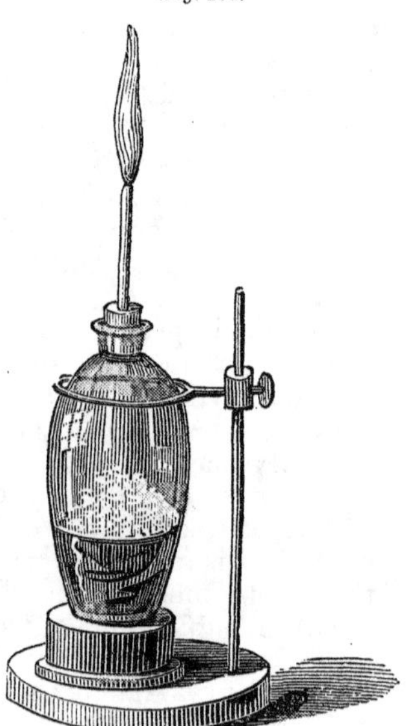

The following experiment proves three facts at the same time—1. The lightness of hydrogen; 2. Its inflammability; 3. That it is not a supporter of combustion. A jar, *a*, *Fig.* 201, is to be filled with hydro-

What are its relations to respiration? How may its combustibility be shown? Describe the philosopher's lamp. How may its inflammability, its non-supporting power, and its lightness be shown?

Fig. 201.

gen at the water trough, and then, being lifted in the air with its mouth downward, a taper placed on a wire is carried into its interior. As the taper passes the mouth of the jar there is a feeble explosion, and the hydrogen, taking fire, burns with a pale flame, but as soon as it is immersed in the gas it is extinguished. It may, however, be relighted, as it is brought out of the jar, at the burning hydrogen, and this may be repeated several times. The combustibility of the gas, and that it is a non-supporter of combustion, are obvious enough. Its lightness is proved by its not flowing out of the mouth of the jar, which it would do at once if it were heavier than atmospheric air.

Fig. 202.

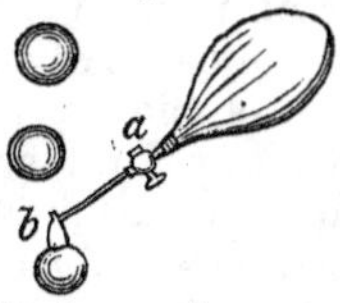

The use of hydrogen for filling balloons depends on its small specific gravity. This property is very distinctly illustrated by filling an India-rubber gas bag with hydrogen, and having attached to the stopcock *a*, *Fig.* 202, a tobacco-pipe, *b*, by dipping the pipe in a solution of soap, bubbles may be blown. These rise through the air with rapidity, and if a light be brought near them they burn with a yellow flame. If the bag be filled with a mixture of hydrogen and air, the bubbles will explode violently.

Fig. 203.

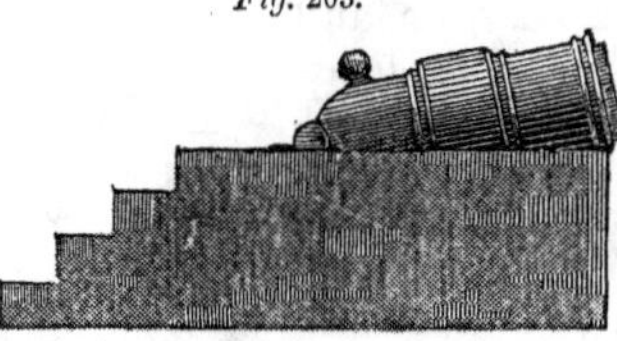

If in a strong brass gun, *Fig.* 203, we place a mixture of hydrogen, 1, and air, 3, and, having inserted the cork tightly, pass a light into the touch-hole, a violent explosion takes place, the hydrogen combining with the oxygen of the air to produce water, *HO*.

Fig. 204.

Musical sounds originate in vibratory movements communicated to the air. If the flame of the philosopher's lamp be covered by the neck of a broken retort, *Fig.* 204, a loud sound is emitted. This arises

To what use is hydrogen applied on account of its lightness? How may this be illustrated on a small scale? Describe the hydrogen mortar. How may musical sounds be produced by hydrogen?

from the circumstance that the hydrogen burns in the tube, giving rise to a series of small explosions which follow each other with rapidity, and these explosions throw the air in the tube into vibration. As the tube is raised or lowered, the explosions occur with different degrees of rapidity, producing sometimes a clattering sound and sometimes a pure musical note.

Whatever may be the circumstances under which hydrogen burns—whether quietly in the philosopher's lamp, or with trivial explosions as in the tube, or with violent detonations, water alone is produced. It may be condensed from burning hydrogen by lowering a cold porcelain plate upon the flame. During the combination of hydrogen and oxygen a very great amount of heat is given out, for the former combines with eight times its weight of the latter, a greater proportion than is met with in the case of any other substance. Advantage is taken of this in the construction of the oxyhydrogen blowpipe, an instrument invented by Dr. Hare, which furnishes us with an efficient means of attaining a high temperature. There are several different forms; in some the gases are mixed in proper proportions in a strong receiver, and set on fire after passing through a Hemming's safety-tube. But it is better to keep them in separate reservoirs, on account of the danger of explosion, and conduct them to a common jet, where they may mix and be burned, as is shown in *Fig.* 205, where the gasometer O contains oxygen and H hydrogen; *a b* are the tubes leading to the jet *c*, where the gases are set on fire. By this instrument platinum can not only be melted, but even vaporized. The intensity of the heat depends, to a great extent, on the fact that, unlike ordinary flames, this is solid—that is, incandescent throughout.

Fig. 205.

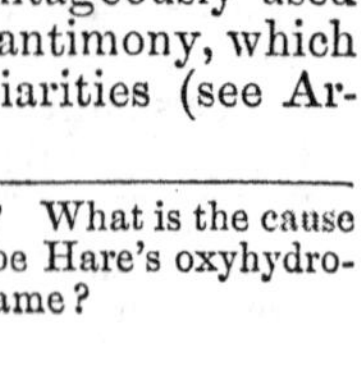

The flame of hydrogen is very advantageously used in detecting the presence of arsenic and antimony, which communicate to it characteristic peculiarities (see Arsenic and Antimony).

What arises from the combustion of hydrogen? What is the cause of the great heat of burning hydrogen? Describe Hare's oxyhydrogen blowpipe. What is the peculiarity of the flame?

In its general relations hydrogen possesses many of the properties of a metal, and has hence been regarded as one of that class in the vaporous condition. Its transparency and gaseous form do not militate against this conclusion, for the vapor of mercury possesses a similar aspect; but it is to be remarked that it is often replaced in combinations by the strongest anti-metal—chlorine.

LECTURE XLII.

WATER.—*Hydrogen Acids.—Water.—Its Composition and Properties.—Crystallized Forms.—Compressibility.—Synthesis of Water.—Analysis of Water.—Its Chemical Relations.—Water of Crystallization and Saline Water.—Acts as a basic, indifferent, and acid body.—Its Impurities.—Hardness.—Mineral Waters.*—PEROXIDE OF HYDROGEN.—*Preparation.—Properties.*

WATER. $HO=9$.

HYDROGEN unites with all the electro-negative substances, and with many of them forms strong acids, as hydrochloric, hydrobromic, hydriodic, hydrofluoric; but with oxygen instead of an acid, a neutral body, water, results. Water contains one atom of each of its elements, combining to form one atom of water. It is therefore a binary compound, and its symbol is HO.

By volume it consists of two of hydrogen with one of oxygen; by weight, one of hydrogen with eight of oxygen. These statements correspond with the first, because the hydrogen atom is twice the volume of that of oxygen, and the weight of an atom of oxygen is eight times that of hydrogen.

Water is a most important and universally diffused substance, covering three fourths of the earth's surface, and entering as a constituent into all animal and vegetable substances and most minerals. An oyster contains 81 per cent., and some of the acalephæ (jelly-fish) 99 per cent. of this ingredient. By its aid the processes

---

To what class of bodies does hydrogen probably belong? When hydrogen unites with electro-negatives, what class of bodies may arise? What is the constitution of water? What are the properties of water?

of decay or oxidation necessary to life are caused to go on with rapidity; while, on the contrary, by desiccation organic substances may be indefinitely preserved.

Water is an inodorous, tasteless fluid, of a slightly bluish-green color, which conducts heat and electricity imperfectly, and refracts light strongly. It freezes at 32° F., and boils at 212° F. if certain precautions be adopted (Lecture XI.). In a capillary tube $\frac{1}{200}$ of an inch in diameter it must be cooled to 1°.4 before freezing, while if a piece of pure ice be heated in a vessel of oil, the heat may be continued till the resulting water has reached 240°, when the whole is converted into vapor with explosion. The specific gravity of water is 1.000, being the standard of comparison of all other liquid and solid bodies. The specific gravity of its vapor, steam, compared with atmospheric air, is 1.6219, and its color is red. Water is a compressible and elastic substance. One cubic inch of it at 60° weighs 252.5 grains, and the cubic foot is so nearly 1000 ounces, that the specific gravity of any substance is very nearly the absolute weight of a cubic foot of it in ounces.

If water be slowly frozen it crystallizes in needles, crossing one another at angles of 60° and 120°, and presenting forms of beautiful symmetry, as seen in *Fig.* 206 (page 236). Similar flower-like shapes may be seen on melting the interior of a block of ice by the aid of a beam of sunlight or the electric lamp.

The compressibility of water is demonstrated and measured by an instrument invented by Oersted, in which pressure can be exerted upon water in a tube by means of a screw. It shows that water is compressed $\frac{1}{22.000}$ part of its volume for each atmosphere of pressure.

Fig. 207.

The constitution of water was first proved by Cavendish and Watt in 1781. It can be illustrated in a variety of ways. Thus, if over a jet of burning hydrogen a cold glass bell be suspended, *Fig.* 207, it becomes soon covered with a misty dew, and if the experiment be prolonged, drops of liquid finally trickle down the sides, and may be

What is the specific gravity of water and steam? What does *Fig.* 206 represent? What is the amount of compressibility of water? How may the composition of water be synthetically shown?

Fig. 206.

caught in a vessel placed to receive them. This liquid is water, which has arisen from the union of hydrogen with the oxygen of the air. It may be slightly acid, from the presence of nitric acid produced by oxidation of the nitrogen of the air.

If in a vessel over the mercurial trough twenty measures of hydrogen are added to ten measures of oxygen, and a small pellet of spongy platinum passed up through the quicksilver, union between the two gases rapidly takes place, so that it is usual, in order to moderate its action, to mix the spongy platinum previously with a little pipe-clay. As the gases unite the mercury rises, until at last they totally disappear. This experiment shows that the constitution of water by volume is 2 of hydrogen to 1 of oxygen.

The composition of water by weight was determined by Berzelius as follows: Let a flask, *a*, *Fig.* 208, con-

Fig. 208.

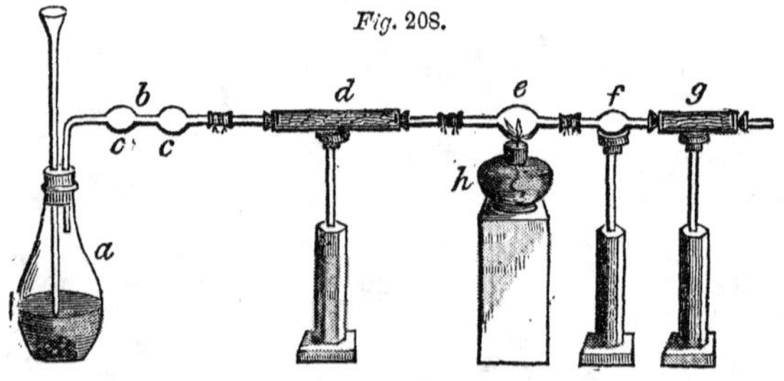

taining zinc and dilute sulphuric acid, be connected by a bent tube, *b*, with another tube, *d*, containing chloride of calcium. The hydrogen which is evolved from the flask consequently deposits any small quantity of water it may be contaminated with in the bulbs *c c*, and then, passing through the chloride of calcium tube, *d*, is perfectly dried. The tube *d* is connected with a tube of hard glass on which a bulb, *e*, is blown. The bulb is filled with a known weight of oxide of copper, which can be raised to a red heat by the spirit-lamp, *h*. As the dry hydrogen passes over the ignited oxide it reduces it, forming with its oxygen water, and leaving pure metallic copper. The water is partly collected in the bulb *f*, and the rest of it is detained by a second chloride of calcium tube, *g*.

What effect has spongy platinum on a mixture of oxygen and hydrogen? Describe the method of Berzelius for determining the composition of water by weight.

If we weigh the tubes *e*, and *f*, and *g*, before and after the experiment, it will be found that for every 8 grains the oxide of copper, *e*, has lost, 9 grains of water have been produced, showing that the constitution of water by weight is 8 of oxygen to 1 of hydrogen.

The composition of water may also be proved analytically by the aid of the Voltaic battery (Lecture XXX.).

Lavoisier determined the composition of water by passing its vapor over fragments of iron made red-hot in a tube. Thus, if from the retort *a*, *Fig.* 209, contain-

Fig. 209.

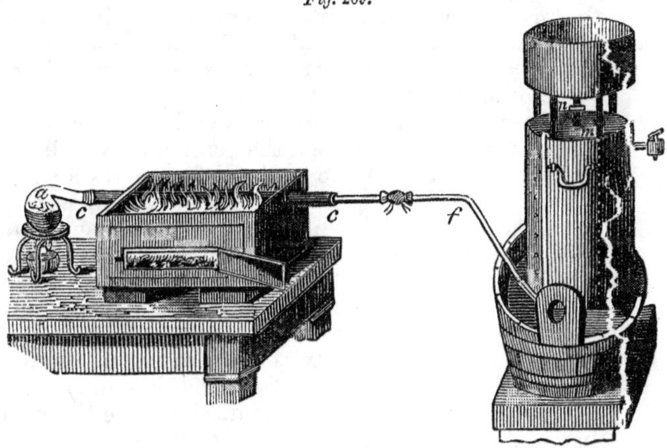

ing water boiling, steam be passed through a red-hot iron tube, *c c*, filled with turnings of iron or iron wire, decomposition takes place, black oxide of iron forming, and hydrogen gas escaping by the tube *f* into the gas-holder *m n*.

The chemical relations of water are of the utmost importance. It exerts a more general solvent action than any other liquid known, holding in solution gaseous and solid substances, acids, alkalies, and salts. As respects gaseous bodies, the quantity which water will take up depends on pressure and temperature. In the case of

How may the analysis of water be effected? Describe Lavoisier's method of analysis. What are the chemical relations of water, its solvent powers, etc.

salts, an increase of temperature generally increases its solvent power. Salt crystals sometimes contain a very considerable quantity of it, as in the case of alum, of which, if a mass be heated, it melts in its own water of crystallization, and after a great quantity of steam is thrown off a dry residue remains. Crystals may contain water in two different states—water of crystallization, which is easily expelled, and saline water, which is with difficulty driven off. In works of chemistry, *Aq* (aqua) signifies the former, and *HO* the latter; thus

$$FeO + SO_3 + HO + 6Aq$$

is the symbol for green vitriol, which is a sulphate of the protoxide of iron, with one atom of saline water and six of water of crystallization. The saline water is only removed by a high temperature, or by being replaced by some other body.

Water unites with many acids with great energy. If mixed with sulphuric acid, the temperature will run up above 300°. With basic bodies the same results may be obtained; as when quicklime is slaked with water, the temperature rises above 570°—more than sufficient to inflame gunpowder. Potassa and soda exhibit similar phenomena. Toward acids water acts as a base; toward bases, as an acid; and toward salts, as an indifferent body.

As found in nature, water is always impure. The water of Loch Katrine, in Scotland, has only 2 grains of solid matter to the gallon, and that of the River Loka, in Sweden, $\frac{1}{20}$ of a grain; but that from the Great Salt Lake has 20,000 grains. Rain-water and melted snow contain the various soluble gases that are in the air; spring, river, well, and mineral waters, the soluble bodies of the strata through which they have flowed. From these they can be purified by distillation. The solid contents in the imperial gallon of some of the principal river waters of Europe is as follows: the Thames, at Greenwich, 28; the Seine, 20; the Rhone, at Lyons, 13; the Danube, at Vienna, 10. Croton water contains 4, Schuylkill 4.3, and Boston (Long Pond) 1.2.

What is meant by water of crystallization and saline water? How is the difference indicated in formulæ? What is the relation of water to acids, bases, and salts? Is water found pure in nature? Give examples of its constitution. What are the solid contents of Thames water, etc.?

The quality termed hardness of water results usually from the presence of a salt of lime or magnesia. Such samples are tested by the aid of a solution of soap in alcohol, the quantity required to produce a permanent froth being ascertained.

Mineral waters are usually divided into four groups, carbonated, saline, sulphurous, and chalybeate, containing respectively in excess carbonic acid, common salt, sulphureted hydrogen, and iron. Many also contain arsenic, as the Vichy and Plombières, waters of France. The water of the Congress Spring, at Saratoga, is composed as follows:

Chloride of Sodium	390	grains	per gallon.
Iodide of Sodium and Bromide of Potassium	6	"	"
Carbonate of Soda	9	"	"
Carbonate of Magnesia	101	"	"
Carbonate of Lime	104	"	"
Carbonate of Iron	1	"	"
Silica and Alumina	1	"	"
	612	"	"

Peroxide of Hydrogen. $HO_2=17$.

This compound, the deutoxide of hydrogen, is a definite compound of oxygen and hydrogen, and not merely oxygenated water. It is prepared by adding a paste of peroxide of barium and water to hydrofluosilicic acid. The liquid is separated by filtration and concentrated in vacuo. It should be kept at a temperature less than 60°.

It is a colorless, sirupy liquid of a disagreeable taste. It bleaches, and is readily decomposed not only by heat, but also by several metals and their oxides, and sometimes with explosion. Peroxide of manganese or lead, for example, entirely resolve it into oxygen and water.

What is meant by hardness of water? How are mineral waters divided? What is the composition of Congress water? How is peroxide of hydrogen prepared? What are its properties?

LECTURE XLIII.

NITROGEN.—*Preparation.—Properties.—Its Indifferent Nature.—Its Combustion and Compounds.—Its Oxygen Compounds.*

ATMOSPHERIC AIR.—*Constitution.—Methods of Analysis.—Eudiometry.—Extent of the Atmosphere.—Relations to Organization.—Pressure of the Air experimentally shown.*

NITROGEN. $N=14$.

NITROGEN gas is readily procured from the atmosphere by burning phosphorus in a bell jar on the pneumatic trough. If a piece of phosphorus be laid in a cup, *Fig.* 210, and set on fire, all the oxygen in the air of the jar, *a*, will be consumed, white flakes of phosphoric acid (PO_5) forming, and these being finally dissolved in the water of the trough, *d*, there is left behind nitrogen contaminated with a trace of the vapor of phosphorus. It may be also prepared by allowing the phosphorus to oxidize slowly at the expense of the air. If the interior of a jar be wetted and iron filings sprinkled over it, they will slowly remove the oxygen from the contained air, leaving the nitrogen, if it is subsequently agitated with water, quite pure. Copper filings with hydrochloric acid may be used in a similar manner.

Fig. 210.

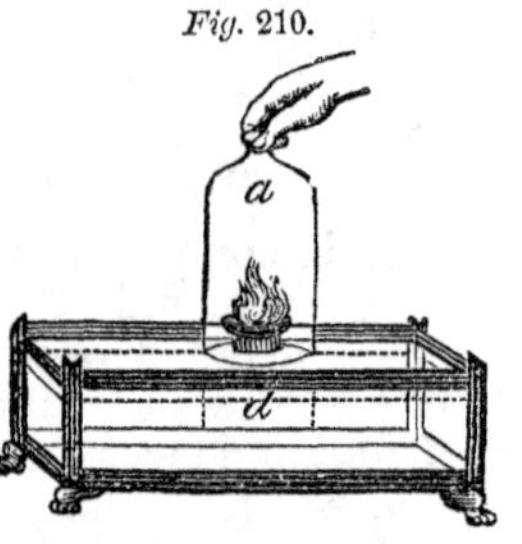

To obtain nitrogen perfectly pure, use, instead of the phosphorus, *Fig.* 210, a porcelain vessel containing pyrogallic acid and a strong solution of caustic potassa. In this method the carbonic acid of the air is also removed, and no impurity but vapor of water remains.

Nitrogen gas (*nitron*, nitre, and *gennao*, to produce) is a colorless, tasteless, and inodorous body, dissolving in water only to the extent of 1½ per cent. by volume.

How is nitrogen prepared by phosphorus? By what other methods may it be prepared? What are the properties of this gas?

It is lighter than atmospheric air, its specific gravity being .967; 100 cubic inches weigh 29.96 grains. It does not support combustion nor respiration, and from the latter circumstance took the name *Azote.* It does not exert any poisonous agency on animals, as is shown by the large proportion in which it exists in the atmosphere.

Under certain circumstances nitrogen undergoes combustion, as when electric sparks are passed through air, and nitric acid (NO_5) formed. Also, when nitrogen is mixed with the detonating compound of oxgyen 1 and hydrogen 2, it may easily be oxidized.

Nitrogen is little disposed to unite with other bodies except when either it or they are in the nascent state. Its compounds are prone to decompose easily, and hence among them we find some of the most remarkable fulminating bodies. The animal and vegetable substances into which it enters as a constituent are characterized by the facility with which they putrefy or oxidize, and, as we shall hereafter find, ferments owe their powers to this element.

Nitrogen unites with oxygen to form five different compounds:

NO, Protoxide of Nitrogen.
NO_2, Deutoxide of Nitrogen.
NO_3, Hyponitrous Acid.
NO_4, Nitrous Acid.
NO_5, Nitric Acid.

With oxygen it also forms atmospheric air; but this is a mixture, and not a compound.

Atmospheric Air.

The mechanical properties and constitution of the atmosphere are, on account of their great importance, first to be described.

The atmosphere consists chiefly of oxygen and nitrogen, in the proportion of 20.8 volumes of the former to 79.2 of the latter. It contains, as an essential ingredient, also a small proportion of carbonic acid, 10,000 parts of air containing from 3.7 to 6.2 parts, or, on an

When does it undergo combustion? Why does it give rise to explosive bodies? Give the compounds of oxygen and nitrogen. Of what is atmospheric air composed?

average, 5 parts of this gas. Besides these, there are variable quantities of the vapor of water, and traces of ammonia, sulphureted hydrogen, and carbureted hydrogen. Certain organic constituents, some of which may be detected by their odor, give it miasmatic properties, causing a variety of diseases. Air is an invisible, elastic substance, only seen to be of a blue color in thick masses, 815 times lighter than water, and is taken as the standard of comparison for the specific gravity of gases. Its specific gravity is therefore =1. One hundred cubic inches of it weigh at the mean temperature (60°), and pressure (30 inches) about 31 grains.

There are many methods by which the analysis of the air can be effected. The eudiometer, *Fig.* 211, consists of a thick glass tube, closed at the upper end and open below, where it dips into a cup or basin of mercury. It is graduated along the side, and has two wires through the upper part, which approach each other, but do not touch. The tube is filled with mercury, is then reversed, and a mixture of equal volumes of air and hydrogen put into it. An electric spark being passed between the wires, the mixture is exploded. The amount of gas left is ascertained by the divisions, and one third of the deficit represents the quantity of oxygen originally present.

Fig. 211.

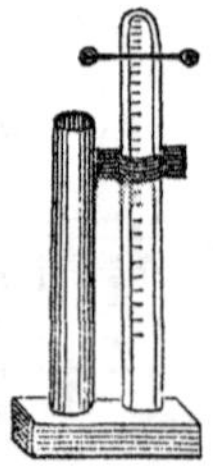

The analysis of air may be accomplished by the aid of pyrogallate of potassa, and subsequently dry potassa. The oxygen, carbonic acid, and aqueous vapor are by these means entirely removed. In such volumetric analyses errors may arise by variations of pressure, temperature, etc., and hence the following method by weight is worthy of attention: Air is deprived of its carbonic acid and water by being made to traverse through tubes containing strong sulphuric acid and caustic potassa. It is then passed through a weighed tube containing metallic copper heated to redness, which deprives it of its oxygen. By connecting the apparatus with a glass vessel in a vacuous state, the nitrogen may

What variable constituents does it contain? What is its specific gravity? How much do 100 cubic inches weigh? Describe the eudiometer. By what other methods may air be analyzed?

be collected. The increase of weight in the copper tube gives the weight of oxygen, while the increase of weight in the glass vessel, that was originally vacuous, gives the nitrogen. Air is thus shown to be composed by weight of

Oxygen......................23.10
Nitrogen......................76.90.

Its composition by volume has already been stated. These proportions remain unchanged, no matter from what part of the globe the air may be taken, nor what the elevation may be.

The earth's atmosphere does not extend indefinitely into space, but terminates at an altitude of about fifty miles; a fact first discovered by Alhazen, a Mohammedan philosopher. It is therefore a mere film on the face of the earth, like the down on a peach, for the globe is nearly 8000 miles in diameter. If a representation of it were placed on a common twelve-inch globe, it would not exceed one sixteenth of an inch in thickness.

Its relations to the world of organization are full of interest. All plants come from it, and all animals return to it, so that it stands as the bond of connection between the two. It is the grave of animal, the cradle of vegetable life.

As we ascend to the more elevated regions the air becomes less dense, for the obvious reason that, as it is a very compressible body, those portions of it near the earth have to sustain the weight of the mass above, and are therefore more dense; but in the higher regions, where the superincumbent pressure is less, the air is more rare, as is shown in the following table:

Height in Miles.	Volume of Air.	Barometric Inches.
0	1	30
2.705	2	15
5.410	4	7.5
8.115	8	3.75
10.820	16	1.875
13.525	32	.9375
16.230	64	.46875
18.935	128	.234375

What is its composition by weight? To what distance does the atmosphere extend? What are its relations to animals and plants? Why does its density decrease with the altitude?

This also shows that the great mass of the atmosphere is within a very short distance of the earth's surface—one half within three miles, and four fifths within eight miles. At different altitudes the temperature varies, being 1° colder for every 350 feet of ascent, partly from the increased capacity for heat, and partly because it is warmed mostly by contact with the earth. The line of perpetual snow is 15,200 feet above the sea at the equator, 3818 at 60° latitude, and only 1000 at 76°. At 85° it has sunk to 117 feet, and nearer the pole actually dips below the soil.

Of the constituents of the air, the oxygen and nitrogen are regarded as fixed, the carbonic acid, water, and ammonia as variable; but every process of respiration and combustion tends to change the amounts. Animals take oxygen and replace it by carbonic acid, while plants do the reverse. A strict balancing is, however, observed between these various operations, so that no change can be detected between the present atmosphere and that existing centuries ago.

Of the mechanical properties of the air, the first to which we have to direct our attention is its pressure, which takes effect equally in all directions, upward, downward, and laterally. Thus, if we take a glass tube, *a*, *Fig.* 212, several feet long, closed at one end and open at the other, and, having filled it with water, place over the mouth a stout piece of paper, *b*, and turn it upside down, the paper will not fall off nor the water flow out. They remain, as it were, suspended on nothing, but, in reality, sustained by the upward pressure of the air. Or if we take a bottle, *a*, *Fig.* 213, with a hole, *b*, half an inch in diameter at the bottom, and, having filled it with water, close the mouth with the finger, it may be held in the air without the water flowing out, though the aperture *b* is wide open. In this instance, again, it is the upward pressure of the air which sustains the liquid.

Fig. 212.

Fig. 213.

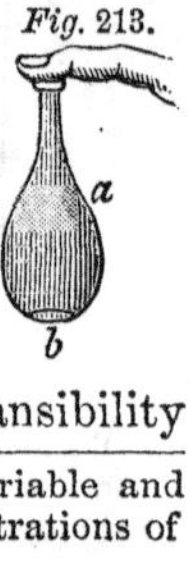

Advantage is taken of the elasticity and expansibility

How does its temperature vary? Which are the variable and which the fixed constituents of the air? Give some illustrations of the upward pressure of the air.

of the air in the construction of the air-pump, an instrument intended for the removal of air from closed vessels. It consists of two cylinders, *Fig.* 214, the pistons

Fig. 214.

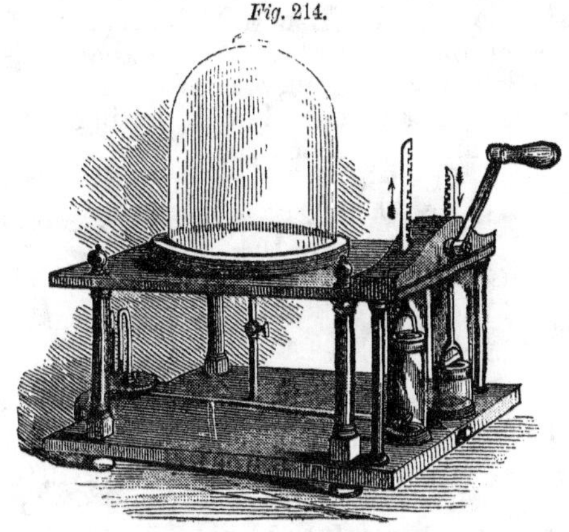

of which may be alternately raised and depressed by a rack and pinion. They communicate by a tube with the interior of a bell-jar or receiver, which stands on the air-pump plate, and makes a tight joint with it because the two are ground flat. On working the handle the air in the bell-jar is, little by little, removed, and eventually only a small fraction of the original amount remains. Let the glass globe *a*, *Fig.* 215, be nearly filled with water, and inverted so that its neck, *b*, dips beneath the water contained in vessel *c*. If the whole be covered with an air-pump receiver, *d*, and the receiver exhausted, the bubble at *a* dilates, and after a time, as the action of the machine continues, fills the entire glass, both bulb and tube. As the pressure is restored by letting air again into the bell-jar, the bubble contracts and eventually regains its original size.

Fig. 215.

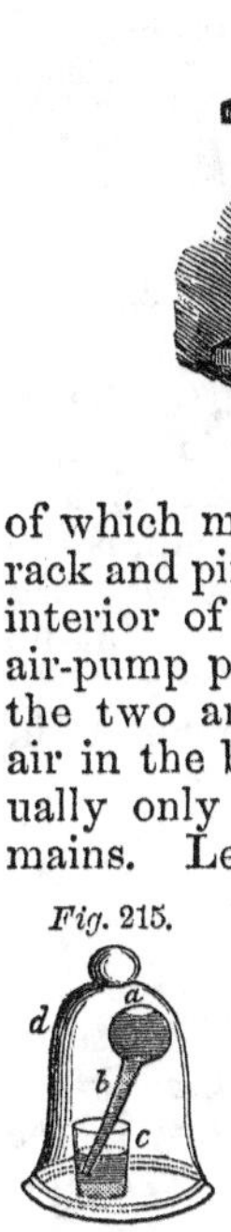

The air-pump enables us to exhibit in a striking manner the chief mechanical properties of the air. Thus, if on the plate of it there be placed a receiver, *a*, *Fig.* 216,

Describe the air-pump. Describe *Fig.* 215. Give illustrations of downward pressure.

Fig. 216.

as soon as the air is exhausted from its interior, the superincumbent pressure retains the glass so firmly in contact that it is impossible to lift it off; but when the air is readmitted it can easily be moved. If within the receiver *a* a smaller one, *b*, be placed, and exhaustion made, while *a* is fixed, *b* can be easily moved by shaking the pump; but, on letting in the air, *b* becomes fixed and *a* loosened.

Fig. 218.

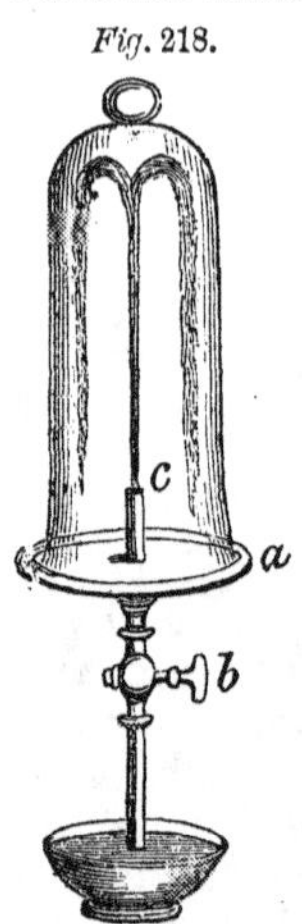

Fig. 217.

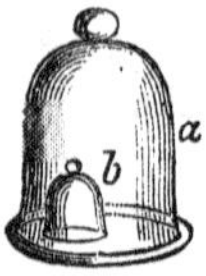

If over the mouth of a jar, *Fig.* 217, placed on the pump, the palm of the hand be laid as the air is exhausted, it is pressed in close contact with the jar, and can only be removed by the exertion of a very considerable force.

On the small plate *a*, *Fig.* 218, furnished with a stopcock, *b*, terminating in a jet, *c*, a tall receiver is placed, and the apparatus being screwed on the air-pump, is exhausted, and the stopcock closed. On being opened when the lower end of the tube dips under water, the water rises to the top of the jar, and forms a fountain in vacuo.

LECTURE XLIV.

ATMOSPHERIC AIR.—*Pressure of the Air.—Exhaustion without an Air-pump.—Determination of the Weight of the Air.—Amount of Pressure.—Elasticity of the Air.—Air in the Pores of Bodies.—Aquatic Respiration.—Preservation of Meats and Fruits.*

THE Magdeburg hemispheres, invented by Otto Guericke, who also invented the air-pump, illustrate in a striking manner atmospheric pressure. They consist of a pair of brass hemispheres, *Fig.* 219, with handles. They

Describe the fountain in vacuo. What are the Magdeburg hemispheres?

Fig. 219.

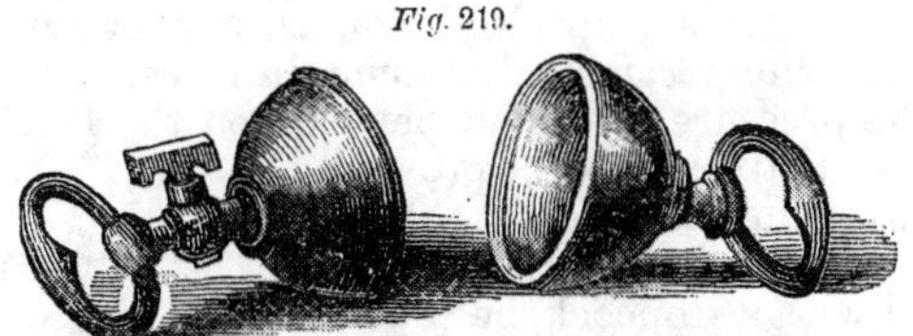

fit without leakage to one another and form a sphere. One of them has a stopcock through which the air may be exhausted; and, on this being done, it will be found impossible to pull them apart, though when the air is readmitted and the pressure restored to the interior, they will fall apart by their own weight.

If over the mouth of an open receiver, *Fig.* 220, a piece of bladder be tied with a waxed thread, when the air is suddenly exhausted the bladder becomes deeply depressed into a spherical concavity by the pressure of the air, and finally bursts inward with a loud explosion.

Fig. 220.

It is upon this principle of atmospheric pressure that the various instruments for cupping act. The simplest method of performing this operation is to place the cupping-glass for a moment over the flame of a large spirit-lamp, and then transfer it rapidly to the skin. As soon as the steam with which the interior of the vessel is filled condenses, a partial vacuum is formed, and the soft parts are pressed into its interior.

For many experiments the air-pump is not required. Thus, if we take a vial, *a*, *Fig.* 221, and fit to the mouth of it a cork, *b*, through which passes a piece of glass tube, *c*, drawn to a narrow jet at one end, but open at the other, by placing the finger over the opening and introducing it into the mouth, the air, by the action of the tongue and muscles of the mouth, may be drawn out to a great extent; and when the exhaustion has been carried as far as possible, by pressing the finger over it the opening may be closed. If now the

Fig. 221.

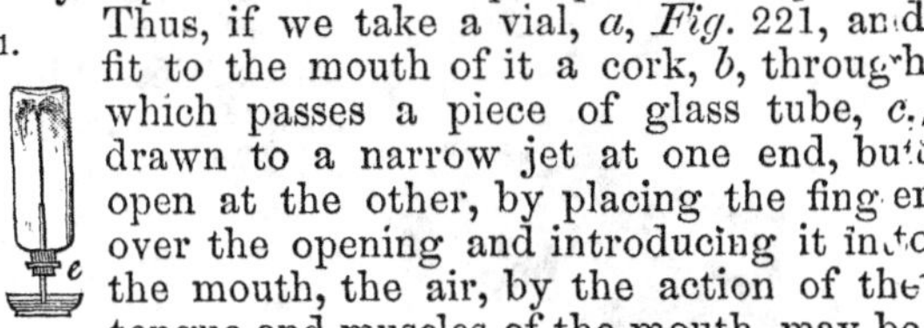

Describe the experiment *Fig.* 220. How is cupping performed? How may exhaustion by the mouth be illustrated?

bottle be turned upside down, as at *e*, in some water, and the finger removed, a fountain in vacuo is formed.

The pressure of the air depends on the fact that it is a heavy body, as may be proved by weighing it directly. Let a light glass flask, *a*, *Fig.* 222, fitted with a stopcock, be counterpoised in a balance, then let the air be exhausted from it, and the loss of weight determined. On opening the stopcock and again admitting the air, it will regain its original weight. A flask containing 100 cubic inches will in this manner lose 31 grains in weight, and that, therefore, is the weight of that amount of air.

Fig. 222.

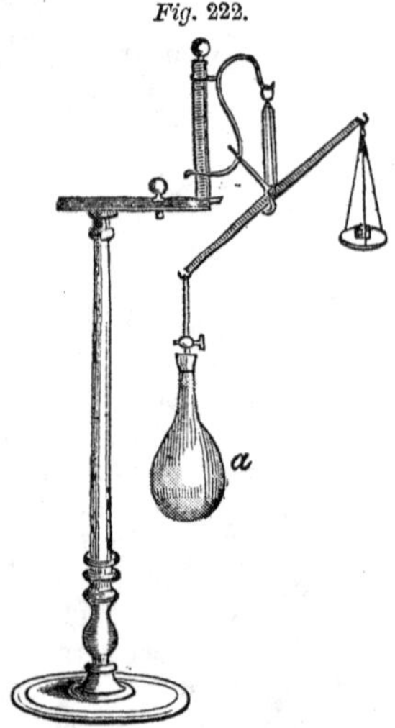

Fig. 223.

Atmospheric air is used as the standard of comparison of the specific gravities of other gases. The process for the determination is very simple. A pear-shaped flask, *F*, *Fig.* 223, furnished with a small stopcock, is screwed to the air-pump and exhausted of air. After being weighed it is attached to the graduated jar *G*, filled with air or any other gaseous body, and, the stopcocks being opened, it fills. It is then reweighed,

How may the weight of air be directly ascertained? How may the relative weight of other gases be determined?

and will be found to have increased several grains. The amount that has passed into the flask is known by reading off at the graduation.

There are several different methods of stating the amount of the mean pressure of the air; thus we say that it is equal to 15 pounds on the square inch, or to a column of mercury 30 inches long, or to a column of water 33 feet long. Upon the body of a man the atmosphere presses with a weight of 30,000 pounds, but it is not felt because the internal air presses outward with similar force. Humboldt exposed his body to a variation of pressure equal to 31,000 pounds without suffering any inconvenience.

That air is a highly elastic substance can be readily shown. Under a bell-jar, *Fig.* 224, let there be placed a half-blown bladder, the neck of which is tied. As the air is removed from the bell the bladder distends, but on restoring the pressure it becomes as flaccid as it was before, showing that the included air expands and contracts as the pressure upon it varies.

Fig. 224.

This may be still more strikingly shown by taking a small India-rubber bag, *Fig.* 225, the mouth of which is closed tightly, and using it instead of the bladder. On rarefying the air in the receiver the bag begins to dilate, and may be extended to several times its original dimensions, as shown by the dotted line; but when the pressure is restored it returns to its former size.

Fig. 225.

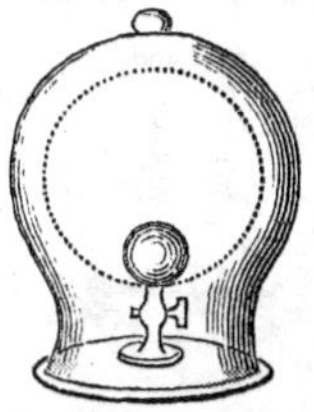

Nor does the expansion take place with an inconsiderable force. If a flaccid bladder be compressed by leaden weights, *Fig.* 226, as soon as the pressure is removed it will push up the weights. Nor does it lose its elasticity by being long confined. Some of the old chemists kept air compressed in copper globes for months, and

Fig. 226.

How may the pressure of the air be stated? What is the pressure on the body of a man? How may the elasticity of air be illustrated? How may it be shown by an India-rubber bag? Give an illustration of the amount of this force.

found that, as soon as liberated, it regained its original size.

By taking advantage of the expansion of air, and by reducing the pressure, its existence in the pores of many bodies may be shown; thus if we place an egg, *Fig.* 227, an apple, or other such objects in a glass of water, and exhaust the air from the covering bell-jar, we shall see bubbles of air in countless numbers escaping through the water. A glass of ale, *Fig.* 228, placed in an exhausted receiver, foams from the escape of carbonic acid gas; and spring or river water, examined in the same way, is found to contain three or four per cent. of gas.

Fig. 227.

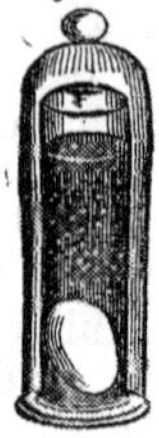

Fig 228.

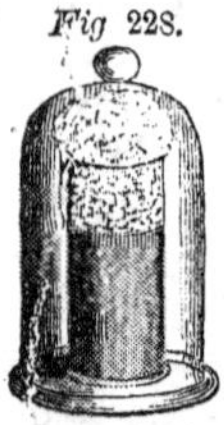

This last fact is of great importance, for it is by the aid of this air that all water-breathing animals exist. Fish do not respire water, but the air contained in it. This air differs from the atmosphere, in containing more oxygen — 33 per cent. instead of 21. The cause of the difference is that oxygen is more soluble in water than nitrogen. Nevertheless, the respiration of aquatic animals is never so perfect as that of air-breathers, on account of the limited supply, and hence such animals are always cold-blooded. When fish are placed in water under an exhausted receiver, *Fig.* 229, they soon die. They are unable to descend to the bottom except by violent exertion, because the air in their swimming-bladder expands and buoys them up. The highest fishes, as the whale, which has a temperature of 98°, breathe the atmosphere by lungs, coming periodically to the surface for that purpose.

Fig. 229.

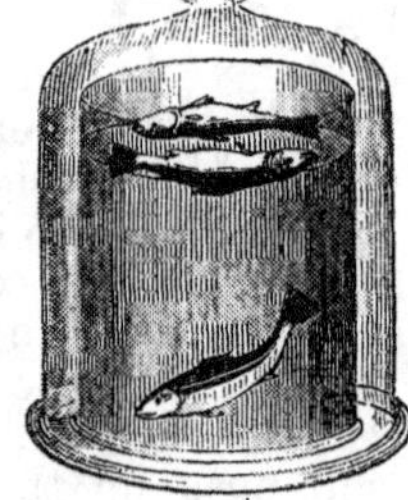

The necessity of air to the support of combustion may be demonstrated by the length of time that a candle

How may the presence of air be detected in the pores of solid bodies? How may it be shown that air exists in water? Of what use is the air in water? How does it differ from atmospheric air? How do fishes under an exhausted receiver act?

will burn in a jar full of air, and in the same jar rendered vacuous. In the latter case it dies out at once, the smoke descending to the bottom of the jar, owing to the rarefaction of the surrounding medium.

Substances prone to decay, such as meats and fruits, may be preserved for a long time in vessels void of air. The process is illustrated in *Fig.* 230. The fruits are placed in a jar, closed by a sound cork covered with sealing-wax. A small hole is made through the cork, and the air exhausted. When the vacuum is as complete as may be, the hole is closed by melting the wax in a converging beam of sunlight. On the large scale, the things to be preserved are inclosed in tin cans, which are sealed by soldering, except a pinhole that is left through the cover. The tins are immersed in boiling water, and when steam issues from the pin-hole it is closed by a drop of melted solder.

Fig. 230.

From the foregoing experiments and considerations, it is demonstrated that the prime fact in pneumatics is, that air has weight; from this arise its pressure and varying density at different altitudes. Its elastic force must also be equal to the pressure upon it; if it were less, the air would compress; if greater, dilate.

How may the necessity of air to combustion be shown? How may meats, etc., be preserved? What facts in pneumatics have we demonstrated?

LECTURE XLV.

Atmospheric Air.—*Construction of the Barometer.—Cause of its Phenomena.—History of its Invention by Torricelli.—Pascal's Experiment.—Illustrations of the Nature of Pressure.—Variability of Pressure.—Disturbances in the Composition of the Air.—Are corrected by the Winds and Diffusion.—Illustrations of Diffusion.—The Air is a Mixture.—Marriotte's Law.*

If we take a tube of glass, *a b*, *Fig.* 231, more than 30 inches long, sealed at one end and open at the other, and, having filled it with quicksilver, invert it in a cup of that metal, *c*, the mercury will not flow out of the tube, but will remain suspended at a height of from 28 to 30 inches, a vacuous space being left at the upper part. A scale, *d*, divided into inches and decimal parts, the zero being at the level of the mercury in the cup, completes the instrument. It is termed a barometer.

Fig. 231.

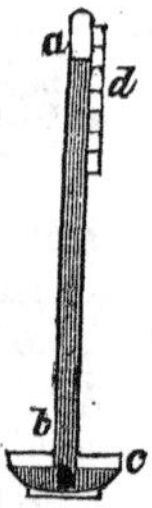

The cause of the suspension of the mercury in the tube is the pressure of the air. This may be demonstrated by placing over the barometer a tall air-pump jar, *Fig.* 232, and exhausting. It will be found that, as the pressure is reduced, the column of mercury falls; and when the pressure is restored, it rises again to the original point.

Fig. 232.

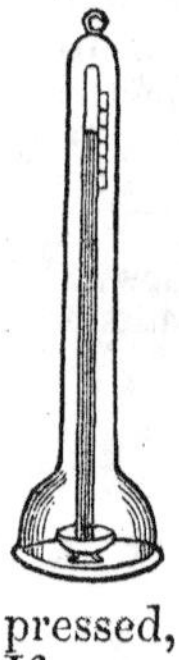

The same fact may be proved in another manner. If a tube upward of 30 inches in length, the upper end of which is closed by a piece of bladder, be filled with mercury and inverted in a cup, *Fig.* 233, the bladder will be deeply depressed, the pressure of the air being borne by it. If now a pinhole be made through the bladder so as to allow the air to press on the top of the mer-

Fig. 233.

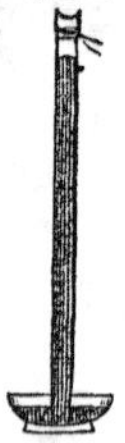

Describe the barometer. What are the causes of suspension of the mercury? How may this be proved? Describe *Fig.* 233.

cury, the column descends to the level of that in the cup below.

The barometer was invented by Torricelli in 1643. Some plumbers working for the Duke of Florence found that it was impossible to make a pump that would raise water more than about 30 feet. This fact, coming to the knowledge of Torricelli, caused him to suspect that water rose in those machines in virtue of the pressure of the air, and not from Nature's abhorrence of a vacuum, as was at the time supposed. If the limit to which water can be raised by suction is reached when the weight of the column of liquid equilibrates the weight of the air, a heavier fluid will be raised to a less height. Accordingly, a pump ought to raise quicksilver only as many inches as water feet, for the respective weights are as 1 to 13½. Torricelli found, by means of a pump fixed to a long glass tube, that such was the case.

That it is the pressure of the air that supports the mercurial column is definitely proved by Pascal's experiment of taking a barometer up a high mountain. As the height of air above the instrument is decreased, the column of mercury also decreases in height. Advantage is taken of this fact to determine the height of mountains.

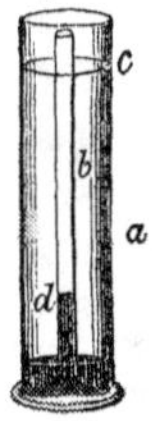

The principle of the barometer may be illustrated by substituting for the pressure of the air the pressure of a column of water. If some quicksilver be put at the bottom of a deep jar, *a*, *Fig.* 234, and a long tube, *b*, plunged into it, on pouring water into the jar, for every 13½ inches in depth poured in, the quicksilver will rise one inch, the mercurial column counterpoising the column of water just as it does the column of air in the case of the barometer.

Fig. 234.

Mr. Boyle discovered that the pressure of the air is variable, the mercurial column sometimes falling to 27 inches, sometimes rising to 30. The range is commonly estimated at 2.5 inches. It is less in the tropics. These changes are irregular, and depend on

---

Who invented the barometer? What were the circumstances of the invention? What was Pascal's experiment? How may the phenomena of the barometer be illustrated by the pressure of a water column? What are the irregularities of the barometer?

meteorological events. A sudden change in the barometer is regarded by seamen as indicating an approaching storm. There are also regular variations during the day, the column rising twice in 24 hours. In winter the first maximum is about 9 A.M., the minimum at 3 P.M.; the second maximum at 9 P.M.

The pressure of the air and its variations can also be measured by an instrument called the Aneroid Barometer, which consists of an elastic metal box partly exhausted of air. By the aid of multiplying levers, the variations in the size of this box are conveyed to an index which plays upon a divided scale, and enables the observer to read the amount. It must be graduated by a standard mercurial instrument.

Many causes tend to give rise to local disturbances in the air. In its lower strata, combustion and respiration are tending to diminish the oxygen and increase the carbonic acid. On the contrary, vegetation, more particularly at the equator, diminishes the carbonic acid and increases the oxygen. But, notwithstanding these disturbances, and the fact that the constituents of the air are of different specific gravities, its composition is nearly the same every where. This is owing partly to the winds and partly to the principle known as the diffusion of gases, which may be illustrated as follows: If two vials, *h c*, *Fig.* 235, are rinsed out, *h* with ammonia and *c* with hydrochloric acid, each becomes filled with the vapor of the liquid used. On placing them mouth to mouth, as in the figure, dense white fumes of sal ammoniac (NH_4Cl) appear simultaneously in both.

Fig. 235.

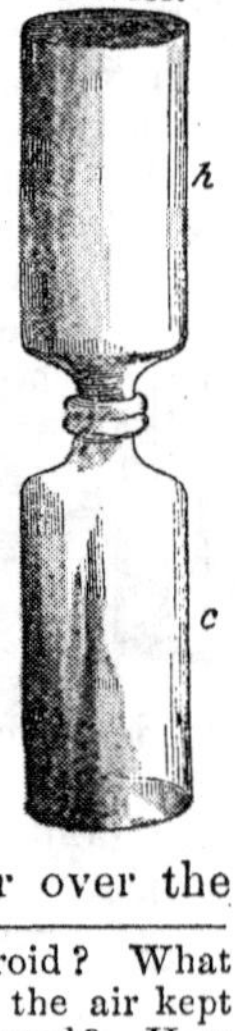

The same effect will take place even though barriers intervene, as may be shown by taking a porous earthenware cup, *a a*, *Fig.* 236, such as is used in Grove's battery, and cementing into its mouth a tube, *b*. A wide-mouthed bottle, *c c*, being placed as a cover over the

What are the diurnal variations? What is the Aneroid? What tends to change the composition of the air? How is the air kept uniform in composition? How may diffusion be illustrated? How may it be shown to occur through barriers?

Fig. 236.

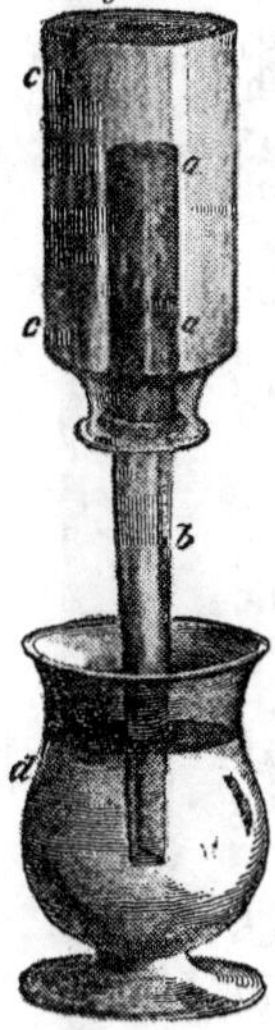

porous cup, it may be filled with hydrogen by displacement. If the end of the tube be put in water in a cup, *d*, the water will rush up the moment the bottle is removed. The hydrogen diffuses out through the porous cup.

Even India-rubber will allow the same results to be shown. Let a bottle, *b*, *Fig.* 237, be taken, and the mouth having been closed with a sheet of India-rubber, let it be exposed under a jar of carbonic acid. It swells up, and assumes a dome shape, the acid having diffused into the bottle. Or, if the bottle be filled with carbonic acid, as at *a*, and exposed to the air, the carbonic acid will diffuse out, and the India-rubber be depressed.

Fig. 237.

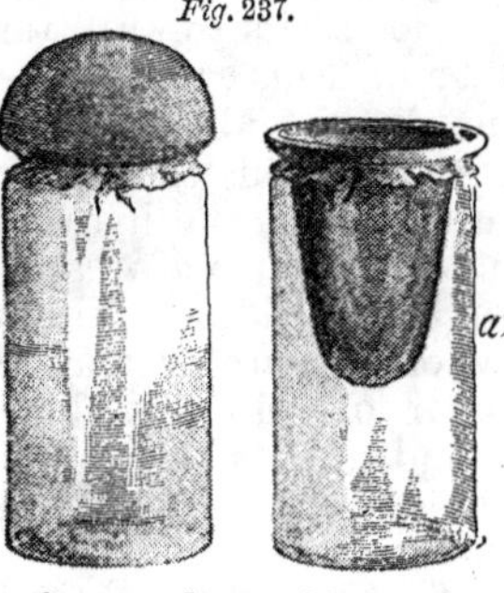

By substances which can not be regarded as having any pores at all, the same phenomena are exhibited. In *Fig.* 238, *a a* is a bottle rinsed out with ammonia. By means of a tube, *b b*, a soap-bubble, *c*, is blown in it. If some of the air from the bubble be drawn into the mouth it will be found to have at once acquired an ammoniacal taste; or, if a rod dipped in hydrochloric acid be presented to the projecting end of the tube *b*, copious white fumes arise.

Fig. 238.

Dr. J. W. Draper showed, by the aid of the apparatus

Describe *Fig.* 237. How may diffusion through poreless media be shown?

Fig. 239, that sulphureted hydrogen will diffuse into atmospheric air, though resisted by a pressure of 750 pounds on the square inch. It consists of a strong tube hermetically sealed at one end, through which a pair of platinum wires, *b c*, pass. The other end, *a a*, is closed by a sheet of India-rubber. The gauge-tube, *d*, which indicates the pressure, carries at its top a little cup to contain acetate of lead, or any other reagent. On connecting *b* and *c* with a Voltaic pile, the water which fills the tube to *e e* is decomposed, and oxygen and hydrogen accumulate in the tube, exerting a pressure greater and greater as the decomposition is prolonged. The whole apparatus may then be subjected to a jar of sulphureted hydrogen. It will be found to pass through with facility, coloring the acetate of lead solution black or deep brown.

Fig. 239.

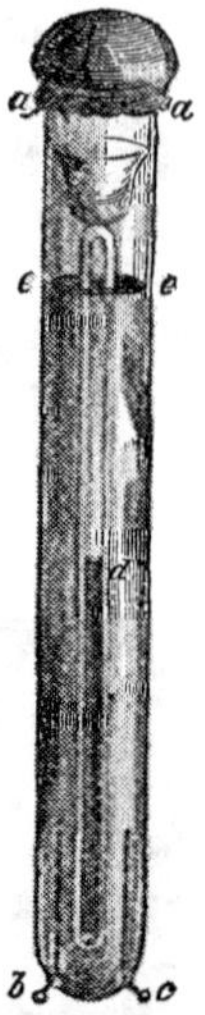

That the atmosphere is a mixture, and not a compound, is proved by its easy decomposibility, its refractive power, and by the fact that its constituents retain their properties unchanged. The oxygen and nitrogen may be determined as already described: the carbonic acid by potassa or lime-water, the aqueous vapor by the method for the dew-point.

Atmospheric air being thus an elastic and compressible body, it is necessary to determine the law of its change of volume under changes of pressure. This is known as Marriotte's law, and applies to many other gases. *The volume of a gas is inversely as the pressure upon it.* This law is of the utmost importance in gaseous chemistry. It is illustrated by the instrument *Fig.* 240, in which *a b* is a bent tube, open at *a* and closed at *b*. The branch *a* may be several feet long, and *b* six inches. A small quantity of mercury is poured into the tube so as to occupy the bend, and shut up a column of air between *d* and *b*. If the tube be filled

Fig. 240.

How may diffusion against pressure be demonstrated? What proofs are there that the atmosphere is a mixture? What is Marriotte's law? How may its truth be proved?

with mercury to the height of 30 inches, the pressure of this column is exerted on the air in *b*; and as there are now the weight of two atmospheres—that of the ordinary atmosphere and that of the mercurial column—it is compressed into half its former volume, *c*. If we bring upon it three atmospheres, it compresses to one third; four, to one fourth. And the law of course holds good for diminution of pressure; reduce the pressure one half, the volume doubles, etc.

LECTURE XLVI.

COMPOUNDS OF NITROGEN AND OXYGEN.—*Protoxide of Nitrogen.—Preparation and Properties.—Constitution.—A Supporter of Combustion.—Physiological Effects.*
Deutoxide of Nitrogen.—Preparation and Properties.—Constitution.—Relations with Free Oxygen.
Hyponitrous Acid.—Preparation and Properties.

PROTOXIDE OF NITROGEN. $NO=22$.

IF nitrate of ammonia be exposed to a temperature of 350° in a retort, *Fig.* 241, it is decomposed, being resolved into water and the protoxide of nitrogen. The former condenses in the neck of the retort, the latter rises into the jar. If whitish fumes are evolved, they indicate that the temperature is too high. The decomposition is very simple.

Fig. 241.

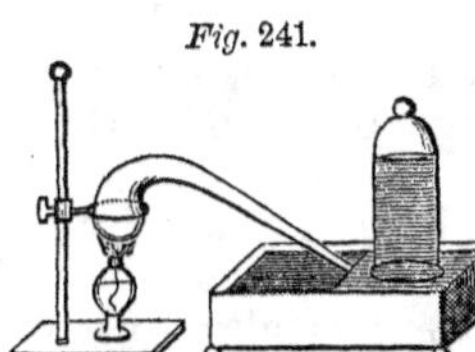

$$NO_5+NH_3=2(NO)+3(HO).$$

One atom of the salt yields two atoms of the protoxide of nitrogen and three of water. One ounce of the nitrate produces two gallons of the gas.

The protoxide of nitrogen is a colorless transparent gas, having a sweetish taste. It is soluble in water, that liquid dissolving about its own volume of the gas, but giving it up on being boiled. Its specific gravity is 1.527. A hundred cubic inches weigh 47.08 grains; it

---

How may protoxide of nitrogen be made? What are its properties?

is therefore half as heavy again as air, and may be collected by displacement, the specific gravity being the same as that of carbonic acid. It may be liquefied at 45° by a pressure of fifty atmospheres, and at 150° below zero freezes into a transparent crystalline solid. In the liquid form it is colorless, and boils at −125°. A drop of it falling on the skin produces the effect of a burn. Mercury sinks in it and freezes into a solid. The liquid protoxide mixed with bisulphide of carbon produces the lowest temperature yet attained, −220°. If the liquid be forced into the air from a jet, a part freezes into a solid, the same occurring when it evaporates in vacuo. It is composed by atom of one of nitrogen and one of oxygen, and by volume of two of nitrogen united to one of oxygen, condensed into two volumes—a constitution like that of water. As it contains half its volume of oxygen, it supports combustion actively; a spark on the wick of a taper introduced into it bursts into a flame, and phosphorus burns with great brilliancy.

Protoxide of nitrogen, or laughing-gas, as it is popularly called from the following circumstances, possesses remarkable physiological properties. When breathed it produces a transient intoxication, owing to the rapid oxidation that is set up throughout the system. It is even more active than pure oxygen, because it is more readily soluble in the blood, and therefore causes a more rapid combustion of the tissues. The individual under its influence has a great flow of ideas, an irresistible tendency to laugh and undergo great muscular exertion, and not infrequently becomes pugnacious. This state is succeeded by one of depression, on account of the accumulation of carbonic acid in the blood, the lungs not being able to remove it as fast as produced. The respiration of the protoxide is not unattended with danger, as it may contain chlorine or deutoxide of nitrogen, or may dangerously stimulate an excitable constitution.

Deutoxide of Nitrogen. $NO_2=30$.

The deutoxide or binoxide of nitrogen is made by the

How may it be liquefied? What are the properties of liquefied protoxide? Why does it support combustion? What are its relations to respiration? Why does it intoxicate? What is the cause of the subsequent depression? How is deutoxide of nitrogen made?

action of dilute nitric acid on slips of copper. The necessary apparatus is represented in *Fig.* 242. Fresh supplies of nitric acid are occasionally to be added through the funnel, *b*, when the action slackens. It is a colorless gas, and may be collected over water, as that fluid only absorbs one twentieth of its volume of it.

Fig. 242.

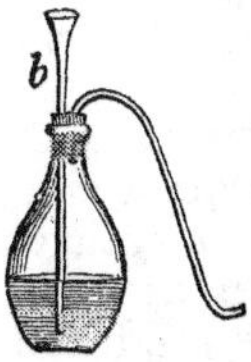

It is composed of equal volumes of nitrogen and oxygen, united without condensation. Its specific gravity is 1.0365; 100 cubic inches weigh 32.10 grains. It does not support combustion, a lighted taper, ignited camphor, or sulphur being extinguished; but if phosphorus in intense ignition be placed in it, the combustion is increased in activity. The same is true of potassium or sodium. The vapor of bisulphide of carbon, mixed with the deutoxide and set on fire, causes the evolution of a most intense light, and the production of carbonic and sulphurous acids, with the liberation of nitrogen.

$$3NO_2 + CS_2 = 2SO_2 + CO_2 + 3N.$$

Iron, arsenic, some sulphides and sulphites, and protochloride of tin, decompose it, protoxide of nitrogen being liberated.

The most remarkable quality of deutoxide of nitrogen is its action on mixtures containing oxygen, as, for example, atmospheric air. It at once produces red fumes of nitrous acid, the deutoxide taking up two atoms of oxygen. For this reason it has been used to effect the analysis of air, certain precautions being adopted. The deutoxide must be added in a small steady stream to the air; red fumes are produced, which are removed by the water of the pneumatic trough. The residual gas is less in volume than the air and deutoxide; one fourth of the deficit represents the volume of oxygen originally present.

Solutions of the protosulphate and protochloride of iron dissolve this gas, a greenish-black liquid being formed. It escapes, however, in a vacuum, leaving the

---

Does it support combustion? What is its action on gaseous mixtures containing oxygen? How may it be used to determine the amount of oxygen? What are its relations to protosulphate and protochloride of iron?

iron salt unchanged; heat only partially expels it. The deutoxide yields in the spectroscope the red band of nitrogen, and near it one derived from the oxygen.

Hyponitrous Acid. $NO_3=38$.

This substance, now frequently called nitrous acid, may be made by mixing four volumes of dry deutoxide of nitrogen with one of dry oxygen over mercury. There arises a green fluid, colorless however at zero, which gives off an orange vapor. It may be produced from the action of 8 parts of nitric acid on 1 of starch, the evolved gases being dried by chloride of calcium, and then conducted into a tube cooled to zero. It is doubtful whether the acid has yet been produced pure; it is generally contaminated with nitrous or nitric acid. Water decomposes it into nitric acid and deutoxide,

$$3NO_3=NO_5+2NO_2.$$

It forms a class of salts called the hyponitrites or nitrites.

LECTURE XLVII.

Compounds of Nitrogen and Oxygen.—*Nitrous Acid.—Preparation and Properties.—Changes of Color by Heat.—Organic Compounds with Nitrous Acid.*

Nitric Acid.—Its Discovery.—Sources in Nature.—Artificial Sources.—Preparation and Properties.—Anhydrous Nitric Acid.—Purification, Tests for.—Its Salts.

Nitrous Acid. $NO_4=46$.

This acid is also called hyponitric acid and peroxide of nitrogen. It may be made from the union of one volume of dry oxygen with two of dry deutoxide of nitrogen, the mixture being cooled to 20°. It also arises in the earthen-ware cup of Grove's battery from the deoxidation of nitric acid. It is most conveniently prepared by distilling in a retort, *a*, *Fig.* 243, dry nitrate of lead at a high temperature, and receiving the product into a tube, *b*, artificially cooled by a freezing mixture, *c*.

How may hyponitrous acid be made? What are its properties? What is the action of water on it? How may nitrous acid be made?

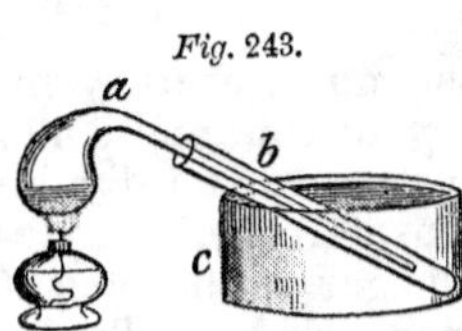

Fig. 243.

The nitrous acid condenses as a colorless liquid, which becomes yellow as the temperature rises. Its specific gravity is 1.45. It crystallizes at 16°, and boils at 82°. It can only be preserved in hermetically-sealed tubes. The vapor is interesting optically; when its temperature is low it is colorless, at 32° it is pale yellow, at 60° deep orange, and finally becomes by farther heating almost black. If the gas be examined by the aid of a prism or spectroscope, a great number of lines are found in the spectrum of light that has passed through it (page 98). As the temperature is caused to rise, these increase in breadth and number to such an extent that eventually no light at all can pass.

Nitrous acid gas, when once mixed with atmospheric air, can with difficulty be condensed into the liquid form. It is irrespirable and of a suffocating odor. Nitrous acid is decomposed by water,

$$3NO_4 = 2NO_5 + NO_2,$$

nitric acid and deutoxide of nitrogen arising at higher temperatures, while at lower ones nitric and hyponitrous acids are formed, $2NO_4 = NO_5 + NO_3$. The vapor of nitrous acid is absorbed by nitric acid, communicating to it colors which vary with the specific gravity of the nitric acid. At 1.5 it is deep orange, at 1.4 yellow, at 1.3 greenish blue, at 1.15 colorless.

Though nitrous acid does not combine without decomposition with alkaline bases, it forms some remarkable combinations with organic bases. Pyroxyline, or gun-cotton of the explosive variety, contains 5 equivalents of this acid, photographic pyroxyline 4 equivalents. That having 3 equivalents of nitrous acid forms an opaline collodion when dissolved in ether and alcohol; that with 2 equivalents is soluble in water.

Nitrous acid is a powerful oxidizer, sulphur, phosphorus, and the metals decomposing it with the evolution of nitrogen.

What are its properties? How does the color of its vapor change by heat? What is seen on examining the spectrum of light passed through it? What is the effect of water on nitrous acid? What are the compounds of nitrous acid with organic bases?

NITRIC ACID. $NO_5=54$.

Nitric acid was discovered in the ninth century by the alchemists. Until the discovery of this and some of the other powerful acids, chemistry can hardly be said to have existed. The Egyptians, Greeks, and Romans had no acid stronger than vinegar. The constitution of nitric acid was determined synthetically by Cavendish. He formed it by passing electric sparks through a mixture of 7 volumes of oxygen and 3 of nitrogen, in contact with a solution of potassa. Nitrate of potassa was obtained.

Nitric acid exists to a small extent in rain-water, and in this case either arises from the effect of the electric flashes upon the atmosphere, or from the oxidation of ammonia in the air. The nitrates of potassa and soda exist naturally in the East Indies and in North and South America. They may be formed artificially by the oxidation or decay of organic matter in contact with basic bodies. In this way the nitre used for gunpowder is produced on a large scale in Europe.

In most of these cases the nitric acid arises from the oxidation of ammonia produced during putrefaction.

$$NH_3+O_8=NO_5+3HO.$$

Common nitric acid is made by distilling equal weights of nitrate of potassa or soda and sulphuric acid. On the large scale the process is conducted in iron vessels, but in the laboratory glass vessels are used.

If a less quantity of sulphuric acid be employed, the nitric acid is of an orange color, from the presence of nitrous acid; and, in addition, the soluble bisulphate of potassa is not formed, but a sparingly soluble sulphate, and the retort may be lost. The decomposition is as follows:

$$KO,NO_5+2(HO,SO_3)=KO,HO,2SO_3+HO,NO_5.$$

The hydrated nitric acid thus produced is a colorless liquid, which boils at 247° if its specific gravity be 1.42, the boiling point being higher if the proportion of water be larger. It freezes at −40°, but, when diluted with half its bulk of water, at −2°. It is decomposed into

When was nitric acid discovered? How did Cavendish form it? Why does nitric acid exist in rain-water? From what sources is nitrate of potassa produced? How may nitric acid arise from ammonia? How may nitric acid be made? What are its properties?

oxygen and nitrogen by passage through a white-hot tube; at a lower temperature, nitrous acid, water, and oxygen arise. In the light it suffers decomposition and turns yellow, on account of the nitrous acid which dissolves in it; it may be freed from that acid by boiling in a glass vessel. From its property of tinging animal substances yellow, it is useful in dyeing. Its action on many metals and combustible bodies is very violent, from its great oxidizing powers. Poured on phosphorus it produces an explosion. If some pieces of iron are placed in a glass under a bell-jar, *Fig.* 244, the vapors of nitrous acid are given off with effervescence. It is often necessary to add a little water to start the action. Ignited charcoal thrown upon strong nitric acid burns vigorously.

Fig. 244.

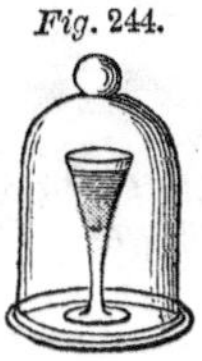

Nitric acid was, until 1840, regarded as a hypothetical body, the strongest aqua fortis of a specific gravity of 1.52 containing one equivalent of water. Its formula is therefore

$$NO_5+HO,$$

though its molecular constitution is regarded as being

$$H, NO_6.$$

The anhydrous acid is formed by the action of chlorine on dry nitrate of silver, heated in a tube at first to 300°, and eventually to 150°. It crystallizes in colorless rhombic prisms, which fuse at 85°, and boil with decomposition at 113°. The solution of these crystals in water causes a rise in temperature, and they then gain the acid properties which they did not possess in the solid state.

Nitric acid may be purified by distillation, the first parts which come over containing chlorine and nitrous acid, and the last parts, containing sulphuric acid, being rejected.

This acid may be detected by the addition of sulphuric acid and a crystal of protosulphate of iron, a brown color being produced; or by its action on copper filings, with the evolution of red fumes. On boiling a nitrate with hydrochloric acid and a piece of gold leaf, the gold

---

On passing through a heated tube, what happens to it? Why is nitric acid yellow? What is its action on the skin and metallic bodies? How may anhydrous nitric acid be made? How may nitric acid be purified? How may it be detected?

is dissolved, forming a yellow solution. The gold will be precipitated as a purple powder by protochloride of tin.

The nitrates deflagrate when burned with combustible matter, as may be shown by igniting a mixture of nitre and sugar. From the solubility of all its salts, nitric acid can not be determined by precipitation. The salts are mostly neutral, though some of the metallic ones are basic.

LECTURE XLVIII.

SULPHUR.—*Sources in Nature.—Its Three Forms.—Properties of Sulphur.—Its Vapor.—Oxygen Compounds.*

Sulphurous Acid.—Preparation and Properties.—Collection by Displacement.—Bleaching Powers.—Liquefaction.—The Sulphites.

SULPHUR. $S=16$.

THE sulphur of commerce is derived either from volcanic regions or from the distillation of metallic sulphides. Iron pyrites contains 54 per cent. of sulphur, contaminated, however, with arsenic. It also exists largely in the sulphates of lime, baryta, etc., and in many organic substances.

It is found in three forms in commerce—roll sulphur, flowers of sulphur, and milk of sulphur. The first receives its name from being cast into cylindrical moulds; the second is derived from the first by sublimation; the third is obtained by precipitation from the tersulphide of potassium or protosulphide of calcium with hydrochloric acid.

Sulphur commonly exists as a solid of a yellow color, and of a specific gravity of 1.99, having no taste, but a peculiar odor. It volatilizes at 180°, and melts at 226° into a yellow liquid. If the temperature be raised to about 450° it changes to a dark brown color, and becomes so viscid that the vessel may be turned upside

When do the nitrates deflagrate? Under what forms does sulphur naturally occur? What are its commercial forms? Describe the properties of sulphur.

down without the sulphur flowing out. At about 800°, if out of contact with the air, it boils, producing an amber-colored vapor. If cooled in water after having been melted at a low temperature, it solidifies into ordinary sulphur; but if heated to near 600° and then suddenly cooled, it becomes elastic like India-rubber, and may be drawn into threads. In this state it may be used for taking casts of coins, etc., because it slowly returns to the hard state. Sulphur presents six different allotropic conditions, in the form of black sulphur being insoluble in bisulphide of carbon, which dissolves the other varieties.

When rubbed on flannel it becomes highly electrical, assuming the negative state. It was formerly used for electrical machines, before the powers of glass were discovered. A roll of it held in the hand crackles, the crystals separating from one another. It is a bad conductor of heat and electricity, and crystallizes under two different forms; it is therefore dimorphous. One of the forms is an acute rhombic octahedron, the other an oblique rhombic prism. When heated to 560° in the open air it takes fire, burning with a blue flame, and gives off a suffocating odor, that of sulphurous acid gas. It is wholly insoluble in water, and but slightly in alcohol, ether, and chloroform. One hundred parts of bisulphide of carbon will dissolve seventy-three parts when warmed.

The vapor of sulphur has the high specific gravity of 6.64. One hundred cubic inches, at the ordinary temperature and pressure, would theoretically weigh 205.44 grains. In it metallic bodies burn like in oxygen gas.

Fig. 245.

If a gun-barrel containing a piece of sulphur, and closed at the muzzle, be heated red-hot, the ignited jet of sulphur vapor issuing from the touch-hole will cause a bunch of iron wire to take fire and burn brilliantly, *Fig.* 245. A very important applica-

What changes occur in it during melting? What electrical condition does it assume? What is its conducting power? Why is it called dimorphous? What is produced by its combustion? What phenomenon is shown by its ignited vapor?

tion, in the arts, of sulphur, is made in the process called vulcanization, which consists in dipping India-rubber in melted sulphur, and then subjecting it to a temperature of about 300°. The rubber retains about two per cent. of sulphur, and gains the property of resisting its usual solvents, and of retaining perfect plasticity and pliancy through a range of many degrees. Silicate of magnesia is often mixed with the rubber, to give a smooth surface. The sulphides may be used instead of sulphur.

Sulphur has a very great range of affinities, combining with most metallic substances in several different proportions, with hydrogen, and with oxygen. With the latter it furnishes seven compounds:

SO_2, Sulphurous Acid.
SO_3, Sulphuric Acid.
S_2O_2, Hyposulphurous or Dithionous Acid.
S_2O_5, Hyposulphuric or Dithionic Acid.
S_3O_5, Trithionic Acid, or Acid of Langlois.
S_4O_5, Tetrathionic Acid, or Acid of Fordos and Gelis.
S_5O_5, Pentathionic Acid.

Of these, the first three are the most important.

SULPHUROUS ACID. $SO_2=32$.

This acid may be formed by burning sulphur in oxygen or in air; in the latter case the gas is mixed with nitrogen. The combustion may be conducted under a bell-jar, the sulphur being placed on a stand. A better process is to partially deoxidize sulphuric acid by heating it with mercury, an oxide of mercury forming, which is converted into a sulphate by the excess of sulphuric acid. It may also be produced by the action of sulphuric acid on charcoal, copper filings, or sulphur. A mixture of three parts of black oxide of manganese and one of sulphur yields it. It must be collected either over mercury or by displacement, unless a solution in water is wanted.

Sulphurous acid is usually a transparent colorless gas, of a sour taste, and suffocating, sulphurous odor. It is

---

Of what use is sulphur in the arts? What are the oxygen compounds of sulphur? How may sulphurous acid be made? What other processes are there for its manufacture? What are the properties of sulphurous acid?

entirely irrespirable, and extinguishes flame at once, being for this reason employed to put out fires in chimneys, a handful of sulphur being burnt at the bottom of the flue. Its specific gravity is 2.2112; one hundred cubic inches weigh 68.48 grains. If a stream of it, which has been cooled by flowing from the generating flask, *a*, *Fig.* 246, through a bent tube, *b*, immersed in cold water, be carried to the bottom of a jar, *c*, the gas will displace the atmospheric air, floating it out of the vessel. This process, called the method of displacement, is useful in collecting gases soluble in water.

Fig. 246.

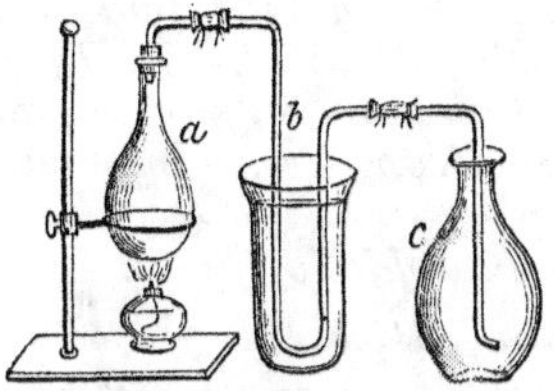

A taper put in a jar of sulphurous acid gas is extinguished at once. If the jar be inverted over water, the gas rapidly dissolves, the liquid taking up about fifty times its volume. Alcohol absorbs 115 volumes. Vegetable colors submitted to it are bleached, but not permanently, as in the case of chlorine, where the coloring matter is destroyed, its hydrogen going to form hydrochloric acid. The colors may be restored by an acid or alkali. Sulphurous acid will support the combustion of potassium or sodium vividly.

This acid gas very readily takes the liquid form if it be cooled by a freezing mixture to 14°. It has then a specific gravity of 1.45, and evaporates so quickly as to produce a very intense cold, by which mercury may be frozen, or water congealed in a red-hot capsule. Sulphurous acid suffers no change at a red heat unless hydrogen be present, when water is formed and sulphur deposited. Oxygen, in presence of water, slowly turns it into sulphuric acid.

Sulphurous acid forms with bases a series of salts—the sulphites. They are easily decomposed by chlorine, nitric acid, and other oxidizing agents, passing into the condition of sulphates; they can also reduce the metallic oxides.

What is the method of displacement? What is its solubility? How does its bleaching power compare with that of chlorine? How may it be condensed? What are the properties of this liquid? What salts does sulphurous acid form?

LECTURE XLIX.

COMPOUNDS OF SULPHUR AND OXYGEN.

Sulphuric Acid.—Anhydrous Sulphuric Acid.—Its Properties.—Nordhausen Oil of Vitriol.—Its Preparation and Constitution.—Common Sulphuric Acid.—Method of Preparation.—Properties.—Impurities.—Tests for.

Hyposulphurous Acid.—Hyposulphite of Soda.

SULPHURIC ACID. $SO_3=40$.

THIS compound is the most important of all acids. By its aid nitric, hydrochloric, and many other acids are prepared. It is also largely consumed in the preparation of carbonate of soda from sea-salt, and of chloride of lime.

There are several varieties of sulphuric acid, differing from one another in the amount of water that they contain. 1st. There is anhydrous sulphuric acid, or sulphuric anhydride (SO_3), which is prepared by heating Nordhausen oil of vitriol to 290°, when a white crystalline substance like asbestos distills over. It fumes in the air, melts at 66°, and boils at 110°. It has an intense affinity for water, hissing like a hot iron when placed in it. The acid properties of this substance are very slight; it shows but little tendency for combination, and does not form true sulphates.

2d. Saxon or Nordhausen oil of vitriol, SO_3+SO_3, HO, is prepared by distilling protosulphate of iron (green vitriol) which has been exposed to a heat sufficient to remove its seven atoms of water. If this dry powder be placed in a retort and exposed to a high temperature, there distills over an oily liquid, hence called oil of vitriol. It is a dihydrate—that is, contains two atoms of acid and one of water. It completely dissolves sulphate of indigo.

3d. Common sulphuric acid, SO_3, HO, called commer-

What is the formula of sulphuric acid? What varieties of sulphuric acid are there? How is the anhydrous acid made? What is the process for making Nordhausen oil of vitriol? What is its composition? How is commercial sulphuric acid made?

cially oil of vitriol. It is made by burning sulphur or pyrites in a regulated current of air, and conducting the sulphurous acid into chambers lined with lead, into which steam and nitrous acid, produced from the action of sulphuric acid on nitre, are admitted. The sulphurous acid takes oxygen from the nitrous acid, reducing it to deutoxide of nitrogen; but that, in turn, takes oxygen from the atmospheric air that is present, and becomes again nitrous acid. The deutoxide acts as an oxygen carrier. The bottom of the chamber, being covered with water, becomes gradually saturated with sulphuric acid, when it is drawn off and concentrated in leaden and then platinum boilers. Its specific gravity is eventually 1.845. It is a dense oily liquid, freezing at —30°, and boiling at 650°. Sulphuric acid of a specific gravity 1.78 freezes at 40° in large crystals.

The affinity of sulphuric acid for water is very intense. If a tube, *b*, containing sulphuric ether, be stirred in a glass, *a*, *Fig.* 247, in which a mixture of sulphuric acid 3 parts, and water 1 part, has been made, the temperature will rise to 300°, and the ether boil. On the same principle, it is useful for removing water from gases, and, as is shown in Lecture XII., that liquid may be frozen on account of the rapidity with which sulphuric acid absorbs its vapor. Organic substances are charred by the action of this acid, which removes the constituent water, and sets the carbon free.

Fig. 247.

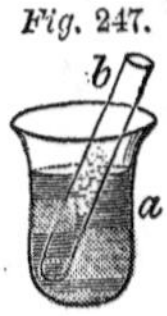

Sulphuric acid is not found pure in commerce, containing sulphate of lead, derived from the lining of the chambers in which it is made, and frequently arsenic, selenium, tin, and nitrous acid. The dark color it presents is due to carbonaceous matter. The acid is tested for by chloride of barium or nitrate of baryta, the white sulphate of baryta being insoluble in water or acids. It reddens black woolen materials, but the stain is removed by ammonia.

In addition to the above hydrates of sulphuric acid, there are two others:

4. Bihydrate, SO_3, $2HO$, boils at 435°, specific gravity 1.78.
5. Terhydrate, SO_3, $3HO$, " 348°, " " 1.63.

What are the properties of sulphuric acid? How may its affinity for water be shown? What are its usual impurities? How is it detected?

HYPOSULPHUROUS ACID. $S_2 O_2$.

This acid is only known in the combined state. On attempting to separate it from its salts, it decomposes into sulphur and sulphurous acid,

$$S_2 O_2 = S + S O_2.$$

The hyposulphite of soda is of great use in photographic operations, from its power of dissolving the compounds of silver. If any trace of the salt is left in a paper proof, it will eventually cause it to become yellow and fade away.

The other compounds of sulphur and oxygen possess but little interest.

LECTURE L.

SULPHUR AND PHOSPHORUS.—*Sulphureted Hydrogen.—Preparation and Properties.—Uses as a Test.—Sulphur Waters.—Persulphide of Hydrogen.*

SELENIUM.

PHOSPHORUS.—*Made from Bone-earth.—Properties.—Shines in the Dark.—Inflammability.—Allotropic Phosphorus.—Compounds with Oxygen.*

SULPHURETED HYDROGEN. $HS = 17$.

Fig. 248.

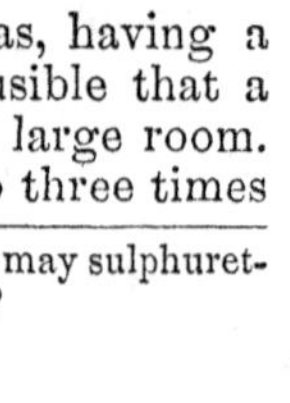

THIS gas may be prepared by the action of hydrochloric acid on sulphide of antimony, in the apparatus *Fig.* 248, or of dilute sulphuric acid on sulphide of iron. It must be collected over either warm or salt water. If made by sulphide of iron, the action is as follows:

$$FeS + S O_3, HO = HS + Fe O, S O_3.$$

It is called also hydrosulphuric acid and sulphydric acid.

Sulphureted hydrogen is a colorless gas, having a fetid odor like rotten eggs. It is so diffusible that a very small quantity will taint the air of a large room. It is absorbed by water, that fluid taking up three times

---

What are the uses of hyposulphurous acid? How may sulphureted hydrogen be prepared? What are its properties?

its volume at 60°. In this form it is rapidly decomposed by the contact of air, the hydrogen forming water with the oxygen, and the sulphur depositing. The specific gravity of sulphureted hydrogen gas is 1.1747; one hundred cubic inches weigh 36.38 grains. It is inflammable, and may be burnt from a jet, as in *Fig.* 249. If the access of air is unlimited, sulphurous acid and water arise; if limited, water is produced and sulphur deposited. It reddens litmus slightly, and combines with metals to form sulphides. For this latter reason it is very valuable in analytical operations, many of the sulphides being insoluble and highly colored: antimony gives an orange precipitate, arsenic a yellow, lead a brown, manganese a flesh-colored. It tarnishes silver, the metal passing through various shades of yellow and orange to blackness. It is liquefied by a pressure of 17 atmospheres at 50°, the specific gravity being 0.9. When cooled to −122° it solidifies into a white substance.

Fig. 249.

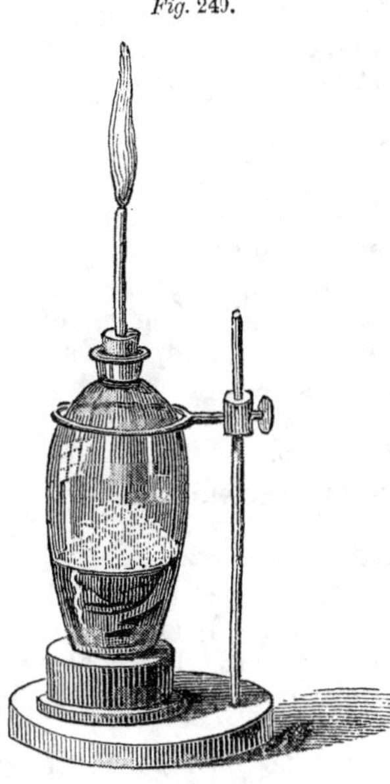

The action of sulphureted hydrogen is illustrated by writing on a sheet of paper with a solution of acetate of lead. The letters are invisible until exposed to a current of the gas, when they become black.

Sulphureted hydrogen is a natural constituent of some mineral waters, as at Sharon, in New York, and at the Virginia Sulphur Springs. It is also found in the air of sewers and in putrefying animal matter, and has been supposed to be the cause of miasmatic fevers. It is

---

What is its solubility? What arises from its combustion? What precipitates does it give with metallic oxides? At what points does it liquefy and solidify? Does it exist naturally?

very poisonous when respired, even when dilute, causing nausea, headache, and faintness.

There is another compound of sulphur and hydrogen, the persulphide of hydrogen, the composition of which is supposed to be HS_5. It is a heavy yellow liquid, of a specific gravity 1.76.

SELENIUM. $Se=40$.

This substance is found in certain varieties of iron pyrites. It resembles sulphur in many respects, and has a reddish-brown color and dim metallic lustre. It tinges flame of a light blue color, and gives off an offensive odor. With oxygen it forms three compounds:

SeO, Oxide of Selenium.
SeO_2, Selenious Acid.
SeO_3, Selenic Acid.

With hydrogen it unites to form seleniureted hydrogen.

PHOSPHORUS. $P=32$.

This substance, so named from shining in the dark, was discovered in 1669 by Brandt. It is now produced from phosphate of lime or bone-earth, but is also found in other animal matters, more particularly in the brain and nervous tissue.

The process for production is to burn the bones, grind them, and digest them in dilute sulphuric acid for six hours, steam being passed into the mixture to hasten the changes. The liquid is then strained, evaporated to dryness, fused, and mixed with one fourth of its weight of charcoal. It must then be distilled at a white heat in a stone-ware retort, the neck of which dips beneath warm water. A part of the phosphoric acid is deoxidized by the charcoal, carbonic oxide escaping and phosphorus coming over. It is purified by melting and straining through chamois leather. In Great Britain about six tons are annually used in the manufacture of matches, one pound making 600,000 matches.

Phosphorus is tasteless, transparent, and colorless.

What are its relations to respiration? What other compound of sulphur and hydrogen is there? Describe selenium. From what is phosphorus derived? Describe the process for its production. To what use is phosphorus applied in the arts? What are its properties?

Exposed to light it turns red, even in a vacuum, owing to undergoing a molecular change. In general appearance it resembles wax. In the air it slowly oxidizes, smoking, and exhaling an electrical odor. At 32° it is brittle; at 110° it melts; at 570° it boils in close vessels. In the air it takes fire at 120°, burning with the evolution of anhydrous phosphoric acid. Its specific gravity is 1.826. Phosphorus is so poisonous that a few grains will destroy life, and those engaged in the manufacture of matches frequently suffer from necrosis of the lower jaw-bone.

Phosphorus requires to be kept under water in order to avoid oxidation, and must also be handled carefully. A few pieces placed between brown paper and rubbed take fire, and it will also inflame if sprinkled with lamp-black or powdered animal charcoal. Placed on dry wood, flannel, feathers, or other non-conducting substances, it will ignite, if in thin slices. In chlorine, or the vapor of bromine or iodine, it burns spontaneously.

If phosphorus is suddenly cooled from the fused condition it undergoes a change, becoming passive. This allotropic modification may also be produced by distilling it in an atmosphere of nitrogen or carbonic acid. For the purposes of commerce it is prepared by keeping phosphorus heated to 450° for three or four weeks in an air-tight iron vessel. It may then present a black, gray, or scarlet color, and will not take fire under a temperature of 500°. It is insoluble in bisulphide of carbon, does not shine in the dark, shows no disposition to unite with sulphur, and will not oxidize in the air.

Phosphorus and oxygen form four compounds:

P_2O, Oxide of Phosphorus.
PO, Hypophosphorous Acid.
PO_3, Phosphorous Acid.
PO_5, Phosphoric Acid.

Is phosphorus poisonous? Give examples of its combustibility. How may passive phosphorus be made? How many compounds of phosphorus and oxygen are there?

LECTURE LI.

COMPOUNDS OF PHOSPHORUS AND OXYGEN. — *Oxide of Phosphorus, Preparation of. — Hypophosphorous and Phosphorous Acids.—Phosphoric Acid.—Preparation. — Compounds with Water. — Properties of these Acids and their Salts.—Phosphureted Hydrogen.—Three Compounds of Phosphorus and Oxygen.—Spontaneous Inflammability of Phosphureted Hydrogen.* — CHLORINE. — *Existence in Nature. — Preparation.—Liquefaction.—Relations to Combustion and Respiration.*

OXIDE OF PHOSPHORUS. P_2O.

WHEN phosphorus is burned in air, the red residue is this body. It may be formed in quantity by passing a stream of oxygen from the tube *a*, *Fig.* 250, upon phosphorus under hot water in a glass, *b*. A brilliant combustion takes place, phosphoric acid and the oxide resulting. The former is dissolved by the water, and the latter, when washed with bisulphide of carbon, is left in a state of purity.

Fig. 250.

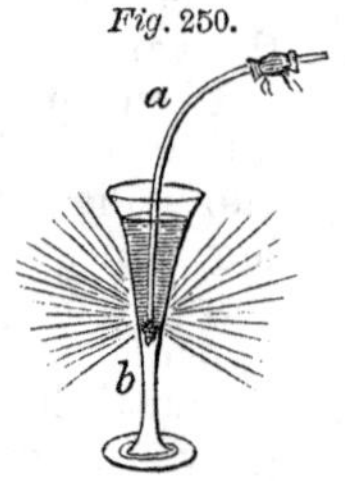

HYPOPHOSPHOROUS ACID, PO,

is prepared by acting on phosphide of barium with water, and treating the solution with sulphuric acid as long as any precipitate falls. It is a powerful deoxidizing agent. Some of the hypophosphites are useful in medicine.

PHOSPHOROUS ACID, PO_3,

is formed by the combustion of phosphorus in a limited amount of dry air. It is then seen as a dry white powder. It may be obtained in solution by setting a number of sticks of phosphorus, inclosed in glass tubes, around a funnel placed in the neck of a bottle. It has

How is oxide of phosphorus made? How are hypophosphorous and phosphorous acids made?

powerful deoxidizing properties, taking the oxygen from sulphuric acid, and causing it to deposit sulphur.

Phosphoric Acid. PO_5.

The anhydrous form originates when phosphorus is burned in dry air or oxygen, *Fig.* 251. It condenses in white snowy flakes, which hiss like hot metal when dipped in water. It may be made by the action of nitric acid on phosphorus. It is very deliquescent, and scarcely shows any acid properties before uniting with water. It may be detected by its yellow precipitate with molybdate of ammonia.

Fig. 251.

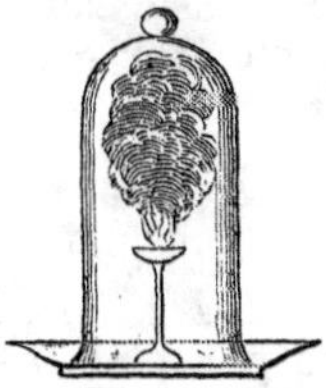

Phosphoric acid unites with water in three proportions, producing

Monobasic	Phosphoric	Acid . . .	PO_5, HO
Bibasic	"	" . . .	PO_5, $2HO$
Tribasic	"	" . . .	PO_5, $3HO$

These also have been called metaphosphoric, pyrophosphoric, and common phosphoric acids.

The first, the protohydrate of phosphoric acid, is produced when any of the watery solutions of phosphoric acid are evaporated to dryness, a body called glacial phosphoric acid resulting. It gives a white granular precipitate with nitrate of silver, and coagulates albumen. Its salts contain one atom of base to one of acid. By boiling with water it goes into the tribasic form.

The second, the bihydrate of phosphoric acid, can be made by heating common phosphoric acid to 417° for some time. It neither precipitates silver nor coagulates albumen, though its salts yield with silver a flaky precipitate. It also turns to tribasic acid by boiling with water.

The third, the terhydrate of phosphoric acid, may be obtained from phosphate of lime by the action of sulphuric acid, sulphate of lime being formed, or by boiling anhydrous phosphoric acid in water. It neither precipitates silver nor coagulates albumen, but its salts give canary-yellow precipitates with nitrate of silver.

How is anhydrous phosphoric acid made? What compounds does it yield with water? How is the protohydrate produced? How are the bihydrate and terhydrate made? What are their properties?

These hydrogen acids of phosphorus give rise to a very complex series of salts, according to the extent to which the hydrogen is replaced by metallic bodies. The monobasic acid can only yield one class of salts, in which all its hydrogen is replaced by a metal; but the bibasic can yield two series, according as the metal replaces one or both atoms of base; the tribasic can yield three series, according as one, or two, or three of the hydrogen atoms are displaced.

PHOSPHURETED HYDROGEN. $PH_3=35$.

Phosphorus forms three compounds with hydrogen: a gas, PH_3; a liquid, PH_2; and a solid, P_2H. The first is best known, and is made by boiling phosphorus in a strong solution of potassa in a retort, *Fig.* 252, the neck

Fig. 252.

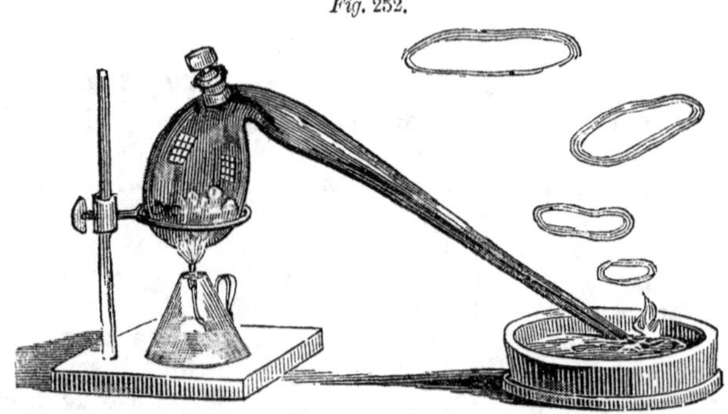

of which dips beneath the surface of water. As the bubbles of gas break on the water they take fire, burning with a bright yellow light, and there ascends through the air a ring of smoke, which dilates as it rises, and exhibits a rotating motion in its parts. The gas may also be made by putting phosphuret of calcium in water.

Phosphureted hydrogen is a colorless gas, having a smell like garlic or putrid fish. Water takes up one eighth its volume. The spontaneous combustibility results from an admixture of the vapor of the liquid

---

How many series of salts can each class yield? Describe the preparation of phosphureted hydrogen. What are its properties?

phosphuret; phosphoric acid and water arise. Its specific gravity is 1.185.

Phosphorus forms also compounds with nitrogen, chlorine, bromine, iodine, and sulphur.

CHLORINE. $Cl = 35.5$.

Chlorine was discovered by Scheele in 1774, and was originally called oxymuriatic acid. It derives its name from its greenish color. It is not found free in nature, but exists in abundance in common salt, the chloride of sodium, a material which gives salinity to the ocean.

Chlorine may be obtained by the action of sulphuric acid on common salt and black oxide of manganese, or better by heating a mixture of hydrochloric acid and black oxide of manganese. The action in the latter case is as follows:

$$Mn\,O_2 + 2HCl = Mn\,Cl + 2HO + Cl.$$

One atom of peroxide of manganese and two of hydrochloric acid give one atom of chloride of manganese, two of water, and one of chlorine. Half the chlorine is given off as gas, and half remains in combination.

The apparatus for its production is seen in *Fig.* 253,

Fig. 253.

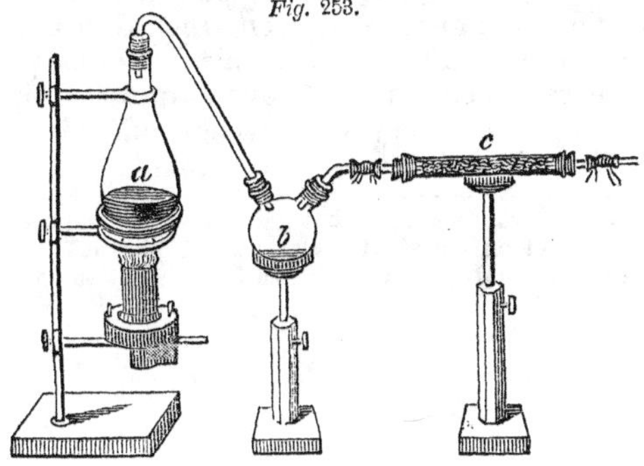

where *a* is the retort, with the generating materials, connected with a small receiver, *b*, to retain part of the water which the gas may bring over; this, again, is connected with a chloride of calcium tube, *c*, which ef-

When was chlorine discovered? In what substances does it occur? How may it be formed? Describe the apparatus for its production.

fects the perfect drying of the gas. As chlorine is very soluble in cold water and acts on mercury, it can neither be collected at the pneumatic nor mercurial trough. It may, however, be gathered over warm water, or a saturated solution of common salt, or by displacement.

Chlorine is a greenish-yellow gas that may be liquefied by a pressure of four atmospheres, or by cooling to $-106°$; it has not been solidified. It forms with water a crystalline hydrate, having the composition $Cl+10HO$. On inclosing these crystals in a bent tube and heating them, the resulting chlorine will be liquefied by its own pressure.

A taper immersed in chlorine burns for a short time, emitting volumes of black smoke, *Fig.* 254, which are due to the fact that the hydrogen of the fatty compound is alone uniting with the gas to produce hydrochloric acid, while the carbon, which has but slight affinity for chlorine, is set free. Powdered antimony or brass leaf immersed in this gas becomes incandescent, and consumes, a chloride resulting. Phosphorus takes fire in it at ordinary temperatures, and burns with a pale flame. The odor of chlorine is pungent, and, even when dilute, irritating to the mucous membrane of the air passages, producing a hoarseness which may last several days. Mixed with aqueous vapor and very much weakened, it is said to be of advantage in pulmonary complaints.

Fig. 254.

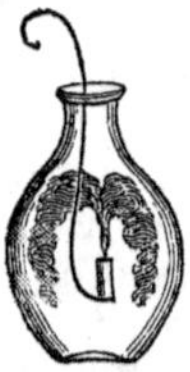

What are its properties? What is its action on a burning taper? How does it act on certain metals and phosphorus? What is its effect on the animal system? Which is the most valuable property of chlorine?

LECTURE LII.

CHLORINE, CONTINUED. — *Bleaching and Disinfecting Powers. — Combustion of Hydrocarbons. — Tests for Chlorine. — Chlorine Water. — Oxygen Compounds. —Hypochlorous, Chlorous, Hypochloric, Chloric, and Perchloric Acids.—Bleaching with Chloride of Lime. —Chloride of Nitrogen an explosive Compound.— Hydrochloric Acid. — Preparation in the Gaseous and Liquid Conditions.*

THE bleaching properties of chlorine render it of the greatest value in the arts. Previous to its introduction, woven fabrics, for example, were bleached by exposure to the sunshine and moisture, a process demanding a length of time and a large open space. The same operation can now be performed in a few hours in a confined apartment. This property may be illustrated by pouring a solution of litmus or indigo through a funnel, *a*, into a flask, *b*, containing chlorine, *Fig.* 255. The decoloration takes place at once. A solution of chlorine in water may also be used. Chlorine is employed by physicians for disinfecting the air of foul rooms, as those in which hospital gangrene has been prevalent. Such effluvia contain hydrogen, which unites with the chlorine, and the noxious compound is decomposed. Some have supposed these properties to be due to the evolution of nascent oxygen. It should be disengaged slowly, if patients are in the room, by the action of dilute sulphuric acid on chloride of lime.

Fig. 255.

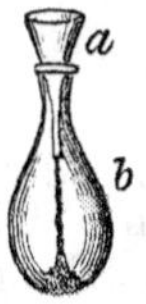

The peculiarities of chlorine as a supporter of combustion are well seen when a piece of blotting-paper, *Fig.* 256, saturated with turpentine, is placed in a jar of the gas. It takes fire with the evolution of clouds of carbon smoke. This phenomenon depends on the intense affinity that chlorine has for the electro-positive bod-

Fig. 256.

What is the most valuable property of chlorine? How may it be illustrated? How does it act on effluvia? What is the cause of the smoke when hydrocarbons are burned in chlorine?

ies, though it unites with carbon with reluctance. A green wax taper, the wick of which carries a spark, will rekindle in chlorine, and continue to burn.

Free chlorine may be detected by its smell, its bleaching action on vegetable compounds, and its white curdy precipitate with nitrate of silver. This compound chloride of silver, which changes to a dark color in the light, is used for the production of photographs on paper, very large quantities being consumed in that application. Chlorine water is made by agitating water in a bottle of chlorine, the mouth of the bottle from time to time being opened under water. This solution decomposes in the sunshine, oxygen gas being liberated and hydrochloric acid formed by the decomposition of the water. The specific gravity of chlorine is 2.487, and 100 cubic inches weigh 77.04 grains.

Chlorine unites with oxygen, producing

ClO, Hypochlorous Acid.
ClO_3, Chlorous Acid.
ClO_4, Peroxide of Chlorine, Hypochloric Acid.
ClO_5, Chloric Acid.
ClO_7, Perchloric Acid.

Hypochlorous Acid. $ClO = 43.5$.

This acid is often obtained by acting on the red oxide of mercury, suspended in water, with chlorine. The gaseous acid is procured by placing the aqueous solution in a tube inverted over mercury, and passing dry nitrate of lime into it. It is deeper-colored than chlorine, a more powerful bleaching agent, and oxidizes vigorously. The warmth of the hand causes it to explode, as does also the direct sunshine. Water dissolves 200 times its volume of this gas. The specific gravity is 3.04, and 100 cubic inches weigh 94.16 grains.

The most common bleaching compound is the chloride or hypochlorite of lime. The composition seems to be $CaO, ClO + CaCl$. The articles to be bleached are saturated with an aqueous solution of this substance, and then washed in dilute sulphuric acid. The quan-

How may chlorine be detected? How may chlorine water be made, and what are its properties? Name the oxygen compounds of chlorine. How is hypochlorous acid made, and what are its properties? What is the constitution of bleaching powder?

tity of chlorine in chloride of lime is determined by the process called chlorimetry. The best method is to ascertain how much arsenious acid, $As\,O_3$, can be raised to the state of arsenic acid, $As\,O_5$, by a weighed quantity of the chloride of lime.

Chlorous Acid. $Cl\,O_3=59.5$.

This gas is of a greenish color, and may be made by mixing arsenious acid with chlorate of potassa and diluted nitric acid, and distilling in a water-bath. Water dissolves six times its volume. It explodes at 130°.

Peroxide of Chlorine, $Cl\,O_4=67.5$,

is made by acting on chlorate of potassa with concentrated sulphuric acid at a temperature not exceeding 100°. It is a yellow gas, exploding at 140°, and soluble in water to the extent of 20 volumes. If into a glass, *a*, *Fig.* 257, containing water, some crystals of chlorate of potassa and fragments of phosphorus are placed, and sulphuric acid is poured upon them through a long funnel, *b*, chlorous acid is liberated, and a brilliant combustion under water ensues, the water becoming yellow. The gas may be readily liquefied; its specific gravity is 2.33.

Fig. 257.

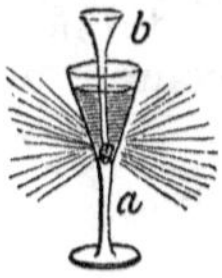

Chloric Acid. $Cl\,O_5=75.5$.

This acid, which only exists in combination with one atom of water, is made by decomposing the chlorate of baryta by sulphuric acid, which must not be added in excess. This solution is to be evaporated in vacuo, when a sour, yellowish, sirupy liquid is produced. It sets on fire substances containing carbon and hydrogen, and has the bleaching power. One of the salts, the chlorate of potassa, is of use in chemistry on account of the facility with which it yields up oxygen. A few grains of this salt ground in a mortar with sulphur explode violently. When mixed with sugar, it is inflamed by a drop of sulphuric acid.

What is chlorimetry? How is chlorous acid made? How is peroxide of chlorine made? In what manner may phosphorus be burned under water? How is chloric acid made? What are the properties of chloric acid?

Perchloric Acid, $Cl\,O_7 = 91.5$,

is obtained by distilling perchlorate of potassa with its own weight of sulphuric acid, mixed with one quarter as much water. At 280° a dense white vapor passes over. It is a colorless, oily, corrosive liquid, exploding like chloride of nitrogen on contact with combustible substances, such as charcoal, ether, etc. It fumes in the air, and can not be kept, even in the dark, without exploding.

Chlorine and Nitrogen. NCl_3 or NCl_4.

This compound, one of the most explosive at present known, is formed when a solution of sal ammoniac is exposed to chlorine. A leaden basin should be placed under the jar to collect the oily liquid. Dulong lost an eye, and Davy was severely wounded by the explosions of this substance. Its specific gravity is 1.65, and it may be distilled at a temperature of 160°. The contact of oily matter causes a detonation.

Chlorine and Hydrogen.
Hydrochloric Acid. $HCl = 36.5$.

This acid, called also muriatic acid, is prepared by placing in a flask one part of salt and two of sulphuric acid. The action may be aided by a spirit-lamp. The gas is conducted through a tube into a bottle containing water. The end of the tube should dip but a short distance beneath the surface of the water, so that if the water should tend to regurgitate, it may be arrested by a suitable bulb, and atmospheric air allowed to pass into the flask. The bottle should be surrounded by ice-water, as a large amount of heat is extricated during the process of solution, and the water increases in volume from one to two thirds. The action is

$$NaCl + 2(SO_3, HO) = HCl + (NaO, 2SO_3, HO);$$

that is, one atom of chloride of sodium and two of sulphuric acid yield one atom of hydrochloric acid and one of bisulphate of soda.

Hydrochloric acid gas may be obtained by heating the liquid thus obtained, and collecting it by displace-

How is perchloric acid made, and what are its properties? What are the properties of the chloride of nitrogen? How is hydrochloric acid made? How may the gas be procured?

ment. It is a transparent colorless gas, having powerful acid properties, and very absorbable by water, that fluid taking up 500 times its volume. It fumes in moist air. If a dry Florence flask be filled with it by displacement, and the mouth of it then opened under water, the water rushes up violently, owing to the quick absorption of the gas. The specific gravity of the gas is 1.2783; one hundred cubic inches weigh 39.59 grains. It contains equal volumes of its constituents, united without condensation.

LECTURE LIII.

CHLORINE CONTINUED.—*Liquefaction of Hydrochloric Acid.—Its Production by Light.—The Chlorine-Hydrogen Photometer.—Action of Hydrochloric Acid on Metallic Oxides.—Solution of Hydrochloric Acid.—Its Properties.—Tests for Nitro-hydrochloric Acid.*
IODINE.—*Method of Preparation.—Properties.—Reactions.—Its Photographic Relations.—Hydriodic Acid.—Compounds of Iodine with Oxygen, Nitrogen, and Chlorine.*

HYDROCHLORIC ACID, although gaseous at ordinary temperatures and pressures, is liquefied by a pressure of forty atmospheres. At 50° it is colorless, and less refractive than water.

The pure gas may be obtained by the direct union of chlorine and hydrogen under the influence of flame, the electric spark, or light. In the dark the gases do not combine, but if a beam of sunlight be thrown upon a flask containing a mixture of equal volumes, a violent explosion results, and the vessel is shattered to pieces.

It was found by Dr. Draper that in this striking experiment the action is due to the chlorine, which, on account of its color, absorbs the indigo ray, and changes from the passive to the active state. It may also be rendered active by spongy platinum, and chlorine which has been exposed by itself to the sunshine unites more

---

What are its properties? How may its solubility be shown? How may hydrochloric acid gas be liquefied? What is the action of sunlight on a mixture of chlorine and hydrogen? To which of these bodies is the action due?

readily with hydrogen. Dr. Draper invented a photometer based on these phenomena (page 91), and with it determined many most valuable photo-chemical facts (Philosophical Magazine, December, 1843).

When hydrochloric acid is brought in contact with metallic oxides, both are decomposed, a metallic chloride and water resulting.

$$MO+HCl=MCl+HO,$$

or

$$M_2O_3+3HCl=M_2Cl_3+3HO;$$

that is, one atom of a metallic protoxide with one of hydrochloric acid yields one of protochloride and one of water. In the case of a sesquioxide, one atom with three of hydrochloric acid gives one of metallic sesquichloride and three of water.

Fig. 258.

The constitution of hydrochloric acid, and its action on metallic oxides, are strikingly shown by taking a flask, *Fig.* 258, filled with it, and pouring a fine stream of peroxide of mercury in through a funnel. The chloride of mercury, corrosive sublimate, forms at once, and drops of water condense on the sides of the flask.

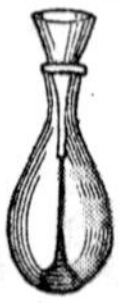

Liquid hydrochloric acid, or spirit of salt, as it is called from its origin, is the most commonly used form. It has, when very concentrated, a specific gravity of 1.21, boiling at 112°, and freezing at −60°. It contains 42.4 per cent. of the acid gas. The boiling point is highest, 230°, when the specific gravity is 1.094, and the liquid contains 20 per cent. of dry gas. The strong acid is weakened and the weak strengthened by boiling.

The commercial acid is generally yellow, partly from chloride of iron and partly from the particles of cork and lute that may have fallen into it. It may also contain sulphuric acid, chlorine, sulphurous acid, tin, or arsenic. From the latter it is separated by distillation over sulphide of barium.

The tests for hydrochloric acid are first the dense white fumes of sal ammoniac, the chloride of ammonium, that it yields with free ammonia. If two bottles that have been rinsed out, one, *c*, with the acid, and the

---

What is the action of hydrochloric acid on metallic oxides? What is the action of hydrochloric acid on peroxide of mercury? What are the properties of liquid hydrochloric acid? What are its impurities? What are the tests for it?

Fig. 259.

other, *h*, with ammonia, are placed mouth to mouth, *Fig.* 259, they are filled with a white cloud very quickly. Second, with nitrate of silver this acid gives a curdy white precipitate of chloride of silver, which is soluble in ammonia. Hydrochloric acid gas is distinguished from chlorine by the absence of bleaching power and its acid qualities, litmus water being used as the test.

Nitro-Hydrochloric Acid

is also called nitro-muriatic acid and aqua regia, and is formed by adding to hydrochloric acid one third of its volume of nitric acid. It possesses the power of dissolving the noble metals, forming chlorides, a property due to the evolution of nascent chlorine. Nitrous acid and water are also set free. Heat accelerates the action, but may cause the loss of some of the chlorine.

Iodine. $I=126$.

This element was discovered in 1811, and is named from the violet color of its vapor. It is made from *kelp*, which is the ash produced by the burning of sea-weed, but is also found in some saline springs, in certain Mexican silver ores.

It may be obtained by lixiviating kelp, and evaporating till no more crystals are produced. The mother-liquor is then treated with sulphuric acid, and subsequently heated with peroxide of manganese in a leaden retort, *a b c*, *Fig.* 260, the iodine distilling over into the receivers, *d*.

It is a solid substance, of a bluish-black color, with a metallic lustre, communicates to the skin a fugitive yellow stain, and smells like a sea-beach. It is very volatile, producing a pale vapor at 60°, and crystallizes in rhombic plates and octahedra. The specific gravity is 4.946. At 220° it melts, and boils at 350°, giving off violet fumes. The specific gravity of the vapor is 8.7066.

Describe the experiment *Fig.* 259. What are the preparation and properties of nitro-muriatic acid? From what source is iodine procured? What is the method of its preparation? What is its appearance?

Fig 260.

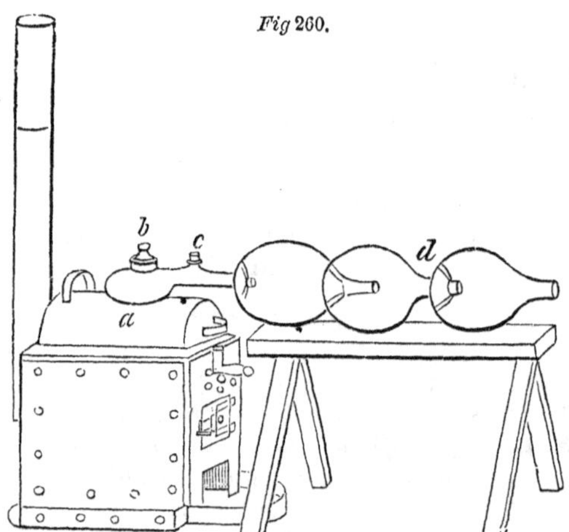

It is one of the heaviest gaseous bodies known; 100 cubic inches weigh 269.64 grains.

Fig. 261.

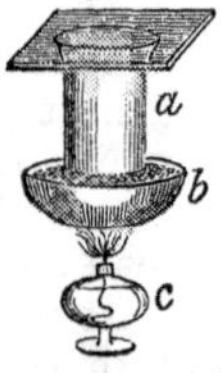

a
b
c

Iodine supports combustion like chlorine. A jar, *a*, *Fig.* 261, containing a few grains of it, placed in a small sand-bath, *b*, and warmed by a spirit-lamp, *c*, may be easily filled with its dense vapor, the atmospheric air floating out before it. A lighted candle plunged in this vapor burns slowly, but phosphorus spontaneously ignites. In the same manner, phosphorus, placed with a few grains of solid iodine in a capsule covered by a jar, *Fig.* 262, takes fire, with the evolution of phosphoric acid, vapor of iodine, and iodide of phosphorus.

Fig. 262.

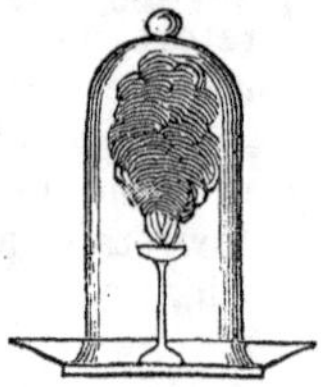

Iodine is but sparingly soluble in water, that liquid taking up, if pure, only $\frac{1}{7000}$ part of its weight, and becoming pale brown. It decomposes in the sunshine, iodic and hydriodic acids arising. If the water contain iodide of potassium, chloride of ammonium, nitrate of ammonia, or hydriodic acid, it will dissolve iodine freely. Alcohol, ether, chloroform, and

---

What are its relations as respects combustion? In what fluids is it soluble?

sulphide of carbon also dissolve it abundantly. The last fluid is valuable in researches on heat, as it only permits the dark rays to pass.

Iodine gives very characteristic reactions. Iodide of potassium with acetate of lead yields a yellow precipitate, which, on cooling after being boiled, assumes a crystalline appearance like flakes of gold leaf. With chloride of mercury a scarlet biniodide is produced. If dried and sublimed in a tube, the yellow crystals which form possess the property of turning red when touched. With a solution of starch, free iodine, or an iodide acidified with nitric acid, yields a blue color; the solution becoming colorless if heated, but the color returning on cooling, providing the temperature has not been carried to the boiling point. If a potato be cut and tincture of iodine poured on the surface, innumerable blue specks make their appearance, corresponding to the position of granules of starch.

Iodine is most valued for its photographic relations. The iodide of silver is one of the most sensitive compounds at present known. The iodides of potassium, ammonium, cadmium, etc., are mixed with collodion or albumen, and, a film of the mixture being spread over glass, is subjected to the action of a solution of nitrate of silver in the dark. After being impressed by the light received through a camera lens, a solution of protosulphate of iron or pyrogallic acid is poured upon the plate. The parts acted on by light immediately receive a dark deposit of metallic silver. The superfluous iodide of silver is removed by a solution of cyanide of potassium or hyposulphite of soda. In the daguerreotype the film of iodide is formed by exposing a polished silver plate to the vapor of iodine.

Hydriodic Acid.—$HI = 127$.

This gas may be made by dissolving in a solution of iodide of potassium as much iodine as it will hold, adding pieces of phosphorus and distilling. A colorless fuming gas is sent over. Its specific gravity is 4.3878; 100 cubic inches weigh 135.89 grains. It extinguishes

What are its reactions? How does it act with a solution of starch? What is its most valuable application? How is hydriodic acid made? What are its properties?

flame and is not inflammable. It liquefies under pressure, and becomes a transparent colorless solid at $-60°$. A solution of hydriodic acid in water may be made by passing sulphureted hydrogen from a flask through water in which iodine is suspended. The acid forms, and sulphur is deposited:

$$I+HS=S+HI.$$

Hydriodic acid has the general relations of hydrochloric acid, and is, like it, very soluble in water. With nitrate of silver it gives a yellow precipitate of iodide of silver.

With oxygen, iodine forms two acids, iodic (IO_5) and periodic (IO_7HO). The former is obtained by boiling iodine with the strongest nitric acid. With nitrogen it gives an iodide (NHI_2), which is prepared by putting powdered iodine in aqua ammonia. The brown powder which forms is dangerously explosive. If placed while moist on paper, it will, on drying, blow up flies that may walk upon it. Chlorine forms two instable compounds—ICl, a liquid; and ICl_3, a solid.

LECTURE LIV.

BROMINE.—*Sources.—Properties.—Compounds.*—FLUORINE.—*Sources.—Hydrofluoric Acid.—Etching on Glass.*—CARBON.—*The Carbon Group.—Allotropic Forms of.—Preparation of Charcoal and Lampblack.—Diamond.—Absorbent Power of Charcoal.—Oxygen Compounds of Carbon.—Carbonic Oxide.—Preparation and Properties.—Chloro-carbonic Acid.*

BROMINE, $Br=78$,

occurs in sea-water, and also in certain brine-springs, both in America and Europe. It is found, among minerals, associated with silver. From its solutions it may be obtained by evaporating the water until the chloride of sodium has crystallized out, and passing a current

---

How may a solution of it be made? What are the oxygen compounds of iodine? What are the properties of iodide of nitrogen? From what source is bromine obtained? Describe the method pursued.

of chlorine gas through it. The solution turns yellow, and on being agitated with ether the bromine is dissolved out. The ether is agitated with potassa, bromate of potassa and bromide of potassium forming. On ignition, oxygen is expelled, and the whole converted into the latter salt, from which the bromine may be distilled by the aid of peroxide of manganese and sulphuric acid.

It is a liquid of a deep reddish-brown color, a disagreeable odor, whence its name, solidifying at —7°. 6, and boiling at 145°. Its specific gravity is about 3., that of the vapor 5.39; 100 cubic inches weigh 166.92 grains. If breathed it produces the effect of a severe cold, which may last for days. It bleaches like chlorine if aqueous vapor be present, but not if dry, whence it is supposed that the bleaching is produced by nascent oxygen derived from the decomposed water. It extinguishes flame, and combines, with explosion, with potassium and phosphorus; antimony burns in it. In its general relations it resembles chlorine.

With oxygen it forms a compound, bromic acid, $Br\,O_5$; and with hydrogen, hydrobromic acid, HBr. The latter is prepared by heating a mixture of phosphorus, bromine, and bromide of potassium with water. At —100° it is a clear colorless liquid, and becomes solid at —124°; 100 cubic inches of the gas weigh 84.53 grains.

The bromide of silver is used as a photographic agent, in combination with the iodide.

Fluorine, $F=19$,

is found abundantly in nature, in combination with calcium, as fluor spar. It occurs also in the topaz, in some kinds of mica, in sedimentary rocks, in teeth, and fossil bones; these last sometimes contain 10 per cent. of fluoride of calcium. Cryolite, found in Greenland, is a fluoride of aluminum and sodium. Fluorine has not with certainty been isolated, though it is stated to be a yellowish-brown gas. It attacks glass and platinum; and though vessels of fluor spar have been substituted

What are the properties of bromine? What other element does it resemble? Describe hydrobromic acid. For what is bromide of silver used? In what forms does fluorine occur?

for those, yet the body obtained seems only to have been a mixture of chlorine and hydrofluoric acid.

It possesses an intense affinity for electro-positive bodies, and gives rise to a series of compounds resembling those of chlorine, iodine, and bromine. It does not unite with oxygen or carbon.

Hydrofluoric Acid, $HF=20$,

is made by decomposing fluoride of calcium by sulphuric acid in a vessel of platinum or lead, the vapors being conducted into a receiver kept at a low temperature. The action is

$$CaF+HO,\ SO_3=CaO,\ SO_3+HF.$$

It is a clear liquid, fuming in the air, boiling at 68°, and having a specific gravity of 1.06. Its attraction for water exceeds that of oil of vitriol, and it produces a malignant ulceration of the skin.

If a piece of glass be coated over with a thin film of beeswax, and letters or other marks made through the wax to the glass with a pointed tool, on setting it over a tin vessel in which, from a mixture of fluor spar and sulphuric acid, hydrofluoric acid vapor is escaping, the glass is corroded away, or etched in the uncovered parts. The liquid acid may also be employed, but the letters are then not so visible.

Carbon. $C=6$.

The carbon group of metalloids comprises three bodies—carbon, boron, and silicon. They are remarkable for being, in the crystalline state, very hard; in the amorphous state, insoluble and non-volatile. Carbon is the principal constituent of the organic, and silicon of the inorganic kingdom.

Carbon occurs under many different allotropic conditions. 1. Diamond, which crystallizes in octahedrons, is transparent, incombustible except in oxygen gas, and the hardest body known; hence its use in cutting glass. 2. Gas carbon, which, unlike diamond, is a good conductor of electricity, and is opaque. 3. The various

---

Are its properties known? How is hydrofluoric acid made? What remarkable quality does it possess? What is the carbon group, and what are its peculiarities? State the allotropic modifications of carbon.

forms of charcoal, anthracite, and coke. 4. Plumbago, which has a metallic lustre, is opaque, and so soft and unctuous that it is used to relieve the friction of machinery and for writing on paper. 5. Lampblack, a powerful absorbent of light and heat, and possessing such strong affinity for oxygen that it can take fire spontaneously in the air.

Other forms might be cited; these, however, are enough to establish the fact that this simple body furnishes varieties which differ more strikingly from each other than many different metallic bodies. It is no doubt owing to the many different states in which carbon exists that its compounds, though containing the same proportions of the same ingredients, yet vary so much in properties.

Charcoal is made by the ignition of wood in close vessels, the volatile materials being dissipated and the carbon left. The nature of the process may be illustrated by taking a slip of wood, *b*, *Fig.* 263, and placing the burning extremity in a test-tube, *a*. This retards the access of the surrounding air, and, as the combustion proceeds, a cylinder of charcoal is left.

Fig. 263.

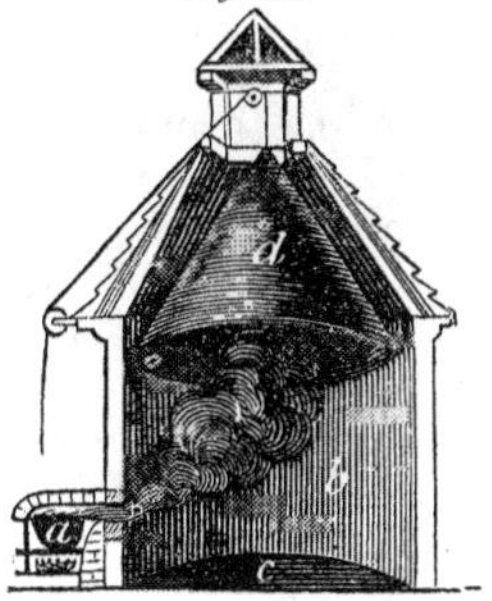

Lampblack is formed on a similar principle. In the iron pot *a*, *Fig.* 264, some pitch or tar is made to boil, a small quantity of air being admitted through apertures in the brick-work. Imperfect combustion takes place, the hydrogen alone burning, the carbon being carried as a dense cloud of smoke into the chamber *b c* by the draught. In this there is a hood or cone of coarse cloth, *d*, which may be raised or lowered by a pulley. The sides of the chamber are covered with leather, and on these the lampblack collects. One of its principal uses is in making printer's ink.

Fig. 264.

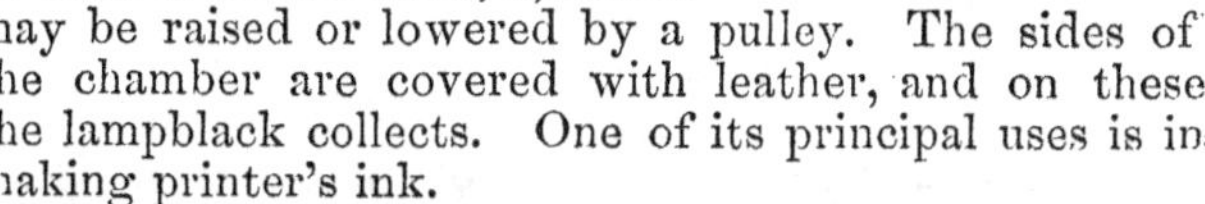

Diamond is the purest form of carbon. Its specific

What effect do these modifications have on its compounds? How is charcoal made? How is lampblack made?

gravity is 3.5. It exhibits a high refractive and dispersive action upon light. The largest diamond known was the Koh-i-Noor, which weighed 900 carats when found; in 1852 it was cut into a brilliant of 162½ carats. The Pitt diamond, one of the crown jewels of France, is probably the finest in the world. Its estimated value is $2,400,000; the Koh-i-Noor is valued at $3,000,000. Diamonds have been found in North Carolina.

Charcoal possesses, in consequence of its porous structure, the quality of absorbing many times its own volume of different gases. One cubic inch of newly-made charcoal will take up of

Ammonia	90	Olefiant Gas	35.
Hydrochloric Acid	85	Carbonic Oxide	9.42
Sulphurous Acid	65	Oxygen	9.25
Sulphureted Hydrogen	55	Nitrogen	7.50
Protoxide of Nitrogen	40	Light carbureted Hydrogen	5.
Carbonic Acid	35	Hydrogen	1.75

The temperature of the charcoal rises as the gas is condensed. Freshly-burned charcoal, put in a mixture of oxygen and sulphureted hydrogen, will cause a violent explosion. It also possesses the power of removing foul effluvia. Ivory black or animal charcoal, which is made by the ignition of bones in close vessels, has the valuable quality of removing organic coloring matters, as is shown by filtering a solution of indigo or brown sugar through it.

In all its forms carbon is infusible, but when burnt in air or oxygen they all give rise to carbonic acid. It combines with several of the metals to form carburets, those of iron being best known. With oxygen it gives several compounds:

CO,	Carbonic Oxide.
CO_2,	Carbonic Acid.
$C_2O_3, 3HO$,	Oxalic "
C_3O_4,	Mesoxalic "
$C_7O_7, 3HO$,	Rhodizonic "
C_5O_4, HO,	Croconic "
C_4O_3, HO,	Mellitic "
C_5O_3	Pyromellitic "

Of these, the first two are of most interest, and will alone be described here.

What are the properties of diamond? Give the absorbing power of charcoal for various gases. What quality has ivory black? What are the oxygen compounds of carbon?

CARBONIC OXIDE, $CO = 14$,

is produced when carbon is burned in a limited supply of oxygen, or when carbonic acid is passed over red-hot iron or over red-hot carbon. In these cases the actions are

$$CO_2 + C = 2(CO),$$
$$CO_2 + Fe = CO + FeO.$$

In the first, carbonic acid unites with one atom of carbon and yields two of carbonic oxide; in the second, it loses one atom of oxygen to the iron and yields one of carbonic oxide. It may also be prepared by heating oxalic acid with oil of vitriol in a flask, *a*, *Fig.* 265, the decomposition giving equal volumes of carbonic acid and carbonic oxide. The evolution is caused by the abstraction of water from the oxalic acid by the sulphuric acid.

Fig. 265.

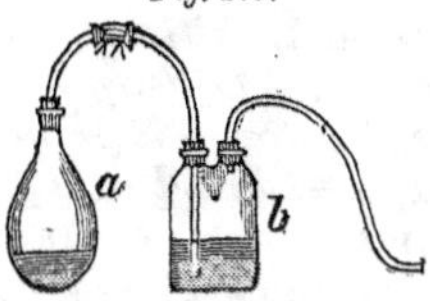

$$C_2O_3 + 3HO + 3(SO_3, HO) = 3(SO_3, 2HO) + CO_2 + CO.$$

The acid may be separated by passing the mixture through a bottle, *b*, containing potassa-water, and the oxide collected over water; but the best process is to heat one part of prussiate of potassa with ten of oil of vitriol in a retort: the carbonic oxide comes over in a state of purity.

Fig. 266.

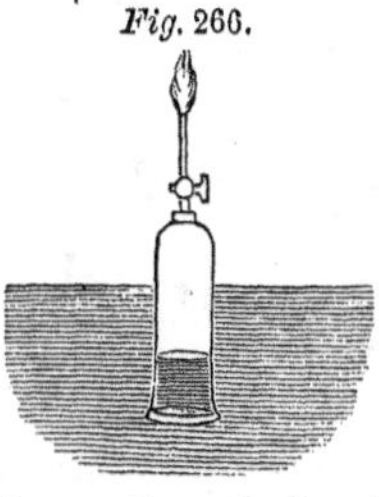

As obtained by any of these processes it is a colorless gas, which may be kept over water, in which it is only soluble to the extent of 6 per cent. It is without odor, and is a narcotic poison. A jet of it, *Fig.* 266, burns in the air with a blue flame, yielding carbonic acid. Its specific gravity is 9.674; 100 cubic inches weigh 29.96 grains. It has never been liquefied. The combustion of this gas produces the blue flame seen on a coal fire. It is evolved by vegetables, more particularly by aquatic plants. Carbonic oxide is a compound radical, giving origin to a series of bodies.

When is carbonic oxide produced? How may it be prepared from oxalic acid? From what other substance may it be made? What are its properties?

Chloro-carbonic acid or phosgene gas (CO, Cl) is formed, under the influence of light, from equal volumes of chlorine and carbonic oxide. It is an acid body of a pungent odor, and is decomposed by water. It is regarded as carbonic acid, in which one atom of oxygen is replaced by one of chlorine.

LECTURE LV.

CARBONIC ACID.—*Prepared by Decomposition.—Results from Combustion.—Properties.—Density.—Relations to Combustion and Respiration.—Solubility in Water.—Produced in Animals.—Liquid and Solid Carbonic Acid.—Light* CARBURETED HYDROGEN. *—Fire-damp.—Marsh Gas.—Artificial Production. —Coal Gas.*

OLEFIANT GAS.—*Preparation.—Properties.—Action with Chlorine.*

CARBONIC ACID. $CO_2=22$.

CARBONIC ACID is commonly prepared by the action of dilute hydrochloric acid on chalk, or any carbonate of lime, the action being

$$CaO,\ CO_2 + HCl = CaCl,\ HO + CO_2;$$

that is, one atom of carbonate of lime and one of hydrochloric acid yield one atom of chloride of calcium and one of water, and one atom of carbonic acid gas is set free. The process may be conducted in a flask, as in *Fig.* 267, the gas being evolved so rapidly that it may be collected over water, though that liquid absorbs its own volume at the ordinary pressure.

Fig. 267.

Carbonic acid is abundantly formed in many processes. It is the result of the complete combustion of carbonaceous bodies; is evolved during the respiration of animals and in alcoholic fermentation. It was called *fixed air* by the old chemists, because a constituent of limestone.

What are the properties of chloro-carbonic acid? How is carbonic acid made? From what natural processes does it arise? Why was it called fixed air?

It is a colorless and transparent gas at common temperatures, with a faint smell and slightly acid taste. It is irrespirable, and acts in a diluted state as a narcotic poison; even air, containing one-tenth of its volume, produces a marked effect; the atmosphere contains one part in 2000. Its specific gravity is 1.527; 100 cubic inches weigh 47.087 grains; it may therefore be collected by displacement. For the same reason, it collects in the bottom of wells and pits, and often suffocates workmen who descend into such places. It does not support combustion; a lighted taper lowered into a jar partly filled with it is at once extinguished. It may be poured from one vessel to another; and if a jar of it be poured on a candle, the light is at once put out. Its density and other qualities may be well illustrated when it is formed by the action of fuming nitric acid on carbonate of ammonia, a smoky cloud marking its position and movements. The Grotto del Cane owes its peculiarity of asphyxiating dogs to the accumulation of this gas in its basin-shaped floor.

Carbonic acid reddens litmus water, but the blue color is restored by boiling, the acid being driven off by the heat. It is soluble in water, which, under pressure, takes up five or six times its volume, constituting the soda-water of commerce. The effervescence of Champagne is due to its escape, and natural waters usually contain more or less of it. The solubility may be shown by agitating it with water in Hope's eudiometer, *Fig.* 268, or by passing it through Nooth's soda-water machine, *Fig.* 269.

Fig. 268.

Fig. 269.

A common test for the presence of carbonic acid in wells is to lower a lighted candle, and if its flame be extinguished it is inferred that the gas is present; but it does not follow that a man may safely descend into such places though a candle may continue to burn; the air may even then contain twenty per cent. of the gas.

What are the properties of carbonic acid? What are its relations to combustion? What results from its great specific gravity? What is soda-water? What is a test for this gas?

If through a tube the breath be made to pass into lime-water, a deposit of carbonate of lime renders the water milky; or if the breath be conducted through litmus water, the color changes to red; the air thus expired from the lungs contains three or four per cent. of carbonic acid. A man throws out about eight ounces of carbon as carbonic acid every day.

Under a pressure of thirty-six atmospheres, or by being cooled to $-106°$, carbonic acid condenses into a liquid four times more expansible by heat than atmospheric air. Thilorier's condensation apparatus is shown in *Fig.* 270. It consists of two iron cylinders—A, em-

Fig. 270.

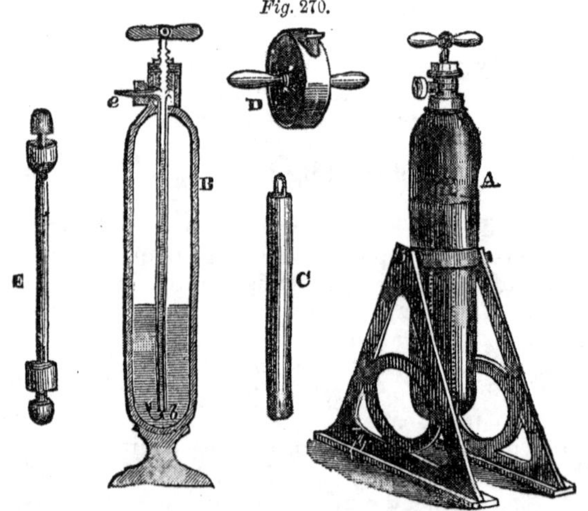

ployed as a retort to generate the gas; B, as a receiver. A is charged with a mixture of carbonate of soda and water; a brass tube, C, containing oil of vitriol, is introduced into it, and the head is screwed on. A is then inverted, and carbonic acid is generated under great pressure. A tube, E, is next made to connect A and B, the latter vessel being immersed in ice. The liquefied gas distills over into B. A tube, *b*, descends nearly to the bottom of B, and terminates above in a fine nozzle,

How can its existence in the breath be proved? How may it be liquefied? Describe the apparatus.

e. As soon as the stopcock at the top of B is opened the liquid carbonic acid is forced out at *e* by the pressure of its vapor, it evaporates rapidly, and, in so doing, produces great cold. A portion of the gas is solidified, and if the jet open into a box, D, the solidified acid collects as a flocculent, snowy-white deposit. Mixed with ether, it produces a temperature of —166°.

Carbonic acid has acid properties in but a feeble degree. It contains its own volume of oxygen, and hence its density, the weight of carbon added not altering the volume of the oxygen in which it is burned. The common test for its presence is lime-water, which deposits carbonate of lime, a white powder. It is entirely absorbed by potassa; and by taking advantage of this fact the most perfect vacuum may be formed.

Carbon and Hydrogen.

These substances unite, producing many compounds, some of which are solid, some liquid, and others gaseous. They are of course all combustible bodies, but the description of most of them belongs to organic chemistry.

Light Carbureted Hydrogen, $CH_2=8$,

occurs abundantly in coal mines, and forms with air explosive mixtures; it is also formed during the putrefaction of vegetable matter under water; on stirring the mud of ponds bubbles of this gas escape—hence the name marsh gas. It is also ejected from petroleum borings, and serves to heat the engine-boilers. The village of Fredonia has been lighted for many years by a well discharging carbureted hydrogen. It may be obtained artificially by heating acetate of potassa with hydrate of baryta. It is a colorless gas, burns with a yellow flame, producing carbonic acid and water, and is, when pure, fatal if respired. Its specific gravity is .5528; 100 cubic inches weigh 17.12 grains. It is the fire-damp of coal mines; choke-damp, which comes after its explosion, being carbonic acid. The gas is decom-

How is solid carbonic acid formed? What is the test for carbonic acid? What is the common property of the compounds of carbon and hydrogen? Where is carbureted hydrogen naturally formed? What are its properties? What are fire-damp and choke-damp?

posed explosively by chlorine in the light, but is not acted on in the dark.

Coal gas contains this gas principally, associated with coke, tar, water, carbonic oxide, carbonic acid, olefiant gas, sulphureted hydrogen, hydrogen, ammonia, and nitrogen. Gas is purified by washing with water, condensation by cold water, and by hydrate of lime. Its specific gravity is about .65, though, if the coal be distilled at a higher temperature, it may only be .345, owing to the decomposition of olefiant gas. The same volume has then much less illuminating power. In London 5,000,000,000 cubic feet are used in a year, producing as much light as 10,000,000,000 of tallow candles.

Olefiant Gas, $C_2H_2 = 14$,

may be made by heating one part of alcohol with four of sulphuric acid in a flask, *a*, *Fig.* 271. The vapor of ether which comes over with it may be removed by causing the gas to pass through a small bottle, *b*, containing sulphuric acid, before being collected at the trough.

Fig. 271.

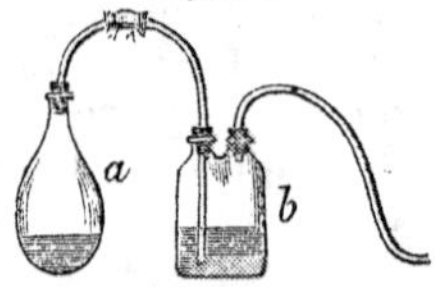

Olefiant gas is transparent and colorless, burns with a beautiful flame, forms an explosive mixture with oxygen, giving rise by its combustion to carbonic acid and water. Its specific gravity is .9674; 100 cubic inches weigh 29.96 grains. If mixed with an equal volume of chlorine an oily liquid condenses, from which olefiant gas receives its name. This is the chloride of olefiant gas, or Dutch liquid, which will be described under Organic Chemistry. With twice its volume of chlorine, if it be set on fire, hydrochloric acid is formed, and carbon deposited as a dense black smoke.

What is the composition of coal gas? How is olefiant gas prepared? What are its properties? From what does it derive its name?

LECTURE LVI.

CYANOGEN.—*Preparation.—Liquefaction.—An Electro-negative Compound Radical.—Bisulphide of Carbon.—Refractive and Dispersive Powers.*
BORON.—*Preparation.—Boron Diamonds.—Boracic Acid.—Nitride of.—Fluoride of.*
SILICON.—*Three Allotropic States.—Silicic Acid.—Fluosilicic Acid.—Compounds of Nitrogen and Hydrogen.—Amidogen.—Ammonia.—Ammonium.—Theory of Berzelius.*

CYANOGEN, *Cy*, OR BICARBURET OF NITROGEN. $C_2N=26$.

Carbon unites with nitrogen, forming a bicarburet when these substances are in the nascent state, and in presence of a base. It may be obtained by exposing the cyanide of mercury to heat, or by heating a mixture of six parts of ferrocyanide of potassium and nine of corrosive sublimate.

It is a colorless gas, having a peculiar odor. It burns with a beautiful purple flame, dissolves to the extent of 4½ volumes in water and 23 volumes in alcohol, condenses into a liquid by a pressure of 3.6 atmospheres at 45°, as may be shown by heating with a lamp cyanide of mercury in a bent tube, *Fig.* 272; the tube being closed at both ends, cyanogen accumulates in the cool extremity. Though a compound body, it has all the properties and characters of a powerful electro-negative element. Its specific gravity is 1.796; 100 cubic inches weigh 55.64 grains. Below −30° it is a transparent solid. A farther description of it and its compounds will be given under Organic Chemistry.

Fig. 272.

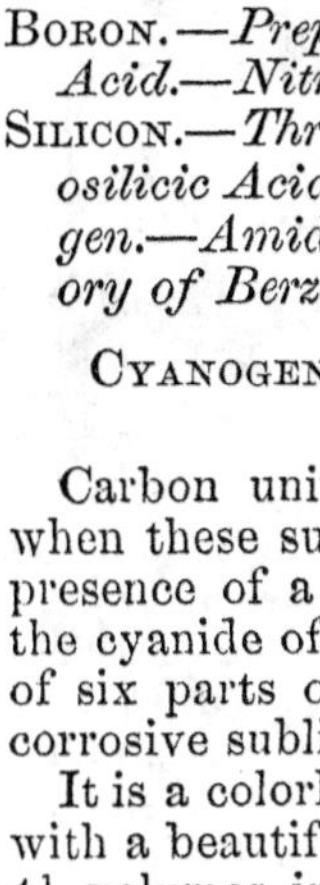

BISULPHIDE OF CARBON, $CS_2=38$,

may be made by passing the vapor of sulphur over charcoal ignited in a tube, and receiving the product in a

How is cyanogen made? What are its properties? How may it be condensed and solidified? How is bisulphide of carbon made?

cold bottle. The apparatus is seen in *Fig.* 273. Into

Fig. 273.

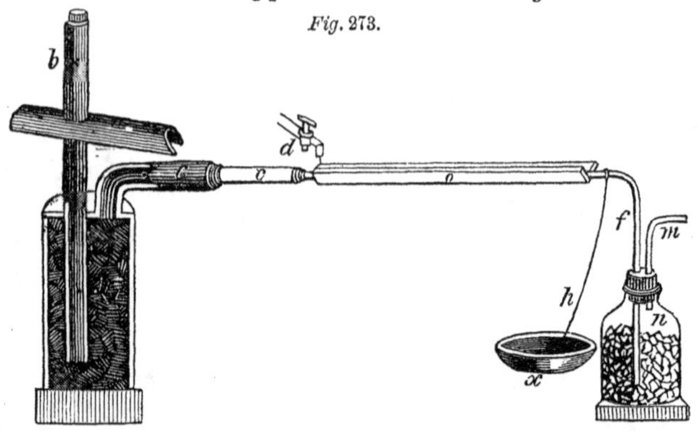

the top of a large bottle two tubes, *b c*, one straight and the other bent, are inserted. The bottle having been filled with charcoal, pieces of sulphur are dropped in through the tube *b* as soon as the bottle is red-hot. The sulphur and carbon unite. The product passes along the tubes *c f*, cooled by a stream of water from the cock *d*, the water being conducted by a string, *h*, into a basin, *x*. The vapor passes into the bottle *n*, which is partly filled with ice, and the incondensible gases pass out through *m*. It is a transparent liquid, specific gravity 1.272, of a very disagreeable odor; has the quality of dissolving sulphur and phosphorus, boils at 100°, and does not freeze at –60°. It is very volatile, and produces an intense cold during evaporation in vacuo, –80° being reached. Its principal point of interest is the powerful refractive and dispersive power it exerts upon light, a property which renders it especially suitable for making prisms to be used in spectroscopes. It is also used as a solvent of sulphur and caoutchouc, in the making of vulcanized India-rubber.

Boron, $B=11$,

is found only in combination with oxygen as boracic acid, from which it may be set free by heating to 300° with twice its weight of potassium or sodium. It may be obtained in the crystalline state by heating to a high

What is its most valuable property? From what is boron derived?

degree 8 parts of aluminum with 10 parts of anhydrous boracic acid, and treating the product with caustic soda, hydrochloric acid, and nitro-hydrofluoric acid. These crystals may present the brilliancy of the diamond, and are so hard that they will scratch its surface. The specific gravity is 2.68. They only take fire under the same circumstances as diamond. The amorphous variety is olive-colored, and burns at 600° in the air into boracic acid. It combines with nitrogen, chlorine, bromine, and fluorine.

Boracic Acid, $BO_3=35$,

exists in the volcanic springs of Tuscany, and, combined with soda, lime, or magnesia, is brought from India, Norway, Sweden, and South America. It may be artificially prepared by dissolving one part of borax in four of hot water, and adding half a part of sulphuric acid. On cooling, the boracic acid is deposited in small crystalline scales, which may be purified by recrystallization. They have a soapy feel, and are soluble in 12 parts of boiling or 50 parts of cold water.

Boracic acid melts at a red heat into a transparent glass, which, combined with oxide of lead or bismuth, has high refractive powers. Its crystals, raised to 212°, lose half of their water. It volatilizes readily when boiled in water, and is soluble in alcohol, the solution burning with a green flame. The experiment may be made in a glass instrument, *a b*, *Fig.* 274, heated by a spirit-lamp, *c*. It is a very feeble acid, and even turns yellow turmeric brown like an alkali. At a red heat it decomposes the sulphates and phosphates. It is employed as a flux for porcelain colors.

Fig. 274.

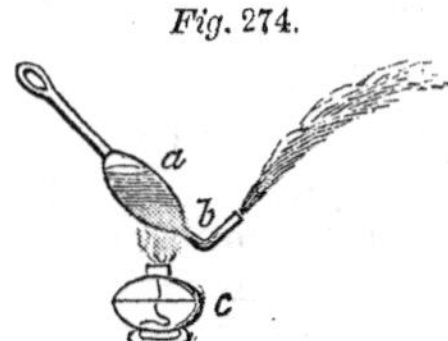

Nitride of Boron, $BN=25$,

is obtained by heating to redness two parts of sal ammoniac and one of borax. It is an amorphous powder, without taste or smell. When calcined with carbonate

What are the peculiarities of its crystals? How is boracic acid prepared? What are its properties? What color does it give to flame? How is nitride of boron made?

of potassa, it decomposes the carbonic acid and forms cyanogen. Heated with steam, it produces boracic acid and ammonia.

TERFLUORIDE OF BORON. FLUOBORIC ACID, $BF_3=68$, is formed when a mixture of fluor spar, boracic acid, and oil of vitriol is heated in a flask. The gas is of a suffocating odor, reddens litmus, extinguishes flame, is dissolved by water to the extent of 700 volumes, and forms dense white fumes with aqueous vapor. It takes water from organic substances, charring them, and can decompose water.

SILICON, $Si=22$,

may be prepared by igniting the silico-fluoride of potassium, *Fig.* 275, with potassium, acting upon the resulting substance with water, which removes the fluoride of potassium, and leaves the silicon as a nut-brown powder.

Fig. 275.

It exhibits three allotropic states, the amorphous, graphitic, and the regular crystalline forms. Aluminum appears to determine the crystalline form of silicon and boron. In decomposing the vapor of chloride of silicon by aluminum in a vessel of hydrogen, silicon may be obtained in hexahedral prisms that can even cut glass.

The amorphous form burns when heated in air, but if previously ignited in close vessels it shrinks in volume, increases in density so as to sink in sulphuric acid, and, passing into another allotropic state, becomes incombustible in oxygen. Silicon can deoxidize carbonic acid, and unites with chlorine, bromine, and fluorine.

SILICIC ACID, $Si\,O_3=46$,

is one of the most abundant bodies in nature, existing under the innumerable forms of the quartz minerals, sands, and sandstones. It is found in every soil, in all waters, is a constituent of many plants, and forms the

What are the properties of fluoboric acid? How may silicon be prepared? What allotropic states does it present? What changes occur in it by heating? What is the constitution of silicic acid?

skeleton of tribes of the lower animals. Rock crystal and flint are pure silicic acid.

It may be obtained for chemical uses as follows: Heat rock crystal to redness and quench it in water. Fuse one part of this with three parts of a mixture of carbonate of soda and carbonate of potassa, dissolve the resulting silicate in water, and decompose with hydrochloric acid. The silicic acid separates as a gelatinous hydrate, slightly soluble in water, which, when washed and dried, yields a white insoluble powder.

Silica is a gritty substance, sufficiently hard to scratch glass. Its specific gravity is 2.66. It combines with the alkalies in excess to form glass. It requires a high temperature for fusion, and is insoluble in any acid except hydrofluoric. At ordinary temperatures the silicates are decomposed by carbonic acid, and it is to this agency that the disintegration of many rocks is due. Soluble glass is made by fusing carbonate of soda with sand and charcoal. The soluble mass is sometimes used as a paint.

FLUORIDE OF SILICON. FLUOSILICIC ACID, $SiF_3=79$, results when silica is dissolved in hydrofluoric acid, or when fluor spar and sand are heated with sulphuric acid. It is a colorless acrid gas, fumes in the air, extinguishes flame, and may be liquefied and solidified. The specific gravity is 3.6; 100 cubic inches weigh 112 grs. Transmitted from the flask which generates it, *a*, *Fig.* 276, through water, it is decomposed, hydrated silica being deposited. To prevent the tube which delivers the gas being stopped up by the liberated silica, some quicksilver, *e*, may be put in the vessel *d*, and the tube dipped into it, so that the bubbles of gas may not come into contact with the wa-

Fig. 276.

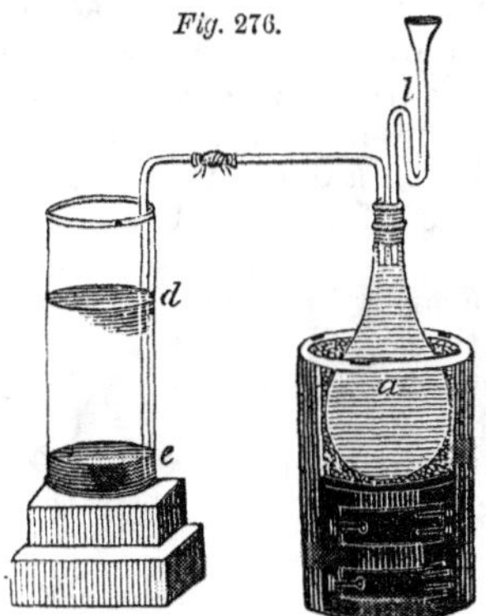

How is it prepared? What are its properties? What is soluble glass? How is fluosilicic acid made?

ter until they have reached the surface of the metal. Sulphuric acid may be introduced through *l*. In the water hydrofluosilicic acid forms. It may be used as a test for potassa, soda, and baryta.

NITROGEN and HYDROGEN yield three compounds:

NH_2, Amidogen.
NH_3, Ammonia.
NH_4, Ammonium.

AMIDOGEN, $NH_2=16$,

is a hypothetical compound radical, the existence of which in several compounds is inferred. On heating potassium or sodium in dry ammonia, one equivalent of hydrogen is set free, and a solid substance remains—the amide of potassium. This, in contact with water, yields potassa and ammonia,

$$K, NH_2+HO=KO+NH_3.$$

Amidogen is an electro-negative compound radical like cyanogen.

AMMONIA. $NH_3=17$.

This substance, called also *volatile alkali*, is an abundant product of the putrefaction of animal matters, and may be obtained from the destructive distillation of horn; hence the name, spirit of hartshorn. It also exists in the air, and is a common product of many chemical reactions. Iron rusting in a damp atmosphere contains a trace of it mixed with the oxide.

Fig. 277.

It may be obtained by heating in a flask, *a*, *Fig.* 277, equal quantities of slacked lime and chloride of ammonium; and as its specific gravity is only .587, it may be collected in a flask or jar, *b*, with the mouth downward, by displacing the heavier air. The action is

$$NH_4Cl+CaO, HO=CaCl+2HO+NH_3.$$

It may also be collected at the mercurial trough, as in *Fig.* 278.

It is a transparent and colorless gas, of excessive pungency, and a strong alkali. It may be liquefied at $-40°$,

What effect has water on it? How many compounds of nitrogen and hydrogen are there? What is amidogen? How does ammonia arise? How may it be procured?

Fig. 278.

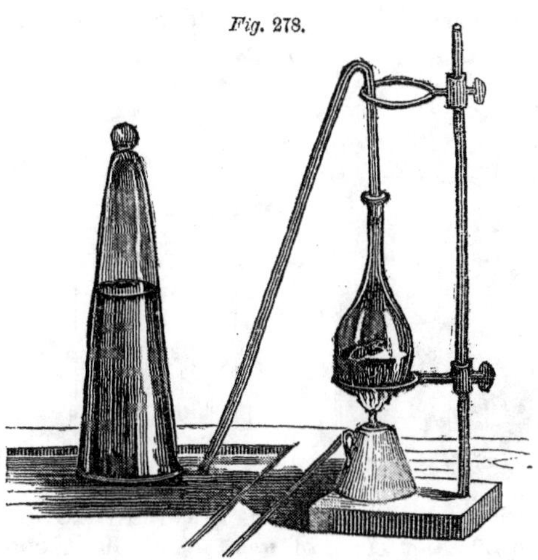

and solidified at −103°, the liquid being most easily obtained by heating in a sealed tube chloride of silver saturated with the gas. It turns turmeric paper brown; is absorbed with great rapidity by water, which at 50° takes up 670 times its volume, a result which may be illustrated by inserting a flask full of it in some cold water, when the water rushes up with sufficient violence to destroy the flask frequently. Ammonia neutralizes the strongest acids, as may be shown by dropping it into litmus water which has been reddened by sulphuric or nitric acid.

It is composed of three volumes of hydrogen with one of nitrogen, condensed into two volumes. It may be recognized by its remarkable odor, and by the formation of white clouds when a rod, *a*, *Fig.* 279, dipped in muriatic acid, is approached to it. One hundred cubic inches weigh 18.19 grains.

Fig. 279.

Its solution in water, aqua ammonia, is prepared by passing the gas evolved from slacked lime and sal ammoniac through Wolfe's bottles,

---

What are its properties? How many volumes of it does water absorb? What is its effect on reddened litmus? What is its composition? How is it detected? How is aqua ammonia made?

as is represeentd in *Fig.* 280. The water will take it

Fig. 280.

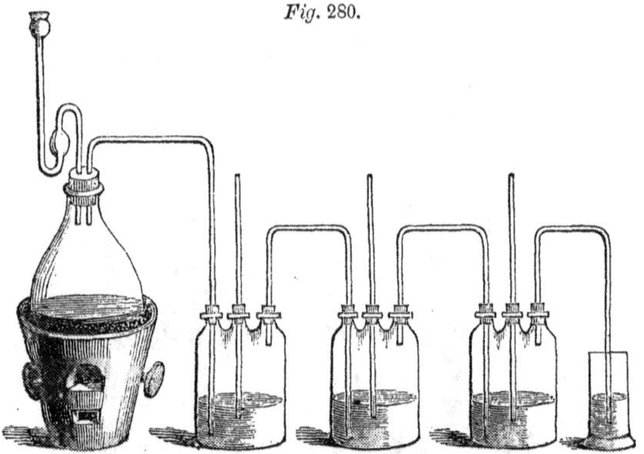

up until its specific gravity is lowered to .875. It then contains 32½ per cent. of gas. At .900 it contains 26 per cent.; at .951, 12.4 per cent.; at .969, 9.5 per cent. This solution is much used for precipitating and neutralizing. It affords the best means of obtaining ammonia, merely requiring to be warmed in a flask, when the gas readily comes off.

AMMONIUM, $Am = NH_4 = 18$,

is a hypothetical body, believed to be of a metallic nature; its symbol is *Am*. It may be combined with mercury by decomposing a solution of an ammoniacal salt by a Voltaic current, the negative pole being in contact with a globule of that metal, or by putting an amalgam of potassium and mercury in a solution of sal ammoniac. Under those circumstances the mercury swells, and eventually becomes of a soft consistency, like butter, preserving its metallic aspect completely. While the mercury increases to thirty times its volume at 100°, it increases in weight only $\frac{1}{2000}$th part. All attempts to separate the ammonium from this amalgam, which crystallizes in cubes at zero, have failed. It decomposes into NH_3 and H.

It is now generally agreed by chemists that ammo-

What is the nature of ammonium? In what state may it be obtained?

nium is the basis of the salts of ammonia, as originally stated by Berzelius. Thus sal ammoniac, called also the muriate of ammonia, is NH_3HCl; this, however, is the same as NH_4+Cl; that is, the chloride of ammonium. But it must be remembered that when dry ammonia and hydrochloric acid gas are brought in contact a white solid is formed. If this is called chloride of ammonium, it would be necessary to suppose that the ammonia has taken hydrogen away from chlorine, whereas ammonia is itself decomposed by chlorine.

In all cases where ammonia forms salts with the so-called oxygen acids, it requires an atom of water, but this water evidently gives it the constitution, not of NH_3+HO, but NH_4+O; the water therefore makes it oxide of ammonium, which will unite with sulphuric, or nitric, or any other acid precisely after the manner of any other metallic oxide. Moreover, the compounds of ammonia with this atom of water are isomorphous with the compounds of potassium.

	Potassium series.	Ammonium series.
Metal..................	K	Am
Oxide..................	KO	AmO
Chloride..............	KCl	$AmCl$
Sulphide..............	KS	AmS
Nitrate................	KO, NO_5	AmO, NO_5
Sulphate..............	KO, SO_3	AmO, SO_3

Of the compounds of ammonium with other bodies, the protosulphide, NH_4S, may be mentioned under the name of hydrosulphate of ammonia. It is used as a test for metals in alkaline solutions. There is also a bisulphide.

What is its relation to the ammonia salts? What is the constitution of the salts of ammonia with oxygen acids? Like what class of bodies does ammonium act? What is hydrosulphate of ammonia?

THE METALS.

LECTURE LVII.

GENERAL PROPERTIES OF THE METALS.—*Definition of a Metal.* — *Color, Specific Gravity, Hardness, Tenacity, and other Properties.* — *Relations to Heat.* — *Compounds with other Bodies and one another.*— *Division into Groups.* — *The Oxides and their Reduction.*— *The Sulphides and their Reduction.*

OF the elementary bodies, by far the larger portion are metallic. By a metal we mean a body which possesses that peculiar manner of reflecting light which is known as metallic lustre; it is also a good conductor of heat and electricity. Of these there are about 55; the number is being continually increased.

Most of the metals are of a white color, but they differ from each other by slight shades, some having a faint blue and some a pinkish shade. There are two that are strikingly colored—gold, which is yellow; and copper, which is red. In specific gravity they differ exceedingly; potassium, sodium, and lithium float in water, and iridium is 21 times as heavy as that liquid.

Many of the metals are malleable—that is, can be extended into thin sheets under the blows of a hammer; others are so brittle that they may be reduced to powder in a mortar; some of them are ductile, and may be drawn into fine wires, the order for ductility not being the same as that for malleability. Thus iron may be drawn into fine wire, but can not be beaten out into such thin sheets as many other metals. Of all metals, gold is the most malleable, and platinum has been drawn into the finest wires, $\frac{1}{30,000}$th of an inch in diameter.

What is the definition of a metal? How many metals are there? What is their color commonly? Which are the colored metals? Which is the lightest, and which the heaviest metal? Of the metals, which is the most ductile, the most malleable, the softest, the hardest. the most fusible?

In hardness the metals differ much—potassium is so soft that it may be moulded by the fingers, but iridium is among the hardest bodies known. In tenacity or strength the same differences are seen. Of all metals, iron is the most tenacious; a wire one tenth of an inch in diameter will sustain a quarter of a ton. The tenacity is, as a rule, decreased as the temperature rises.

In their relations to heat, well-marked distinctions may be traced. Mercury, at ordinary temperatures, is in a melted condition, but platinum can only be fused by the oxyhydrogen blow-pipe or an electrical current. As respects volatility, mercury, cadmium, potassium, sodium, zinc, arsenic, and tellurium may be distilled or sublimed at a red heat. Gold and silver may be vaporized at a very high heat.

The metals unite with electro-negative bodies and with each other. In decomposition by the Voltaic battery they pass to the negative pole, and are therefore called electro-positive. Their compounds with oxygen, chlorine, etc., pass under the name of oxides, chlorides, etc.; their compounds with each other under the name of alloys; or, if mercury be present, amalgams. They also unite with sulphur, phosphorus, and carbon.

Chemical writers usually divide the metals into groups founded upon their relations to oxygen gas. The following simple division will be adopted in this book: 1st. Metals which decompose water at common temperatures. 2d. Metals which only decompose water at a red heat. 3d. Metals which can not decompose water at all.

1st Group.	Indium.	Zinc.
Potassium.	Glucinum.	Cadmium.
Sodium.	Zirconium.	Tin.
Lithium.	Thorium.	
Cæsium.	Yttrium.	3d Group.
Rubidium.	Erbium.	Chromium.
Barium.	Terbium.	Vanadium.
Strontium.	Cerium.	Tungsten.
Calcium.	Lanthanum.	Molybdenum.
	Didymium.	Osmium.
2d Group.	Manganese.	Columbium.
Magnesium.	Iron.	Niobium.
Aluminum.	Nickel.	Ilmenium.
Thallium.	Cobalt.	Titanium.

With what other substances do they unite? Into what groups may they be divided?

Arsenic.	Lead.	Palladium.
Antimony.	Bismuth.	Platinum.
Tellurium.	Silver.	Rhodium.
Uranium.	Mercury.	Iridium.
Copper.	Gold.	Ruthenium.

In addition, there are Dianium, Norium, and Pelopium, the existence of which is uncertain.

The older chemists divided the metals into four classes: 1st. Alkaline, as potassium; 2d. Earthy, as magnesium; 3d. Imperfect, as zinc; 4th. Noble, as gold.

The Metallic Oxides.

Metallic substances unite with oxygen with different degrees of intensity and in very different proportions, many of them giving rise to a complete set of oxides, and producing, 1st. Basic Oxides; 2d. Neutral or Indifferent Oxides; 3d. Metallic Acids.

1st. The basic oxides are commonly protoxides or sesquioxides, which form neutral salts with hydrogen acids with the production of water. To form such salts, for every atom of oxygen in the base there is required one atom of acid. A basic protoxide therefore requires one atom of acid, a sesquioxide three, and a deutoxide two, to form a neutral salt.

2d. The neutral or indifferent oxides contain more oxygen than the basic, and when heated with acids give off the oxygen, a basic oxide resulting.

3d. The metallic acids always contain most oxygen. They may be sesquioxides, deutoxides, teroxides, or quadroxides, and are commonly formed by deflagrating the metal with nitrate of potassa.

Reduction of the Metallic Oxides.

Some of the oxides, as those of mercury, silver, and gold, may be reduced by heat alone, but the greater number require the conjoint action of carbon, which at a high temperature decomposes them, with evolution of carbonic oxide. Among powerful reducing agents may be mentioned the formiates and the cyanide of potas-

What was the old division? What substances do metals yield with oxygen? What is the peculiarity of basic oxides; of neutral oxides; of metallic acids? By what processes may metallic oxides be reduced?

sium, the former acting through the affinity of carbonic oxide for oxygen, and the latter through the affinity of carbon and potassium conjointly. The deoxidation of metals may also be accomplished by reducing agents, such as phosphorous and sulphurous acids, or by the action of other metals; iron, for instance, will precipitate metallic copper from its solutions.

The Voltaic current affords a powerful means of effecting the reduction of metals. By its aid the alkaline metals were discovered. The electrotype, already described, is an example of its action; solutions of metallic salts are readily decomposed by it. Thus, if a glass jar, T, *Fig.* 281, be divided into halves, and a paper diaphragm, D, be introduced between them, the halves being tightly pressed together by the ring B B, if the jar be filled with any metallic solution, such as the sulphate of soda, and the positive and negative wires of the battery dipped in the opposite compartments, after a time the metallic oxide will be found in one of them and the acid in the other, a total decomposition having taken place.

Fig. 281.

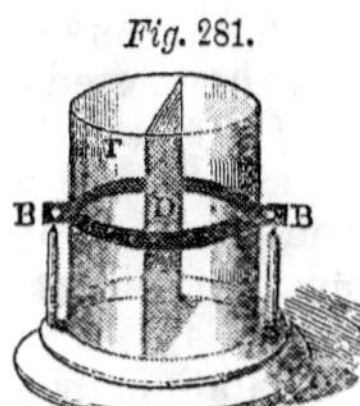

The Metallic Sulphides.

Many of these, as the sulphides of iron, lead, and copper, are found abundantly in nature, or they may be made artificially by heating the metal with sulphur, or by deoxidizing metallic sulphates by charcoal or hydrogen gas, which converts them into sulphides, or by the action of sulphureted hydrogen on their oxides, a metallic sulphide and water being produced. Iron, manganese, zinc, cobalt, and nickel may be precipitated by hydrosulphate of ammonia from alkaline solutions.

The sulphides of a metal are usually equal in number and similar in constitution to its oxides; and as oxygen compounds unite with each other to produce oxygen salts, the sulphides in like manner also unite with each other to produce sulphur salts.

Describe the experiment *Fig.* 281. How may metallic sulphides be made? What relation exists between sulphides and oxides?

Reduction of the Sulphides.

The metallic sulphides may often be reduced by melting them with another metal having a more powerful affinity for sulphur; thus iron filings will decompose sulphide of antimony, sulphide of iron forming and antimony being set free. On the large scale, however, a different process is resorted to; the sulphide, by roasting, is converted into a sulphate, much of the sulphur being expelled during the process as sulphurous or sulphuric acid. The resulting sulphate is then acted on by lime and carbon at a high temperature; the lime decomposes the sulphate, setting free the metallic oxide, which is at once reduced by the carbon, the sulphate of lime turning simultaneously into the sulphide of calcium, and floating on the surface of the metal as a slag.

The metals also unite with chlorine, iodine, bromine, carbon, phosphorus, etc., and some with hydrogen and nitrogen. These compounds will be described in their proper places.

LECTURE LVIII.

Potassium.—*Discovery and Properties.—Preparation. — Relation to Oxygen and Water. — Its Oxides. — Caustic Potassa. — Tests for Potassa. —Its Existence in the Soil and Plants.—Haloid Compounds of Potassium. — Salts of the Protoxide, the Carbonate, Nitrate, Chlorate, etc.*

Potassium. $K=39$.

Potassium (Kalium) was first obtained by Sir Humphrey Davy in 1807. He decomposed its hydrated oxide, potassa, by a Voltaic current. From the positive pole oxygen gas escaped in bubbles, and metallic potassium in globules, together with hydrogen, appeared at the negative.

It was subsequently discovered that the same substance could be decomposed by iron, and also by carbon at a high temperature; and the latter of these sub-

How may the sulphides be reduced? What is the process on a large scale? How was potassium first obtained? What process is now in use?

stances is now exclusively resorted to for the preparation of potassium. The carbonate of potassa is ignited with charcoal in an iron bottle, and the potassium received into a vessel containing naphtha. The productiveness of the operation is greatly interfered with by the circumstance that the carbonic oxide which is evolved, as it cools below a red heat unites with much of the potassium, producing a gray substance, which chokes the tubes and diminishes the yield of the metal. Not more than one fourth of the potassium contained in the carbonate is obtained.

Potassium is a bluish-white metal, which at 32° is brittle, melts at 150°, and boils at a red heat, yielding a green vapor. Its specific gravity is .865; it is therefore much lighter than water, on the surface of which it floats. At 70° it may be moulded with the fingers, being soft and pasty.

It possesses an intense affinity for oxygen, and hence requires to be kept under naphtha, a liquid containing no oxygen. A piece of it thrown upon water, *Fig.* 282, takes fire, and burns with a beautiful pink flame. In the air it speedily tarnishes, as is seen on cutting a mass with a penknife; and even in contact with ice there is decomposition with flame. In these cases the combustion arises from the hydrogen uniting with the oxygen of the air and reproducing water, the potassium simultaneously burning into potassa. Potassium is used for obtaining other metals, as aluminum, magnesium, etc., from their oxides.

Fig. 282.

Potassium and Oxygen.

There are three oxides of potassium—a suboxide, a protoxide, and a peroxide; K_2O, KO, and KO_3.

Suboxide of Potassium, $K_2O=86$,

is formed by heating potassium in a limited amount of air. It takes fire when heated, and is converted by water into potassa, hydrogen being evolved.

What interferes with the productiveness of the operation? What are the properties of potassium? How does it act on the surface of water? Of what use is potassium? How many oxides does it form? How is the suboxide formed?

Protoxide of Potassium, $KO=47$,

is made by heating one atom of potassium with one of hydrate of potassa, $K+KO, HO=2KO+H$.

Hydrated Oxide of Potassium, $KO, HO=56$,

is best procured by boiling two parts of pure carbonate of potassa with twenty of water, and, having previously slacked one part of quick-lime with hot water, the cream which it forms is to be added by degrees, and the whole boiled. The process should be conducted in an iron vessel, to which a lid can be adapted, so as to exclude the air during cooling. The resulting carbonate of lime settles, and the hydrate may be obtained by evaporating the solution rapidly in a silver vessel, pouring out the melted residue on a silver plate, or casting it in the form of small cylinders.

The decomposition which takes place is very simple:

$$KO, CO_2+CaO, HO=CaO, CO_2+KO, HO;$$

that is, the lime takes carbonic acid from the carbonate of potassa, and the oxide of potassium unites with water. The solution may be known to be free from carbonic acid by not effervescing when mixed with the stronger acids.

The hydrate of potassa, *caustic potash*, is a white solid, having a powerful affinity for water, and abstracting it rapidly from the air. Taken between the fingers, it communicates to them a soft feeling, and, if a concentrated solution be used, soon effects a disorganization; it is hence employed by surgeons as an escharotic. It possesses pre-eminently the alkaline qualities, and, indeed, may be taken as the type of that class of bodies; neutralizes the most powerful acids perfectly, and communicates to turmeric paper or solution a brown tint. It turns the reddened infusion of litmus blue, dissolves flint glass, and, possessing an intense affinity for carbonic acid, is used in organic analysis to absorb that gas. Heated in the blow-pipe flame, it gives a charac-

---

How is the protoxide formed? How is the hydrated oxide obtained? Describe the decomposition. What are the properties and uses of caustic potassa?

teristic violet tinge, and in the spectroscope, easily recognized lines.

Potassa in combination occurs in fertile soils, and is essential to the growth of land plants, from the ashes of which its carbonate is abundantly procured. This may be shown by filtering water through the ashes of wood, when the clear liquid will be found to answer all the tests indicating the presence of potassa. It also occurs abundantly in feldspar, and hence is found in clays. The want of fertility in soils is often due to the absence or exhaustion of this body.

The bichloride of platinum gives, with a solution of potassa, a yellow precipitate of the chloride of potassium and platinum. When the amount of potassa is small, it is well to add alcohol at first, in which the double chloride is almost insoluble. Ammonia yields a similar precipitate, but this may be avoided by exposing the substance to a red heat before testing. Perchloric acid with alcohol yields a white precipitate. Tartaric acid, if added in excess, and the mixture stirred with a glass rod, bearing gently on the sides of the vessel, gives white streaks of the bitartrate of potassa wherever the rod has passed over the glass.

Of other compounds of potassium the following may be mentioned:

Peroxide of Potassium, KO_2
Chloride of Potassium, KCl
Iodide of Potassium, KI
Cyanide of Potassium, KCy
Bromide of Potassium, KBr
Protosulphide of Potassium, KS
Tersulphide of Potassium, KS_3
Ferrocyanide of Potassium, K_2, Fe, Cy_3

It also combines with hydrogen in two proportions, producing a solid and a gas; the latter takes fire spontaneously in the air. With ammonia it produces potassiamide, the composition of which is KNH_2.

Of these compounds the most important are, 1. The peroxide of potassium, which is formed by burning potassium in oxygen. It can support combustion, and is

How may the existence of potassa in plants be shown? What are the tests for potassa? Name some of the other compounds. What is potassiamide? What are the properties of the peroxide of potassium?

decomposed by water, evolving oxygen and leaving potassa in solution. 2. The chloride of potassium is analogous to common salt, and may be formed by burning potassium in chlorine. 3. The iodide, much of which is consumed in medicine and photography under the name of hydriodate of potassa. It is prepared by dissolving iodine in a solution of potassa till the liquid begins to appear brown, then evaporating to dryness, and igniting the residue; oxygen is evolved, and iodide of potassium remains; it may then be crystallized from the solution in water, the form assumed being that of cubes. It is very soluble in water and hot alcohol, and will dissolve large quantities of iodine. 4. Cyanide of potassium, which is of great use in photography for dissolving iodide and bromide of silver, will be described in organic chemistry. 5. Bromide of potassium may be made like the iodide, and is used as an ingredient of collodion in photography on glass.

Salts of the Protoxide of Potassium.

Carbonate of Potassa is obtained by lixiviating the ashes of plants. In an impure state it forms the potashes and pearlashes of commerce. It may be obtained pure by igniting the bitartrate with half its weight of the nitrate of potassa. It has an alkaline taste, its solution feels greasy to the fingers, it is very soluble in water, which takes up nearly its own weight, and it is deliquescent.

Bicarbonate of Potassa is formed by transmitting a stream of carbonic acid through a solution of the former salt. It crystallizes in eight-sided prisms with dihedral summits, and has most of the properties of the carbonate.

Sulphate of Potassa is formed by neutralizing the following salt: crystallizes in anhydrous, short, six-sided prisms, terminated by six-sided pyramids, soluble in twelve times their weight of water, and insoluble in alcohol.

Bisulphate of Potassa is a residue of the production of nitric acid. It is soluble in water, but is decomposed

---

What are the properties of the chloride, iodide, cyanide, bromide of potassium? From what is the carbonate obtained? How are the bicarbonate, sulphate, and bisulphate formed?

by an excess of water into the neutral sulphate and acid. It crystallizes in rhombohedrons.

Nitrate of Potassa (Saltpetre) is extracted on the large scale from certain soils in which organic matter is decaying in contact with potassa. In the Mammoth Cave nitrate of lime exists in the soil, and is used to produce nitre by decomposition with wood ashes. It crystallizes in six-sided prisms with dihedral summits; fuses at a heat below redness, with evolution of oxygen gas. Its solubility varies greatly with the temperature, 100 parts of water at 77° dissolving 38 parts, and at 212° 246 parts. This salt enters as an essential ingredient into gunpowder, which is composed of about one atom of nitrate of potassa, one of sulphur, and three of carbon. The sulphur of this mixture accelerates the combustion, while the oxygen of the nitric acid forms carbonic acid with the charcoal. The products, therefore, of the perfect combustion of gunpowder are carbonic acid, nitrogen, and sulphide of potassium. It commonly happens, however, that sulphate of potassa is formed. The proportions of the ingredients of gunpowder are varied for different uses. The powder used for mining, for example, contains more sulphur than that used for fire-arms. Gunpowder must be granulated in order to secure its rapid combustion.

Chlorate of Potassa. When a stream of chlorine is passed into a solution of potassa, the chloride of potassium and the chlorate of potassa result; the latter is deposited in rhomboidal tables. The chlorate of potassa is anhydrous; it dissolves in 18 parts of cold water and 2½ of hot, melts at a red heat with evolution of pure oxygen, deflagrates with combustible bodies, as sulphur, with much violence.

What is the origin and use of the nitrate? What is the composition of gunpowder? How is the chlorate made?

LECTURE LIX.

SODIUM.—*Preparation.—Relations to Oxygen and Water.—Color communicated to Flame.—Its Oxides.—The Hydrated Oxide.—Tests for Sodium.—Haloid Compounds.—Common Salt.—Salts of the Protoxide, Carbonates, Sulphate, Nitrate, Phosphate, etc.*—LITHIUM.—CÆSIUM.—RUBIDIUM.—BARIUM.—*Its Oxides.—Haloid Compounds.—Salts of the Protoxide.*

SODIUM. *Na*=23.

SODIUM (Natrium) may be obtained by the same process as potassium, but is best procured by igniting the calcined acetate of soda with powdered charcoal in an iron bottle; and, as the sodium does not act on carbonic oxide, the operation is much more productive than in the case of the other metal. Like potassium, it has to be kept under the surface of naphtha.

In color sodium resembles silver; its specific gravity is .97; it therefore floats on water. It melts at about 200°, and is volatile like potassium, its vapor being colorless. Thrown upon water, it decomposes it with a hissing sound, and with the evolution of hydrogen, but no flame appears. If, however, the water be hot, or the metal be put in a muslin bag, then a beautiful yellow flame, characteristic of sodium and its compounds, is the result. If a drop or two of water be put on freshly-cut sodium, it takes fire.

SODIUM AND OXYGEN.

With oxygen, sodium forms three compounds—the suboxide, protoxide, and peroxide.

Protoxide of Sodium. *Na O*=31.

This, like the corresponding potassium compound, is produced by oxidizing sodium in dry air. It is a white powder, which attracts moisture from the air, and forms the hydrated oxide of sodium, commonly called caustic soda.

How is sodium obtained, and what are its uses? What are its properties? What oxygen compounds does it yield?

Hydrated Oxide of Sodium, $NaO+HO=40$,

or caustic soda, may be made by the same process as that given for caustic potassa, by using carbonate of soda, and, when the resulting carbonate of lime has settled, evaporating the liquid. The best proportions are, one part of quick-lime to five of carbonate of soda in crystals.

Caustic soda resembles caustic potassa in most of its properties. It is deliquescent, has a strong affinity for carbonic acid, and acts upon animal tissues as an escharotic. Its salts are generally more soluble than the potassa salts, and on this are founded the methods recommended for distinguishing the latter compounds from it. Moreover, the soda compounds communicate to the flame of alcohol, or to the blow-pipe flame, a yellow color, and give in the spectroscope the yellow line D.

Chloride of Sodium. $NaCl=58.5$.

Common salt is obtained abundantly from the sea, in which it exists to the extent of about four ounces to the gallon. It is also found as rock salt, and in brine springs. The springs in Onondaga County, New York, contain one seventh, and the Great Salt Lake one fifth of nearly pure salt in their waters.

Common salt is the type of that extensive class of compounds which have derived the name of salt bodies from it. It crystallizes in cubes, and when in mass is often perfectly transparent, and permits the passage of heat of every temperature through it freely. It melts into a liquid at a red heat, and is not more soluble in hot than cold water. It is extensively used in the preparation of hydrochloric acid and chlorine; immense quantities, also, are annually consumed in the preparation of carbonate of soda, which is made by first acting on common salt with oil of vitriol, so as to turn it into sulphate of soda, and, igniting this with charcoal and carbonate of lime, an impure carbonate of soda is the result, known under the name of black ash, or British

---

How is caustic soda made? What are its properties and uses? What reactions does sodium yield? What is the constitution of common salt? Whence is it obtained? What are its properties? To what uses is it put?

barilla. Common salt is extensively used for the curing of meat. It is also an essential article of food, being decomposed in the animal system, and furnishing hydrochloric acid to the gastric juice, and soda to the bile and pancreatic juice.

The compounds of sodium with bromine, iodine, sulphur, etc., are not of interest.

Salts of the Protoxide of Sodium.

Carbonate of Soda is sometimes obtained by lixiviating the ashes of sea-weeds. Large quantities are now procured from the decomposition of sulphate of soda by sawdust and lime at a high temperature, the carbonaceous matter decomposing the sulphuric acid and generating carbonic acid, which unites with the soda, while the liberated sulphur is partly dissipated and partly unites with the calcium. From the resulting mass, *black ash*, carbonate of soda is obtained by lixiviation. The crystals, as found in commerce, contain generally ten atoms of water; there are two other varieties, the one containing eight atoms and the other one atom of water. Large quantities of the carbonate of soda are also sold in an uncrystallized state under the name of salts of soda. The figure of the crystals of this salt is a rhombic octahedron. They effloresce on exposure to the air. They are soluble in twice their weight of water at 60°, and in less than their own weight at 212°.

Bicarbonate of Soda, or the double carbonate of soda and water, is formed by transmitting a stream of carbonic acid through a solution of the carbonate, and is in the form of a white powder. It is less soluble in water than the former, requiring ten parts of water at 60°. There is a sesquicarbonate which passes in commerce under the name of Trona.

Chlorinated Carbonate of Soda, or Labarraque's disinfecting liquid, is produced by passing chlorine slowly through a solution of carbonate of soda. It is extensively used for the destruction of noxious odors and exhalations.

Sulphate of Soda, or Glauber's Salt, occurs as a nat-

Why is it a necessary ingredient of food? From what source is carbonate of soda obtained? Describe its preparation. How is bicarbonate of soda made? What is Labarraque's solution? What is Glauber's salt?

ural product, and also as the result of the preparation of hydrochloric acid. It is in prismatic crystals of a bitter taste, efflorescing in the air, and becoming anhydrous. Water dissolves more than three times its weight of this salt at 93°, but above that degree it is less soluble. When a solution of three parts of this salt in two parts of water is corked up in a flask while boiling, it may be cooled without crystallization taking place; but if the cork be withdrawn crystallization at once commences, or if it does not, the introduction of any solid matter produces it, and the temperature of the solution at once rises.

Nitrate of Soda is found abundantly in different parts of America, more particularly in Peru and Chili. In the soil it crystallizes in rhomboids. It dissolves in twice its weight of cold water, and, from its deliquescence, can not be used in the manufacture of gunpowder.

Phosphate of Soda (tribasic) is formed by neutralizing phosphoric acid with carbonate of soda: two of the hydrogen atoms are replaced. It crystallizes in oblique rhombic prisms, dissolves in three times its weight of cold water, is of an alkaline taste, and gives a lemon-yellow precipitate with nitrate of silver. By the addition of soda to it a subphosphate is formed, in which all three of the hydrogen atoms of the acid are replaced. By the addition of phosphoric acid to the ordinary phosphate till it ceases to give any precipitate with the chloride of barium, the biphosphate of soda results—a salt very soluble in water. Its crystals are rhombic prisms. In it only one of the hydrogen atoms is replaced.

Microcosmic Salt, or the phosphate of soda, ammonia, and water, is made by dissolving 5 parts of crystallized rhombic phosphate of soda with 2 of crystallized phosphate of ammonia. At a low heat it parts with its water of crystallization, and, the temperature rising, it loses its ammonia and saline water, becoming monobasic phosphate of soda. It is much used in blowpipe experiments.

What are its peculiarities of crystallization? Why can not the nitrate be used for gunpowder? What is the difference between the phosphate, subphosphate, and biphosphate? What is microcosmic salt?

Pyrophosphate of Soda (bibasic) is procured by heating the phosphate. It gives a white precipitate with nitrate of silver.

Metaphosphate of Soda (monobasic) is formed by heating microcosmic salt to redness. It is soluble in water, melts at a red heat, and gives, with dilute solutions of the earthy and metallic salts, viscid precipitates.

Biborate of Soda, commonly called Borax. It is imported in a crude state from the East Indies, and manufactured from the natural boracic acid of Italy by the addition of carbonate of soda. It crystallizes in octahedrons or in oblique prisms, the former containing five, the latter ten atoms of water, all of which is lost by exposure to a red heat, the salt then fusing into a glass. It is of great use in blowpipe experiments.

Hyposulphite of Soda is manufactured on a large scale for photographic purposes, because it can easily dissolve every compound of silver except the sulphide, and that portion of chloride which has been decomposed by light. It may be formed by digesting sulphur in hot sulphite of soda, or by passing sulphurous acid through a solution of sulphite of sodium.

LITHIUM, $Li = 7$,

was discovered in 1817, and may be obtained by decomposing the chloride by an electric current. It is reddish white, softer than lead, and can be pressed into wire. It is the lightest known solid element, the specific gravity being .594, so that it floats on naphtha; it fuses at 356°. Heated in air, it burns with an intense white light, forming lithia, $Li\,O$.

Lithia has only been found to any extent in certain minerals, as spodumene, lepidolite, etc. It is characterized by the crimson-red color it imparts to flame, and by its spectral lines. It yields a series of salts, chloride, nitrate, sulphate, phosphate, carbonate, etc.

CÆSIUM, $Cæ = 133$,

was discovered by Bunsen and Kirchhoff in making a

How are pyrophosphate and metaphosphate of soda formed? From what source is borax derived, and what are its uses? Of what use is hyposulphite of soda? In what minerals does lithium occur, and what are its properties? What is the characteristic of lithium?

spectrum analysis of the Durckheim mineral water, which contains about three grains of the chloride per ton. Its name is derived from the two grayish-blue lines peculiar to its spectrum. It decomposes water, cæsia forming and hydrogen being liberated. The hydrate of cæsia is soluble in alcohol, corrodes platinum like lithia, and is volatile at a high temperature. Among the known salts are the chloride, carbonate, nitrate, and sulphate.

Rubidium, $Rb=85$,

is named from two intensely red lines which its spectrum has near the end of the less refrangible rays. It is commonly associated with cæsium, and is found to a slight extent ($\frac{1}{400,000}$) in sea-water. It may be obtained by electrolysis from its chloride. It decomposes water, forming an oxide, $Rb\,O$, which is a powerful base, producing a series of salts. These salts, as well as those of cæsia, resemble those of potassa closely, in giving, with tartaric acid and chloride of platinum, precipitates.

Barium. $Ba=69$.

The existence of barium was first proved by Davy in 1808. He isolated it by passing a Voltaic current through mercury in contact with hydrate of baryta; an amalgam formed, from which the mercury was subsequently distilled, leaving the barium as a metal of gray color like cast iron, heavier than sulphuric acid, the specific gravity being 1.5, obtaining oxygen rapidly from the air or water, and giving rise to the production of baryta, $Ba\,O$. Barium may also be made by passing potassium in vapor over red-hot baryta.

Protoxide of Barium, $Ba\,O=77$,

may be obtained by igniting the nitrate of baryta, the decomposition being

$$Ba\,O,\,NO_5=Ba\,O+NO_4+O;$$

that is, one atom of nitrate of baryta yields one of protoxide of barium, one of nitrous acid, and one of oxygen gas. It is a white substance, having a strong affinity

From what does cæsium derive its name? How is rubidium obtained, and why is it so called? How was barium first obtained? What is the process for obtaining the protoxide? What are its properties?

for water, with which it exhibits the phenomenon of slacking, as is the case, to a less extent, with lime, heat being evolved. It has an acrid taste, is soluble in water, and absorbs carbonic acid from the air. The specific gravity is 4., being the heaviest of the earths; hence its name. The soluble salts are poisonous.

Hydrate of Baryta, $Ba\,O,\ HO = 86$,

is formed by slacking the protoxide, and is a white powder, very soluble in hot, but less so in cold water, yielding therefore crystals—hexagonal prisms, when a hot solution cools; these contain ten equivalents of water. The cold solution is used as a test for carbonic and sulphuric acids, with which it forms insoluble white precipitates.

The solution is most easily obtained by calcining the native sulphate with pulverized charcoal, which converts it into the sulphide of barium. To a boiling solution of this body oxide of copper is added till the liquid ceases to blacken a solution of acetate of lead. On being filtered, the solution of hydrate of barytes is procured.

Peroxide of Barium, $Ba\,O_2 = 85$,

is made by igniting chlorate of potassa with baryta, or by passing oxygen over baryta in a red-hot tube. It is used in the preparation of the peroxide of hydrogen and making of oxygen.

Of the other compounds of barium, the chloride is much used as a test for sulphuric acid. It may be made by decomposing carbonate of baryta by hydrochloric acid. The sulphide of barium is made by igniting the sulphate of baryta, heavy spar, with charcoal, which deoxidizes both the sulphuric acid and the baryta. It dissolves in hot water, and from this solution caustic baryta may be obtained by boiling with the oxide of lead or copper, and separating the sulphides of those metals by filtration. By acting upon it with hydrochloric or nitric acid the chloride or nitrate may be prepared.

How is the hydrate formed? How the peroxide? Of what use is the chloride? How is the sulphide made?

SALTS OF THE PROTOXIDE OF BARIUM.

Carbonate of Baryta is found native as the mineral *Witherite*, and may be prepared by precipitating a soluble salt of baryta with an alkaline carbonate. It is soluble in 4300 times its weight of cold water and 2300 of boiling. Its density is 4.33.

Sulphate of Baryta, found native abundantly as *heavy spar*, and from it most of the compounds of barium are prepared. Its density is 4.47. It crystallizes in tabular plates, and is wholly insoluble in water.

LECTURE LX.

STRONTIUM. — *Uses in Pyrotechny.* — *Salts of Protoxide.*—CALCIUM.—*Protoxide of.*—*Sources in Nature.* —*Tests for.* — *Haloid Compounds, Chloride, Fluoride, Sulphide.* — *Salts of the Protoxide, Carbonate, Sulphate, Phosphate, Chloride.*—MAGNESIUM.—*Protoxide and Salts of.*—ALUMINUM.—*Method of obtaining.* — *Sesquioxide.* — *Uses in the Arts.* — *Salts of.* — *Alums.*—GLUCINUM.— ZIRCONIUM.— THORIUM.—YTTRIUM. — CERIUM. — LANTHANUM. — DIDYMIUM. — THALLIUM.—INDIUM.

STRONTIUM, $Sr=44$,

is a fixed metal of a gray color, with a reddish reflection, having a specific gravity of 2.5. It oxidizes on exposure to the air, and decomposes water without flame. The metal is obtained by electrolysis from the chloride, the poles of the battery being of iron. Its natural compounds are the sulphate and carbonate.

Strontium yields a protoxide, which is the basis of a series of salts differing from baryta salts in not being poisonous. The hydrated protoxide is formed by slacking the protoxide with water. Strontia gives in the spectroscope eight characteristic bands of color — six red, one orange, and one blue, but no green.

What are the properties of the carbonate and sulphate of baryta? What are the properties of strontia? What are its spectroscopic reactions?

SALTS OF THE PROTOXIDE OF STRONTIUM.

Carbonate of Strontia, the strontianite of mineralogists, is a rare mineral of a greenish tint.

Sulphate of Strontia, the celestine of mineralogists, is found in Sicily, and is of a blue tint. It is not as heavy as sulphate of baryta, and is soluble in 3600 parts of water.

Nitrate of Strontia forms an ingredient of the red fire of theatres, which is composed of forty parts of it, united with thirteen of sulphur, five of chlorate of potassa, and four of sulphide of antimony. It crystallizes in octahedra, soluble in five parts of cold water and one half its weight of boiling water.

CALCIUM, $Ca=20$,

is obtained by electrolysis from the chloride, and is of a yellowish color, harder than lead, malleable, fusible at a red heat, burning with scintillations when heated in air, chlorine, vapor of bromine, iodine, or sulphur. Its specific gravity is 1.57; it decomposes water without flame, and yields a protoxide, quick-lime or lime.

Lime occurs as a carbonate in the various limestones, marbles, chalks, etc., which form in many countries extensive mountain ranges. Its other salts are very abundant.

From the carbonate, quick-lime may be obtained by exposure to a bright red heat. If the limestone contains silica it may be overburned, a silicate of lime forming, which prevents the lime from slacking. It possesses a strong affinity for water, and unites therewith with a great elevation of temperature, as exhibited in the process of slacking, the heat being sufficient to inflame gunpowder. Exposed to a high temperature, as in the *lime light*, it phosphoresces splendidly. The hydrate which forms when lime is slacked is white; it is soluble to a small extent in water, and it is remarkable that cold water dissolves more than hot—one part of lime requiring 1280 of water at 212°, but only 656 at 32°. Lime-water is colorless, of a partially caustic

---

Describe the carbonate, sulphate, and nitrate of strontia. Describe calcium. How does lime occur? How is quick-lime produced? What is the action of water on it? What is lime-water?

taste; neutralizes acids perfectly, restoring to reddened litmus its blue color. It is used as a test for carbonic acid, with which it gives the white carbonate of lime. Milk of lime is lime-water in which hydrate of lime is mechanically suspended. The hardening of mortars depends partly on the formation of carbonate of lime from the hydrate that they contain, and partly on the production of a silicate of lime. Hydraulic lime, which possesses the property of setting under water, contains silicate of alumina and oxide of iron.

Lime is best detected by oxalate of ammonia, with which it gives the white precipitate of oxalate of lime if the solution be not acid. Lime is valuable in agriculture from causing the decay of organic matter, and decomposing such minerals as yield potassa?

Compounds of Calcium.

Chloride of Calcium, $Ca\,Cl = 55.5$, is formed by dissolving carbonate of lime in hydrochloric acid, and evaporating the solution to dryness. It is exceedingly deliquescent, and in the uncrystallized state is much used for collecting water in organic analysis, and for drying gases. It is very soluble in water and alcohol.

Fluoride of Calcium, $CaF = 39$, called also fluor spar, and frequently found associated with lead. It crystallizes in cubes, octahedrons, etc., of various colors, and is procured of many beautiful varieties in Derbyshire; they may be turned in a lathe. It exists in fossil and recent bones to a small extent, and is the source from which the compounds of fluorine are derived. Chlorophane, one of the varieties, phosphoresces with a pale green light when heated.

Sulphide of Calcium, obtained by igniting the sulphate of lime with charcoal, or passing sulphureted hydrogen over red-hot lime, constitutes Canton's phosphorus, which may also be made by igniting sulphur with oyster-shells. It possesses, when fresh, the curious property of shining in the dark after exposure to the sunlight or electric spark.

What does the hardening of mortar depend on? What is the test for lime? For what is chloride of calcium used? Under what forms does fluoride of calcium occur? What quality does sulphide of calcium possess?

Salts of the Protoxide of Calcium.

Carbonate of Lime is abundantly found in nature, forming whole ranges of mountains. The most valuable white marbles are those which come from Paros, Pentelicus, and Carrara. It occurs pure in the form of Iceland spar in rhomboidal crystals possessed of double refraction. It is dimorphous, and assumes the form of six-sided prisms in the mineral called Arragonite. It is anhydrous, insoluble in water, but in water charged with carbonic acid it is soluble, and is deposited from such a liquid on boiling, or by the diffusion of the carbonic acid into the air. The carbonic acid is expelled from this salt by a red heat and the action of the more powerful acids. Carbonate of lime may be obtained in union with water by boiling hydrate of lime with sugar.

The formation of stalactites and stalagmites in caves depends on the solubility of carbonate of lime in carbonic acid water. As the water trickles from the roof the carbonate is partly deposited in the inverted cone suspended from the roof, called the stalactite, and partly in the cone, the stalagmite, on the floor, as seen in *Fig.* 283.

Fig. 283.

Sulphate of Lime, or *Gypsum*, occurs native both in

What are the forms of carbonate of lime? When does it dissolve in water? What are stalagmites and stalactites?

crystals and in extensive crystalline masses. It contains two atoms of water, and forms selenite. Alabaster is also a sulphate of lime. Anhydrite contains no water. On calcining the hydrous sulphate of lime at a low red heat it becomes plaster of Paris, and has the property of setting into a hard mass when made into a paste with water. It is then used for making plaster casts and for hard-finishing walls. The sulphate of lime is soluble in 500 parts of water, and communicates hardness to it, so that it becomes unfit for washing and cooking purposes.

Phosphate of Lime, or *Bone-earth*, is a tribasic phosphate. It is precipitated when the ash of bones is dissolved in hydrochloric acid, and the solution neutralized by ammonia. It exists native as phosphorite and apatite. Phosphate of lime is an important constituent of plants, and is necessary in the animal economy to repair the waste of the bones. Many plants, as the turnip, can be grown of great size by manuring with ground bones.

Chloride of Lime, or *Bleaching Powder*, is made by exposing hydrate of lime to chlorine. It is a white powder, exhaling an odor of chlorine, and is extensively employed as a bleaching agent, acidified water being used to liberate the chlorine. Chlorimetry is the process for determining the amount of chlorine that chloride of lime contains. It is usually about 30 per cent.

Magnesium. $Mg=12$.

Magnesium was first obtained by Davy in 1808, by passing the vapor of potassium over white-hot magnesia, but was not accurately examined until 1830, when Bussy prepared it by heating anhydrous chloride of magnesium with sodium. It may be procured by electrolyzing the fused chloride of magnesium. It is a silvery-white, ductile, malleable metal, fusible at a red heat, volatile at the same temperature as zinc, and burns with great brilliancy, evolving an intensely white light when heated in air. This light has been used for producing

---

Under what forms does sulphate of lime occur, and what are its uses? Of what value is phosphate of lime? What is bleaching powder? How is magnesium obtained? What are its properties? What are its peculiarities of combustion?

photographs of dark interiors. It is unchanged in dry air, and is only slowly oxidized in damp air. It does not decompose water, but is rapidly dissolved by dilute acids. It burns when heated in chlorine, and in the vapor of bromine, iodine, and sulphur. Its specific gravity is 1.74.

Protoxide of Magnesium. $Mg\,O=20$.

This substance, called also *Calcined Magnesia*, or simply *Magnesia*, may be made by heating the carbonate to low redness; the carbonic acid is driven off, and the magnesia remains as a white powder, insoluble in water, but neutralizing acids completely, and forming with them a complete series of salts.

Magnesia occurs very abundantly in nature, often associated as a carbonate with carbonate of lime, as in dolomitic limestone. It also occurs in fertile soils, and is essential to the growth of certain plants.

It is well distinguished from all the foregoing alkaline earths by the relation of its sulphate. The sulphates of baryta, strontia, and lime form a series of salts, the solubility of which in water is constantly increasing; to these the corresponding magnesia salt may be added; it is very soluble.

Magnesia is precipitated from its sulphate by the caustic alkalies, and by the carbonates of potassa and soda as a carbonate, but not by the carbonate of ammonia in the cold. It may be detected by adding carbonate of ammonia and phosphate of soda in succession, when the phosphate of magnesia and ammonia is precipitated. Heated before the blowpipe, after having been moistened with nitrate of cobalt, magnesia becomes of a pinkish color.

Salts of the Protoxide of Magnesium.

Carbonate of Magnesia is found native, and may be prepared by boiling the sulphate with an alkaline carbonate, diffusing the precipitate in water, and passing a stream of carbonic acid through it; by spontaneous evaporation the carbonate of magnesia is deposited in crystals. The carbonate of magnesia, the *magnesia alba*

What names are given to the protoxide? What is dolomite? How may magnesia be detected? How is its carbonate prepared?

of the shops, is prepared by precipitating the sulphate of magnesia with the carbonate of potassa. It occurs in light white cubical cakes or in powder, and is not a true carbonate, for it does not contain a full equivalent of carbonic acid. It is said to be a compound of one atom of hydrate of magnesia with three atoms of hydrated carbonate of magnesia. It is very slightly soluble in water.

Sulphate of Magnesia—Epsom Salts of commerce—is produced by the action of dilute sulphuric acid on magnesian limestone. Its crystals are small four-sided prisms, soluble in an equal weight of cold and three fourths their weight of boiling water, the solution having a bitter taste. A low heat expels six out of the seven equivalents of the combined water.

Phosphate of Magnesia and Ammonia, one of the varieties of urinary calculus, may be formed artificially when a tribasic phosphate, a salt of ammonia, and a salt of magnesia are mixed together.

Aluminum. $Al=14$.

Aluminum is obtained by passing the vapor of chloride of aluminum over sodium heated in a porcelain tube. Intense ignition ensues, and the reduced aluminum forms metallic globules ($Al_2Cl_3+3Na=2Al+3NaCl$). The mineral *Cryolite*, a double fluoride of aluminum and sodium, has also been used as a source of the metal, and is decomposed when heated with sodium, yielding globules of aluminum imbedded in fused fluoride of sodium.

Aluminum is a white malleable and ductile metal of the hardness of silver. Its specific gravity when rolled is about 2.67, and when cast 2.56; its point of fusion is 1750°. It is very slightly acted on by air or water at common temperatures. When intensely heated in a current of air it suffers only slight oxidation; heated to redness in an atmosphere of steam, it is slowly oxidized. It is readily acted on by hydrochloric acid, which evolves hydrogen and forms chloride of aluminum. Neither sulphuric nor nitric acid affects it at common temperatures, but when boiled in the latter it is oxidized as long as

---

What is Epsom Salt? What is the composition of phosphatic calculus? How is aluminum prepared? What are its properties? Under what circumstances does it oxidize?

the heat is maintained. In concentrated solutions of potassa and soda it is oxidized and hydrogen is liberated. The action is increased when the alkaline solution is heated. It forms alloys with many of the other metals, but does not combine with mercury. It is not affected by sulphur or sulphureted hydrogen. The general characters of this metal are such as to fit it for many useful purposes, but its high price has hitherto limited these applications.

Sesquioxide of Aluminum. $Al_2 O_3 = 52$.

This oxide, called also alumina and clay, occurs naturally under certain forms which are highly prized, as the ruby and sapphire. In a more impure condition it yields the various common clays, which also contain silica or metallic oxides, or other extraneous bodies.

Alumina may be prepared from the sulphate of alumina and potassa, common alum, by precipitating the sulphuric acid by chloride of barium. The sulphate of baryta goes down, and there is left in the solution chloride of potassium and chloride of aluminum. When the mass is dried, water is decomposed; hydrochloric acid is then expelled, and alumina, mixed with the chloride of potassium, remains behind. The latter is to be dissolved away by water, leaving the alumina as a white substance, which, with water, forms a plastic mass, capable of being moulded, and retaining its shape when baked. After ignition it adheres to the tongue, and during the act of drying it contracts considerably in volume, a property which formerly gave rise to the invention of Wedgewood's pyrometer.

The presence of alumina gives to the clays those properties which fit them for the purpose of the potter and brickmaker. Alumina is also used as a mordant to fix the colors of certain dyes upon cloth.

Alumina is precipitated from its solutions by fixed alkalies, which yield a white hydrate of alumina, soluble in an excess of the precipitant. It is also thrown down by alkaline carbonates, and when these precipitations are made in a solution tinged with coloring matter, the

What are the natural forms of the oxide? How may alumina be prepared? Why is it used in Wedgewood's pyrometer? What is a mordant?

alumina carries it down with it. Such colored precipitates pass under the name of lakes; and it is this property of attaching such colors to itself, enabling it to cause their firm adhesion to cloth fibre, which is the cause of its application as a mordant.

Among the purposes to which alumina is applied may be mentioned the manufacture of PORCELAIN, and the different kinds of earthen-ware. The former substance, first made by the Chinese, is very compact and translucent. It consists essentially of clay mixed with a fusible body which binds all its parts together, and is covered with a glaze, which does not terminate abruptly on the surface, but pervades the substance of the mass. In this respect it differs from common earthen-ware. Feldspar, or the silicate of lime, are bodies suitable for communicating this glassy structure.

In the manufacture of porcelain great care is taken to select clay free from iron. It is mixed with powdered quartz and feldspar, and the requisite shape given it either by the potter's wheel or by pressing it into moulds. It is then dried in the air, and more perfectly in a furnace, and, when ignited, forms *biscuit*. This is dipped in the glaze, suspended in water, and becomes covered over with a uniform coat of it. It now remains to dry it once more, and fuse the glaze upon it.

EARTHEN-WARE consists of a white clay mixed with silica. It is glazed with a fusible material containing oxide of lead, and colored of different tints by metallic oxides; for example, blue by cobalt.

Connected with the manufacture of pottery may also be mentioned the manufacture of GLASS, of which there are several varieties, some consisting of silica, potassa or soda, and lime, others containing a large quantity of oxide of lead. If silica be heated with carbonate of potassa and lime, or oxide of lead, carbonic acid is expelled and glass forms. The mass is kept in a fused condition till it is free from air bubbles, and is then cooled until it becomes plastic, so that it may be blown or moulded.

Articles of glass, after they are manufactured, require

How may alumina be recognized? What are lakes? What is the composition of porcelain and earthen-ware? How is glass made? Why must it be annealed?

to be annealed or slowly cooled down. This allows their parts to assume a regular structure, and prevents excessive brittleness.

Soluble Glass is formed when silica is heated with twice its weight of carbonate of soda or potassa. It derives its name from the fact that it is for the most part soluble in water.

SALTS OF THE SESQUIOXIDE OF ALUMINUM.

Sulphate of Alumina is made by dissolving alumina in dilute sulphuric acid. It enters into the composition of the alums.

Sulphate of Alumina and Potassa (*Alum*).—This important salt is prepared from alum slate. It crystallizes in octahedrons, has an astringent taste, reddens litmus paper. It dissolves in about eighteen times its weight of cold, and less than its own weight of boiling water. It contains twenty-four atoms of water, and, when exposed to heat, foams up, melting in its own water, which, being evaporated away, leaves a white porous mass, commonly called burnt alum.

In the same way that the sulphate of potassa unites with the sulphate of alumina, so also do the sulphates of ammonia and of soda, forming respectively the ammoniacal and soda alums. The alumina in the common alum may be replaced also by the sesquioxides of iron, manganese, or chromium, giving iron, manganese, and chrome alums.

GLUCINUM. $G=7$.

Glucinum is obtained by decomposing its chloride by means of sodium. It is a gray malleable metal; specific gravity 2.1. Its fusing point is a little below that of silver; it is not altered by exposure to air, and is difficult of oxidation, even in the flame of the blowpipe.

ZIRCONIUM, $Zr=34$,

is procured oy acting on the potassio-fluoride of zirconium by potassium at a red heat. When cold the product is thrown into water, and the zirconium separates

---

What is soluble glass? What are the properties of sulphate of alumina and potassa? What other alums are there? How is glucinum made? How is zirconium prepared?

in the form of a black powder, having the appearance of plumbago. It is difficultly soluble in the acids, with the exception of the hydrofluoric, which readily dissolves it, evolving hydrogen. Heated in the atmosphere it burns into zirconia.

Thorium. *Th*=60.

By passing a current of dry chlorine over a mixture of thorina and charcoal powder, a crystalline chloride of thorium is obtained, which is easily decomposed by potassium, and the product is thorium. It is of a gray color, metallic lustre, and apparently malleable. It is not oxidized by water, but, when heated in the air, it burns into thorina. It is feebly acted on by sulphuric acid, and scarcely touched by nitric acid; it is not attacked by the caustic alkalies at a boiling heat. Hydrochloric acid dissolves it, with the evolution of hydrogen.

Yttrium, *Y*=32,

is obtained by decomposing its chloride by potassium. It is gray, brittle, and resists the action of air and water.

Yttria is always accompanied by *Erbia* and *Terbia*, the oxides of two metallic bases, Erbium and Terbium. Erbia is pale yellow and terbia pale red, but neither has been adequately examined.

Cerium. *Ce*=46.

By heating chloride of cerium with potassium an alloy is obtained, which evolves hydrogen when put into water, and leaves cerium in the form of a gray metallic powder. Heated in the air it burns into an oxide, and it is soluble in the weakest acids, with the evolution of hydrogen.

Lanthanum, *La*=44,

is associated with cerium. All its salts are said to be colorless. When the oxalate is heated it leaves a white carbonate, which at a higher temperature is converted into a light-brown anhydrous oxide. The white hydrated oxide attracts carbonic acid so rapidly that it can

What are the properties of thorium? What other metals accompany yttrium? How is cerium prepared? What are the properties of lanthanum?

not be completely washed upon a filter without conversion into carbonate.

DIDYMIUM, *Di*=48,

accompanies lanthanum and cerium in the minerals containing the latter metal. The atomic weights attached to these metals are of doubtful accuracy.

THALLIUM, *Tl*=203,

was discovered by Crookes in 1861 on account of the green line it gives in the spectroscope, whence its name. It exists largely in iron and copper pyrites, and in many minerals, as lepidolite. It resembles cadmium, being white and of a high metallic lustre; the specific gravity is about 11.85. It is one of the softest of the metals, a piece of lead scratching it readily; the melting point is 561°; it welds at the ordinary temperature. It oxidizes quickly in the air; does not decompose water, but evolves hydrogen from steam at a red heat.

It forms salts with acids, which are, for the most part, colorless and poisonous. The oxides are TlO and TlO_3. In Marsh's apparatus it gives a stain resembling that of arsenic, but is distinguished from it by exposure to iodine, which turns it yellow; the yellow iodide is insoluble in sulphide of ammonium.

INDIUM, *In*=36,

was discovered by Reich and Richter in arsenical pyrites. It is a lead-gray metal, ductile, and very soft, giving a streak on paper. It is most easily detected by its spectrum, which contains a line of indigo-blue light. Its chloride is very volatile. Heated to a bright red, the metal itself volatilizes.

Where is didymium found? What are the circumstances of the discovery of thallium? What are its properties? How is indium detected?

LECTURE LXI.

MANGANESE.—*Its Seven Oxides.—The Peroxide and its Applications.—Mineral Chameleon.—Acids of Manganese.—Salts of the Protoxide.*—IRON.—*Its Natural Forms.—Reduction on the Great Scale.—Cast Iron.—Wrought Iron.—Steel.—Passive Iron.*

MANGANESE. $Mn=28$.

MANGANESE may be procured by igniting its oxides with a mixture of lampblack and oil in a powerful furnace, the reduction being somewhat difficult. It is a white metal, specific gravity 8.013, requiring a white heat for its fusion, and oxidizing readily in the air. It is remarkable for the number of oxygen compounds which it yields; they are,

$$MnO \ldots Mn_2O_3 \ldots MnO_2 \ldots MnO_3 \ldots Mn_2O_7 \ldots Mn_3O_4 \ldots Mn_4O_7,$$

designated respectively,

Protoxide of Manganese.	Permanganic Acid.
Sesquioxide of Manganese.	Red oxide of Manganese.
Peroxide of Manganese.	Varvicite.
Manganic Acid.	

Of these, the protoxide may be made by passing hydrogen gas over red-hot peroxide of manganese. It is of a green color, is a basic body, and forms a series of salts, of which the sulphate is used in dyeing. It is isomorphous with magnesia and zinc. Sulphide of ammonium yields with it a flesh-colored precipitate, ferrocyanide of potassium a white, and the chloride of soda a dark brown hydrated peroxide. The sesquioxide is made by igniting the peroxide, as will be presently explained. The red oxide and varvicite occur as minerals, but, of the whole series, the peroxide is by far the most valuable.

Peroxide of Manganese, $MnO_2=44$,

is found abundantly as a mineral, and passes in commerce under the name of black oxide of manganese, a

How may manganese be obtained? What are its properties? How many oxides does it furnish? How does the peroxide occur?

name indicating its color. It is insoluble in water, and when exposed to a red heat gives off one fourth of its oxygen, forming the sesquioxide, as stated above, the action being

$$2(Mn\,O_2)\ldots=\ldots Mn_2\,O_3+O.$$

On this fact is founded one of the processes for obtaining oxygen gas. Heated with hydrochloric acid, it yields chlorine, as has been explained. It was formerly called glass-maker's soap, from the circumstance that it removes, when added to melted glass, the stain of protoxide of iron, by turning it into peroxide, and causes the glass to become colorless; but if too great a proportion of peroxide of manganese be used, the glass assumes an amethystine color.

Peroxide of manganese, when ignited with caustic potassa in a platinum crucible, yields a substance known as *Mineral Chameleon*, which is of a green color. Water dissolves from it the *Manganate of Potassa*, which is of a beautiful grass-green, the solution speedily passing through a variety of shades of purple, blue, and red. When mineral chameleon is dissolved in hot water, a red solution is obtained of the

Permanganate of Potassa. The solution of this salt is now much employed in volumetric analysis. It readily parts with its oxygen to organic matter and deoxidizing bodies generally; it loses its color, and the brown hydrated peroxide of manganese is deposited. A standard solution is employed for determining the amount of organic matter in air and water. It is also used as a disinfectant. From the permanganate of baryta a crimson solution of *Permanganic Acid* may be procured by the aid of sulphuric acid.

Among other compounds of manganese, the following may be named:

Protochloride of Manganese,	$MnCl$	$= 63.5$
Perchloride " "	Mn_2Cl_7	$=304.5$
Perfluoride " "	Mn_2Fl_7	$=199$

The protochloride may be made by acting on the peroxide with hydrochloric acid, evaporating to dryness, and fusing at a red heat. On digesting with water, the

Of what use is it? What is mineral chameleon? What are its properties? Of what use is permanganate of potassa? How may the chlorides of manganese be formed?

protochloride dissolves, and any impurity of iron is left in the state of oxide. Then, by crystallizing, the chloride can be obtained in pink crystals. The perchloride is produced when permanganate of potassa, common salt, and sulphuric acid are heated. It is a dark greenish and volatile liquid. The perfluoride is obtained by distilling sulphuric acid, permanganate of potassa, and fluor spar; it is a greenish-yellow gas.

Salts of the Protoxide of Manganese.

Protosulphate of Manganese, formed by dissolving protoxide of manganese in sulphuric acid. The figure of its crystals depends on the temperature at which they were formed. They have a rose-colored tint. It is insoluble in alcohol, very soluble in water, and is used by the dyers to produce a fine brown color.

There is but one sulphide of manganese. It is obtained as a hydrate when manganese is precipitated by sulphide of ammonium (MnS, HO). It is of a flesh-red color.

Iron. $Fe=28$.

Iron sometimes occurs in a native state and as meteoric iron, also as oxide, carbonate, sulphide, etc. It is one of the most abundant of the metals. Much of what is found in commerce is derived from clay iron-stone, which is an impure carbonate containing silica, alumina, magnesia, and other foreign substances. The native peroxide of iron, red hæmatite; the hydrated peroxide, brown hæmatite; the black oxide, or magnetic iron ore, furnish some of the finer varieties of the metal.

From *clay iron-stone* metallic iron is procured by the action of carbonaceous matter and lime at a high temperature. The ore, having been roasted, is thrown into the furnace with coal and lime. If the iron is in the ore as a silicate, the lime decomposes it at those high temperatures, forming a slag of silicate of lime, and the oxide of iron set free is instantly reduced by the carbonaceous matter; the metal, sinking down, protected by the slag, is let off by opening a hole in the bottom of the furnace.

What are the properties of the fluoride? What is the use of the protosulphate? What are the forms under which iron occurs? How is it procured from clay iron-stone?

The quantity produced in Great Britain may be estimated at 3,000,000 tons per annum.

The substance thus produced is not pure iron; it contains carbon and other impurities, and passes under the name of cast or pig iron. It is purified by melting and sudden cooling, which converts it into *fine metal;* this fine metal is then melted under exposure to air, which burns off the carbon as carbonic oxide, and the mass, from being perfectly fluid, becomes coherent. It is now subjected to violent mechanical action, such as hammering or rolling; this forces out or burns off the impurities, increases its tenacity, and it becomes the wrought iron of commerce.

Cast Iron melts readily at a bright red heat, and expands in solidifying; on this depends its valuable application for making castings. Kept under the surface of salt water for a length of time, cast iron becomes converted into a body somewhat like plumbago, due probably to the removal of the iron as a chloride; the carbon which is left behind is sometimes observed, as it dries, to become hot—a phenomenon to be accounted for by its porous state. These facts have been frequently verified in the case of cannon which have lain for years at the bottom of the sea. There are two forms of cast iron, white and gray; the former contains about five per cent. of carbon, the latter three or four. The structure of cast iron is crystalline, as is shown in *Fig.* 284.

Fig. 284.

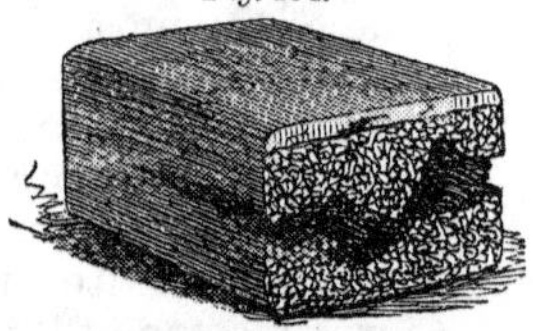

Pure Iron may be obtained by decomposing precipitated peroxide of iron by hydrogen gas, and melting the result. The metal has a bluish color, is more ductile than malleable, and is the most tenacious of all the elements. It becomes very soft at a red heat, and possesses the welding property; on this depends the art of forging it. Its specific gravity is 7.8. It is one of the few magnetic bodies, and, when soft, its magnetism is so transient that it may gain and lose that quality a thou-

What are cast iron and fine metal? How is wrought iron made? What are the properties of cast iron? What changes does it undergo in sea-water? How may pure iron be obtained?

sand times in a minute. The melting point of iron is very high, about 3300°. In the mode of preparing it from cast iron it does not undergo the process of fusion, but its particles are simply welded together. The fibrous structure which wrought iron possesses, *Fig.* 285, is the chief cause of its great tenacity; a wire $\frac{1}{36}$th of an inch in diameter will bear a weight of sixty pounds.

Fig. 285.

Steel, which is a valuable preparation of iron, is made by placing alternate strata of iron bars and charcoal powder in a close box and keeping them red-hot. The process is known by the name of cementation. The iron gains about 1.5 per cent. of carbon. Steel is much more fusible than iron, and becomes excessively hard and brittle by being brought to a red heat, and then suddenly quenched in cold water. When allowed to cool slowly it is quite soft, but various degrees of elasticity and hardness may be given to the hardened steel by the process of tempering, which is effected by again heating it up to a fixed point. Various colors form on the surface, which are an index of the *temper;* at 460°, for example, the color is *straw*, and the hardness is suitable for fine cutlery.

The quality of steel is tested by washing its clean surface with dilute nitric acid, which ought to produce a uniform gray color. If the steel be imperfect and contain veins of iron, they are shown by their difference of color; and this is the cause of the veined appearance of Damascus steel. Case-hardening is an operation performed upon wrought iron, by which it is superficially converted into steel. It is accomplished by heating to redness in contact with charcoal powder, or by the aid of ferrocyanide of potassium.

By placing a piece of platinum in nitric acid of a specific gravity of 1.34, and then bringing an iron wire in contact with it and withdrawing the platinum, the iron

---

What is the difference in structure of cast and wrought iron? How is steel made, and what are its properties? What is the process of tempering? How is the quality of steel tested? What is case-hardening? How may iron be rendered passive?

assumes a passive or allotropic state. It now exhibits no tendency to unite with oxygen, can not precipitate copper from its solutions, and simulates the properties of platinum and gold.

LECTURE LXII.

IRON.—*Oxides of.*—*Three Oxides and Ferric Acid.*—*Tests for Iron.*—*Salts of the Protoxide and Peroxide.*—*The Sulphides.*—NICKEL.—*Its Reduction from the Oxalate.*—COBALT.—*Smalt.*—*Zaffre.*—*Sympathetic Ink.*—ZINC.—*Distillation of.*—*Salts of the Protoxide.*

IRON AND OXYGEN.

IRON burns with rapidity in oxygen gas, as may be proved by igniting a piece of it in wire coiled into a spiral form in a jar of that gas, *Fig.* 286, when it will be found to take fire and burn beautifully. In atmospheric air, under favorable circumstances, the combustibility of this metal may be proved. Thus fine iron filings, sprinkled in the flame of a spirit-lamp, burn with scintillations. Exposed to air and moisture, it slowly rusts. Iron yields four oxides:

Fig. 286.

Protoxide	FeO	= 36
Black Oxide	Fe_3O_4	=116
Peroxide	Fe_2O_3	= 80
Ferric Acid	FeO_3	= 52

Protoxide of Iron. FeO=36.

This oxide exists, united with acids, in an extensive series of salts, from which it is thrown down as a hydrate by alkalies, and is then of a white color, which darkens as it passes into the state of peroxide. Ferrocyanide of potassium gives a white precipitate, and the ferridcyanide a deep blue. Sulphide of ammonium gives a black sulphide of iron. Sulphureted hydrogen and gallic acid give no precipitate.

How may iron be rapidly oxidized? How many oxides does it yield? What are the reactions of the protoxide?

Black Oxide of Iron. $Fe_3O_4=116$.

This oxide, known also as the magnet or loadstone, is found as a mineral. It is a compound of the protoxide and peroxide. The scales of iron found in blacksmiths' forges mainly consist of it. It may also be produced by decomposing the vapor of water by metallic iron in a red-hot tube.

Peroxide of Iron, $Fe_2O_3=80$,

is found in nature as oligist iron, or as a hydrate. It may be produced artificially as a hydrate by precipitation from a solution of persulphate of iron by a caustic or carbonated alkali, or in a pure state by igniting green vitriol; there is then left a red powder, known as *rouge*, used for polishing metals. This oxide is not magnetic; it is the basis of a series of salts, which yield, with alkalies, a brown hydrated peroxide; with ferrocyanide of potassium, Prussian blue; with sulphocyanide of potassium, a blood-red solution; with tannin and gallic acid, a black. This last is of considerable interest, constituting the basis of ordinary ink.

The presence of iron can always be determined by passing it into the condition of peroxide, and applying the foregoing tests.

Ferric Acid, $FeO_3=52$,

is prepared by heating peroxide of iron with four parts of nitrate of potassa. The result is treated with cold water, which yields a red solution of the ferrate of potassa. This slowly decomposes in the cold, and very rapidly when the solution is warm. The ferrate of baryta precipitates when the potassa solution is acted on by a soluble salt of baryta. It is a permanent body, of a crimson color.

Among other compounds of iron, the following may be named:

What is the magnet? What are the natural forms of the peroxide? How may it be prepared? For what is it used? What are its reactions? How is ferric acid made, and what are its properties?

Protochloride of Iron	$FeCl$	= 63.5
Perchloride "	Fe_2Cl_3	=162.5
Protiodide "	FeI	=154
Protosulphide "	FeS	= 44
Sesquisulphide "	Fe_2S_3	=104
Bisulphide "	FeS_2	= 60

Of these, the protochloride is formed by passing hydrochloric acid over red-hot iron: it is white, but forms a green solution in water. The perchloride, in solution, by dissolving peroxide of iron in hydrochloric acid. The protiodide, by boiling an excess of iron filings with iodine, and evaporating; it forms, on cooling, a dark gray mass. Its solution absorbs oxygen from the air. The protosulphide of iron, which is much used for forming sulphureted hydrogen, may be made by heating a mass of iron to a white heat, and applying to it roll sulphur, and receiving the melted globules in a bucket of water; it may also be procured by igniting iron filings with sulphur. The bisulphide occurs abundantly as a mineral of a golden-yellow color, crystallized in cubes or allied forms, and known as *Iron Pyrites.* It frequently assumes the form of various organic remains, being one of the common petrifying agents, but in this state differs essentially from the cubic pyrites both in color and oxidizability, these fossil remains rapidly decaying under exposure to the air, but the other form being unacted on. Besides these, there is a sulphide of iron which is magnetic.

Salts of the Protoxide of Iron.

Carbonate of Iron may be obtained from the sulphate by an alkaline carbonate, falling as a whitish precipitate. It turns brown, however, from the absorption of oxygen. It occurs as a mineral in spathic iron, and dissolves in water containing carbonic acid, forming chalybeate waters.

Protosulphate of Iron—Copperas—Green Vitriol—is prepared largely by the oxidation of iron pyrites, and crystallizes in oblique prisms of a grass-green color. It has a styptic taste, dissolves in twice its weight of cold and three fourths its weight of boiling water. It con-

Name some other compounds of iron. How may these compounds be formed? What is iron pyrites? What forms does it present? How is carbonate of iron formed? How is the sulphate made?

tains seven atoms of water. At a low red heat it becomes anhydrous. In this state it is used for the manufacture of the Nordhausen sulphuric acid. It is employed as a developer in collodion photography.

Salts of the Peroxide of Iron.

Persulphate of Iron may be formed by adding to a solution of the protosulphate of iron half an equivalent of sulphuric acid, and peroxidizing by nitric acid. With water it forms a red solution.

The native oxides of iron are, 1. Magnetic iron ore; 2. Specular and micaceous iron ore, found of singular beauty at Elba; 3. Hæmatite, or red ironstone, comprising several hydrated varieties; 4. Bog ore, found in marshy places and of recent origin.

Nickel. $Ni = 30$.

Nickel may be obtained by igniting its oxalate in a covered crucible, carbonic acid escaping, and the metal being reduced.

$$NiO + C_2O_3 \ldots = \ldots Ni + 2(CO_2);$$

one atom of the oxalate of nickel yielding one of the metal and two of carbonic acid gas.

Nickel is a white metal, requiring a high temperature for fusion. It is magnetic, and has a specific gravity of 8.5. An alloy of nickel and iron forms a principal metallic ingredient in most aerolites, and in the masses of native iron found in various parts of the world, the nickel being from 1.5 to 8.5 per cent. With copper it forms a hard white alloy, used for coinage in the United States. German silver consists of copper 8 parts, nickel 3 to 4 parts, zinc 3½ parts. It unites with oxygen, forming a protoxide and sesquioxide, the former yielding salts of a green color; the latter is an indifferent body.

Salts of the Protoxide of Nickel.

Sulphate of Nickel crystallizes from its solutions with seven atoms of water in slender green prisms, which, when exposed to the sun, change into an aggregate of octahedrons, becoming opaque.

What are its uses? How is the persulphate obtained? What are the native oxides of iron? How is nickel prepared? What are its properties? When does it occur with iron? What change takes place in sulphate of nickel in sunlight?

COBALT, $Co=30$,

is generally associated with iron and nickel, and with them occurs in meteoric iron. Like the preceding metal, it may be obtained by igniting its oxalate in a covered crucible, carbonic acid being disengaged and metallic cobalt left. It is a pinkish-white metal, requiring a high temperature for fusion. Its specific gravity is 8.9. It is magnetic. It forms a protoxide and a sesquioxide, the former being the basis of a class of salts which are chiefly of a pink or blue color. *Smalt* is a silicate of cobalt, and *Zaffre* an impure oxide; the former is used to communicate to paper a faint blue tinge, and the blue color which the oxide gives to glass is taken advantage of in coloring the common varieties of earthen-ware. Cobalt is easily detected upon this principle.

The chloride of cobalt may be made by dissolving the oxide or the metal in hydrochloric acid. It is a pink solution, which turns blue when dried. It forms a beautiful *sympathetic ink*, for letters written with it, especially on paper which has a pinkish tinge, are entirely invisible, but become of a bright blue color when the paper is warmed, the letters again fading as they become cool and moist.

ZINC. $Zn=32$.

Zinc is a very abundant metal, immense quantities of it occurring in the state of New Jersey and in various other places. From zinc blende, which is a sulphide, converted by roasting into an oxide, or from the carbonate brought into the same state by ignition, the metal may be obtained by the process of distillation by descent. The oxide, mixed with charcoal, is introduced into a crucible which has an iron tube passing through a hole in its bottom, as seen in *Fig.* 287, and, the lid being luted on, the temperature is raised to a white heat, and the zinc, distilling over, may be condensed in water.

Fig. 287.

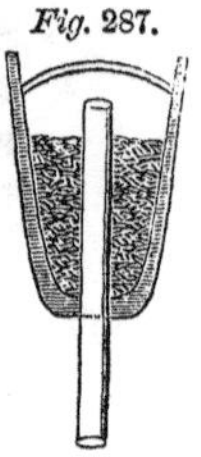

How is cobalt procured? What is smalt? What is zaffre? What are the uses of cobalt? What is sympathetic ink? How is zinc obtained?

Zinc is a bluish-white metal, which melts at about 770°, and, if exposed to a bright red heat in the air, takes fire and burns with a brilliant pale-green flame. Its specific gravity is about 7. At common temperatures it is brittle, but it may be rolled into thin sheets at about 300°, and then retains its malleability when cold. During its combustion there arises from it a great quantity of flocculent oxide, which formerly went under the name of *nihil album*, or philosopher's wool. Among the compounds of zinc may be mentioned

Protoxide of Zinc	$ZnO = 40$
Chloride "	$ZnCl = 67.5$
Sulphide "	$ZnS = 48$

Of these, the oxide is formed, as has been said, during the combustion of zinc. It is also precipitated as a white hydrate from its soluble salts by potassa or soda, soluble in excess of the precipitant. The chloride may be made by the action of hydrochloric acid on metallic zinc. It is used in the arts for soldering under the name of butter of zinc. The sulphide occurs as a mineral under the name of zinc blende.

If pieces of hot iron be dipped into melted zinc, they acquire the appearance of tin plate, for which they are a valuable substitute, inasmuch as the zinced iron is prevented from oxidation and rusting by the electrical relations of the metals. This is commonly known as galvanized iron.

Salts of the Protoxide of Zinc.

Sulphate of Zinc—White Vitriol.—This salt is formed in the process for procuring hydrogen gas by the action of dilute sulphuric acid on zinc. It crystallizes in colorless prisms with seven atoms of water, and is soluble in two and a half parts of cold water. It has a styptic taste, and reddens vegetable blues. There are three different subsulphates of this oxide.

Silicate of Zinc, the electric calamine of mineralogists, remarkable for becoming electric when heated.

What is remarkable in rolling it? What is nihil album? How are the oxide and chloride formed? What is galvanized iron? How is white vitriol made? What is electric calamine?

LECTURE LXIII.

CADMIUM.—*Sources of.*—*Its Volatility.*—TIN.—*Block and Grain.*—*Its Properties.*—*Protoxide and Stannic Acid.*—*Chlorides of Tin.*—*Mosaic Gold.*—*Its Uses.*—CHROMIUM.—*Chrome Iron.*—*Green Oxide and its Uses.*—*Chromic Acid.*—*Salts of the Sesquioxide.*—*Salts of Chromic Acid.*—VANADIUM.—TUNGSTEN.—COLUMBIUM.—NIOBIUM.—ILMENIUM.—NORIUM.—PELOPIUM.—DIANIUM.—MOLYBDENUM.—TITANIUM.

CADMIUM. $Cd=56$.

This metal was discovered in 1817 by Stromeyer. It is contained in certain ores of zinc, and, being more volatile than zinc, passes over with the first portions of metal, from which it may be separated by dissolving it in dilute sulphuric acid, and passing sulphide of hydrogen through the solution; the sulphide of cadmium thus precipitated is then dissolved in hydrochloric acid, and precipitated by carbonate of ammonia. This precipitate, after having been washed and dried, is mixed with charcoal, and reduced in an earthen-ware retort; the cadmium passes over at a dull red heat.

Cadmium, in its physical properties, much resembles tin, but it is rather harder and more tenacious; it crackles when bent. Its specific gravity is from 8.60 to 8.69. It fuses below the temperature required by tin.

Bromide and iodide of cadmium are of use in photography.

TIN. $Sn=50$.

Tin has been known from remote ages, and was obtained at a very early period from Spain and Britain by the Phœnicians. It occurs most abundantly in Cornwall, the mines of which afford about 3000 tons annually. It is also found in Germany, Bohemia, and Hungary; in Chili and Mexico, in Malacca, India, and the island of Banca. The native peroxide is the principal

How does cadmium occur? How is it separated? What are its properties? Where is tin found, and in what form?

ore of tin. It may be reduced by the action of charcoal at a high temperature.

Tin is a white metal like silver. It oxidizes in the air superficially, the action ceasing as soon as a thin crust is formed. At a red heat it oxidizes rapidly, forming *putty powder*, used for polishing metals. It is very malleable, and may be rolled into thin foil. When bent backward and forward it emits a crackling sound. It is very soft; its specific gravity 7.2. It melts at 442°, and burns when raised to a high temperature in the air. Some of its compounds are

Protoxide of Tin	SnO	= 58
Sesquioxide "	Sn_2O_3	=124
Peroxide "	SnO_2	= 66
Protochloride "	$SnCl$	= 85.5
Perchloride "	$SnCl_2$	=121
Protosulphide "	SnS	= 66
Persulphide "	SnS_2	= 82

The protoxide may be made by precipitation from the protochloride by carbonate of potassa. It is to be washed with warm water, and its water finally driven off by a current of carbonic acid gas at a red heat. It is of a black color, is easily set on fire in atmospheric air, passing into the condition of peroxide. Its salts reduce the noble metals to the metallic state, when added to their solutions, and yield with the chloride of gold *the Purple of Cassius.* The peroxide, called also stannic acid, from exhibiting weak acid properties, may be made by the action of nitric acid on tin. It is a hydrate in the form of a white powder, insoluble in acids and water; but if obtained by precipitation from perchloride of tin, it is soluble both in acids and alkalies. Melted with glass, it forms a white enamel.

The protochloride may be made by dissolving tin in warm hydrochloric acid. The solution, when concentrated, deposits crystals of the hydrated protochloride. These are decomposed when heated. The anhydrous protochloride may be made by passing hydrochloric acid gas over metallic tin at a red heat. The perchloride is

---

State the properties of tin. What is the crackling of tin? How is the protoxide made, and how do its salts act? How is stannic acid made? What effect has it in glass? How is the protochloride made?

procured by distilling eight parts of tin with twenty-four of corrosive sublimate. It is a smoking fluid, and was formerly called the *Fuming Liquor of Libavius.* A solution of this substance, much used in dyeing, is made by dissolving tin in nitromuriatic acid, or by warming a solution of the protochloride with a little nitric acid.

Of the sulphides, the first may be formed by pouring melted tin on sulphur, and igniting the powdered result with more sulphur in a crucible. It is a bluish-gray compound. The persulphide is obtained when two parts of peroxide of tin, two of sulphur, and one of sal ammoniac are ignited in a retort. It is a body of a golden-yellow color, formerly called *Aurum Musivum*, or Mosaic gold, in small scales of a greasy feel, and is used for exciting electrical machines, being much more energetic than the common amalgam, though less durable in its power.

Tin furnishes several valuable metallic combinations; *Tin Plate* is sheet iron superficially alloyed with it. The *soft solders* are alloys of lead and tin. *Pewter* is an alloy with antimony.

Moirè Metallique is tin plate which has been superficially acted on by an acid so as to display by reflected light the crystalline texture of the tin.

CHROMIUM. $Cr=26$.

CHROMIUM occurs abundantly near Baltimore as the chromate of iron (*Chrome Iron*), more rarely as the red chromate of lead. The metal may be obtained by the action of charcoal on the oxide at a high temperature, and is of a yellowish-white color. It takes its name from its tendency to produce highly-colored compounds. It is very infusible, and has a specific gravity of about 6. Its compounds to be here described are

Sesquioxide of Chromium	$Cr_2O_3 = 76$
Chromic Acid	$CrO_3 = 50$
Sesquichloride of Chromium..	$Cr_2Cl_3=158.5$

The sesquioxide may be prepared by heating the

What is the Liquor of Libavius? How is Mosaic gold made, and what is its use? What is Moirè Metallique? What natural forms does chromium present? What is the origin of its name? How is the sesquioxide prepared?

chromate of mercury to redness in a crucible. The mercury is driven off, and the chromic acid partially deoxidized, leaving a beautiful grass-green powder, the sesquioxide. It may also be obtained by heating the bichromate of potassa red-hot, and washing the residue in water; also as a hydrate, by boiling a solution of bichromate of potassa with hydrochloric acid, and adding alcohol; the mixture becomes of a green color, and ammonia precipitates the hydrated sesquioxide. It is a weak base, yielding a class of salts of a blue or green color. In the state of hydrate it is soluble in acids, but on making it red-hot it suddenly becomes incandescent, passes into another allotropic state, and is now insoluble. This sesquioxide is isomorphous with the sesquioxides of iron and alumina. In its two allotropic states it yields corresponding classes of salts, one of which is green and the other reddish-green. It is used for communicating a green color to porcelain.

Chromic Acid may be made by adding one volume of a saturated solution of bichromate of potassa to one and a half of oil of vitriol. On cooling, red crystals of chromic acid are deposited. It is isomorphous with sulphuric acid, produces with bases yellow and red salts, is a powerful oxidizing agent, is decomposed by a red heat into the sesquioxide, destroys the color of indigo and other dyes, and may be detected by producing, with the salts of lead, chrome yellow, and by its ready passage, under the influence of deoxidizing agents, into the sesquioxide.

The sesquichloride is procured when chlorine is passed over a mixture of the sesquioxide and charcoal in a red-hot tube. It is a lilac-colored body, which forms a green solution in water. There is also an oxychloride, which may be distilled as a deep red liquid from a mixture of chromate of potassa, common salt, and oil of vitriol. The fluoride, which is a red gas, is obtained by distilling in a silver retort a mixture of chromate of lead, fluor spar, and oil of vitriol. It is decomposed by the moisture of the air, forming chromic and hydrofluoric acids.

What is its use? How is chromic acid made? What is its effect on indigo and dyes? How are the chloride and fluoride obtained?

Salts of the Sesquioxide of Chromium.

Sulphate of Chromium and Potassa (*Chrome Alum*). When the oxide of chromium is dissolved in sulphuric acid and mixed with the sulphate of potassa and a little free sulphuric acid, crystals of chrome alum are deposited in red or blue octahedrons.

Chrome Iron, a compound of the sesquioxide of chromium and the protoxide of iron, is found native, crystallized in octahedrons, and also massive. It furnishes most of the compounds of chromium.

Salts of Chromic Acid.

Chromate of Potassa may be made by igniting chrome iron with one fifth its weight of nitrate of potassa. It crystallizes in small lemon-yellow prisms, and is very soluble in hot water. The crystals are anhydrous.

Bichromate of Potassa may be prepared from the former by adding an equivalent of acetic acid. It crystallizes in prisms of a ruby red. Large quantities are consumed by dyers.

Chromate of Lead (*Chrome Yellow*), obtained by precipitation from either of the foregoing salts by a soluble salt of lead. It is used as a paint.

Dichromate of Lead is formed by adding chromate of lead to melted nitrate of potassa, and dissolving out the chromate of potassa and excess of nitre by water. It is of a beautiful red color.

Vanadium, $V=68$,

occurs in certain lead ores, and may be obtained by decomposing the chloride by a current of dry ammonia in a glass tube heated over a spirit-lamp. It has a silvery lustre, is brittle, and not acted upon by air or water at common temperatures. At a dull red heat it burns into a black oxide. It is not dissolved by sulphuric or hydrochloric acids, but nitric acid and nitro-hydrochloric acid yield with it dark blue solutions. It has three oxides, VO, VO_2, VO_3.

What is chrome alum? What is chrome iron? How are the chromates of potassa made? What is chrome yellow? How is vanadium prepared, and what are its properties?

TUNGSTEN, $W=92$,

is obtained by passing hydrogen over ignited tungstic acid mixed with charcoal. It is very difficult of fusion, hard, brittle, and of an iron-gray color. Its specific gravity is 17.6. It is oxidized by the action of heat and air, and by nitric acid. It communicates valuable properties to steel. Tungsten is also called Wolframium, from Wolfram, which is a tungstate of iron and manganese.

COLUMBIUM, $Ta=184$,

called also Tantalum, was discovered in a North American mineral, *Columbite*, in 1801. It is obtained by heating potassium with the potassio-fluoride of columbium, and washing the mass in water. It remains in the form of a black powder, which, on pressure, resembles iron.

NIOBIUM—ILMENIUM—NORIUM—PELOPIUM—DIANIUM.

Among these rare metals, the two first mentioned have been announced as associated with columbium in some varieties of tantalite, but their distinctive characters have been, as yet, very imperfectly ascertained. Niobium is considered by some chemists to be columbium, while the metal pelopium has no independent existence, the pelopic and niobic acids being identical. Another metal of this series is called Dianium, but Daville states that dianic acid is hyponiobic acid.

MOLYBDENUM, $Mo=48$,

is a whitish, brittle, and very difficultly fusible metal, forming three compounds with oxygen, $Mo\,O$, $Mo\,O_2$, $Mo\,O_3$. A salt of the last, the molybdate of ammonia, is useful as a test for phosphoric acid.

URANIUM, $U=60$,

is a white, slightly malleable metal, unchanged by air and water at common temperatures. Its peroxide, $U_2\,O_3$, is used to give a greenish-yellow color to glass. The nitrate of uranium is of use in photography.

How is tungsten prepared, and what are its properties? How is columbium prepared? What remarks are to be made about the rare metals niobium, etc.? Of what value is molybdenum? What are the properties of uranium?

TITANIUM, $Ti=24$,

exists as titanic acid, TiO_2, in Rutilite, Anatase, and Oysanite. Titanic acid is useful in the coloring of the gums of artificial teeth and in porcelain painting.

LECTURE LXIV.

ARSENIC.—*Preparation of the Metal.—Properties of Arsenious Acid.—Two Varieties of it.—The methods of detecting it.—Process in Cases of Poisoning.—Sulphureted Hydrogen Test.—Marsh's Test.—The Copper Test.—Difficulties arising from Antimony.*

ARSENIC. $As=75$.

ARSENIC is obtained by sublimation in a current of air of the arsenide of cobalt and iron, the vapor condensing as a white oxide. This being mixed with powdered charcoal or black flux, and heated, the metallic arsenic sublimes. The process may be conducted in a tall vial, *Fig.* 288, imbedded in a crucible filled with sand, two thirds of the vial projecting above the heated sand. On this cooler portion the metal condenses. It is also sometimes found in a native state.

Fig. 288.

Arsenic is a metallic body, of a steel-gray color. It is very brittle; its specific gravity is 5.88, and, when slowly sublimed, it crystallizes in rhombohedrons. At 400° it sublimes without undergoing fusion, its melting point being much higher than that of sublimation. Its vapor has a smell of garlic, as may be readily recognized by throwing a little arsenious acid on a red-hot coal. Arsenic prepared by black flux tarnishes, it is said, from containing a little potassium. Among its compounds the following may be mentioned:

Arsenious Acid	$AsO_3= 99$
Arsenic Acid	$AsO_5=115$
Bisulphide of Arsenic	$AsS_2=107$
Tersulphide of Arsenic	$AsS_3=123$
Arseniureted Hydrogen	$AsH_3= 78$

How does titanium exist? How is metallic arsenic obtained? What are its properties? What is its odor?

Arsenious Acid is formed when arsenic is sublimed in atmospheric air. It is a white substance, which, when the process is conducted slowly, crystallizes in octahedrons. Similar octahedral crystals may be obtained by heating arsenious acid itself in a tube to 380°. When the operation has been recently performed and a large mass sublimed, it is a glassy, transparent body, which in the course of time slowly becomes milk-white. The specific gravity of arsenious acid is 3.7. It is nearly tasteless, of sparing solubility in water, the two varieties differing in this respect. By 100 parts of boiling water, 11.5 of the opaque, but only 9.7 of the transparent, are dissolved. This substance passes currently under the name of arsenic. It ought not to be forgotten that the arsenic of chemical writers and that of commerce are very different bodies: the one is black and the other white; the one is a metal and the other its oxide.

Arsenious acid may be detected by several methods:

1st. With ammonia sulphate of copper it gives an emerald-green precipitate—the arsenite of copper, or Scheele's green.

2d. With the ammonia nitrate of silver, a canary-yellow precipitate—the arsenite of silver.

3d. With sulphureted hydrogen, when previously acidulated with acetic or hydrochloric acid, it yields a yellow precipitate, the tersulphide of arsenic, orpiment. This, when dried and ignited with black flux (a mixture of charcoal and carbonate of potassa, obtained by igniting cream of tartar in a covered crucible), yields a sublimate of metallic arsenic.

4th. With the materials for generating hydrogen gas—that is, sulphuric acid, zinc, and water, placed in a bottle—if arsenious acid be present, arseniureted hydrogen is disengaged. When set on fire, it burns with a pale blue flame, emitting a white smoke; and if a piece of cold glass be held in the flame, there is deposited upon it a black spot of arsenic, surrounded by a white border

---

How is arsenious acid made? What is its crystalline form? What is the difference between the arsenic of chemists and that of commerce? What is the action of arsenious acid on ammonia sulphate of copper? With ammonia nitrate of silver? With sulphureted hydrogen? What is the process for detecting it as arseniureted hydrogen?

of arsenious acid. This stain is volatilized on heating the glass. Or if the arseniureted hydrogen be conducted through a tube of Bohemian glass, made red-hot at one point by a spirit-lamp, it is decomposed, and metallic arsenic deposited on the cooler portions beyond the ignited space.

5th. If a solution containing arsenious acid be acidulated with hydrochloric acid and boiled with slips of copper, the metallic arsenic is deposited upon the copper as an iron-gray crust. This is called Reinsch's test.

In cases of poisoning by this substance, it is unsatisfactory to apply, in the first instance, color-giving tests, such as the first, second, and third, as the liquor obtained from the stomach is itself highly colored and turbid. It is therefore desirable to examine that organ and its contents minutely, endeavoring to discover any white granules, or specks, which may be supposed to be arsenious acid, and, if such are found, to examine them separately.

The contents of the stomach, the larger pieces having been divided, are to be boiled in water and strained through a linen cloth. A current of chlorine gas passed through this liquid coagulates and separates much of the animal matter; or, what is more convenient, if the solution be first acidulated with nitric acid, and then nitrate of silver be added, much of the animal matter may be removed. By the addition of a solution of common salt, the excess of the silver salt may be precipitated, and the liquor, being filtered, is then fit for the third or fourth of the foregoing tests.

In the application of sulphureted hydrogen, the liquor having been clarified as just stated, the gas is passed through it until it smells strongly. It is then to be boiled for a short time, to expel the excess of gas, and filtered. The yellow precipitate of tersulphide of arsenic, or orpiment, which is collected, is to be thoroughly dried, and introduced, with twice its bulk of black flux, into the bulb, *a*, of a tube, such as *Fig.* 289, made of hard glass. On the temperature being

Fig. 289.

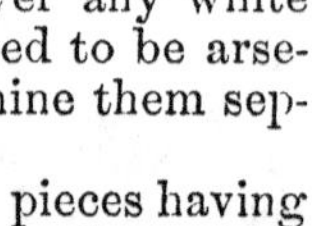

What is Reinsch's test? When are color tests applicable? What is the method of testing a stomach? Describe the sulphureted hydrogen test.

raised by a lamp, metallic arsenic sublimes, forming an iron-black ring round the part *b*. By cutting off the bulb of the tube and heating the black crust gradually, it slowly sublimes toward the colder part, producing a white deposit of arsenious acid in octahedral crystals.

In the application of *Marsh's test*, the liquor, having been cleared either by chlorine or by nitrate of silver, as above described, and mixed with a little sulphuric acid, is to be introduced into an apparatus, A B, *Fig.* 290, composed of a bent glass tube, one arm of which, B, is longer than the other. A piece of zinc is suspended by a thread in the arm A, so that it shall not reach the curved part. The liquid, on being poured into B while the stopcock above A is open, at first fills A, but, as hydrogen commences at once to form, as soon as the stopcock is closed it will accumulate. If the stopcock be opened and a light applied to the jet, the issuing gas will take fire, and, on holding a piece of white porcelain in the flame, the arsenic will accumulate on it as a metallic ring; or the jet may be replaced by a bent tube of hard glass, which is to be kept heated at one point for a length of time. The arsenic will be detected as a black deposit, though it exist to an extremely minute extent.

Fig. 290.

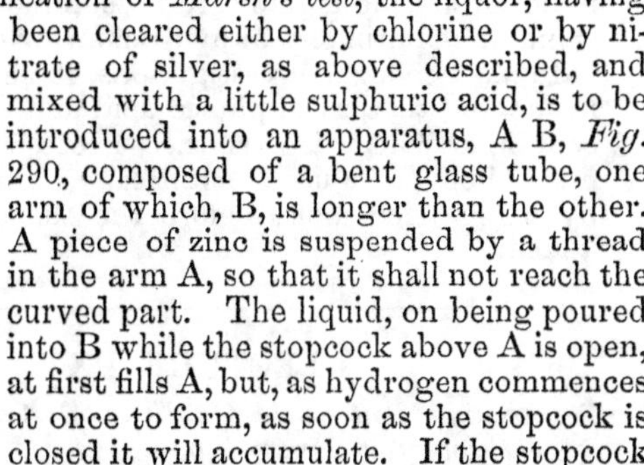

If the liquor, notwithstanding the care taken to clear it, froths when the hydrogen is disengaged, so as to interfere with the results by choking the tube, the gas is best collected under a jar at the pneumatic trough, and may be subsequently examined.

The fifth test, by copper, may be sometimes advantageously applied to collect the arsenic from solutions; the crust upon the copper may be subsequently examined, either by sublimation or otherwise.

It is to be remembered that antimony will yield results closely resembling those of arsenic by Marsh's test; but on heating the glass plate on which the stain

Describe Marsh's test. How may minute quantities be detected by this test? When the liquor froths, what has to be done? When is the copper test advantageously applied? What metal acts like arsenic? How is it distinguished?

has been deposited, if it be arsenic it will totally volatilize away; but if antimony, it will not disappear, but only give rise to a yellow oxide. The antimony stain is readily soluble in hydrosulphate of ammonia, the arsenic stain with difficulty. On evaporating, the sulphide of arsenic is found to be soluble in ammonia, the sulphide of antimony in hydrochloric acid.

In medico-legal investigations, it should also be remembered that, as sulphuric acid and zinc of commerce contain arsenic, it is absolutely necessary that the specimens about to be used be critically examined themselves by being tried alone before the suspected solution is added.

LECTURE LXV.

ARSENIC.—*Antiseptic Quality of Arsenious Acid.—Antidote for Poisoning.—Arsenic Acid.—Isomorphous with Phosphoric Acid.—Realgar and Orpiment.—Arseniureted Hydrogen.*—ANTIMONY.—*Reduction of.—Oxides, Chlorides, and Sulphides of.—Antimoniureted Hydrogen.—Detection of Antimony.*—TELLURIUM.—COPPER.—*Reduction of.—Use of Oxide.—Detection of.—Salts of Protoxide.*

ARSENIOUS ACID possesses a remarkable antiseptic quality, and hence often preserves the bodies of persons who have been poisoned by it. Advantage is also taken of this fact by the collectors of objects of natural history in preserving their specimens.

The antidote for poisoning by arsenic is the hydrated sesquioxide of iron. It may be made by adding carbonate of soda to the chloride of iron. It should be given in the moist state, mixed with water. After being once dried it loses much of its power. It produces an inert basic arsenite of the peroxide of iron.

Arsenic Acid is found in nature in union with various bases. It may be made by acting on arsenious acid with nitric acid, with the addition of a little hydrochloric acid, and evaporating till the nitric acid is expelled. The re-

---

Why must the sulphuric acid and zinc be examined? Of what use is the antiseptic quality of arsenic? What is the antidote for arsenic? How is it prepared? How is arsenic acid prepared?

sulting acid contains three atoms of water, and is isomorphous with tribasic phosphoric acid. The arseniates yield, with nitrate of silver, a dark red precipitate of the tribasic arseniate of silver. It should not be forgotten, in medico-legal inquiries respecting arsenic, that the arseniate of lime may naturally replace phosphate of lime in bone-earth, and this acid substitute the phosphoric in other parts of the system.

The bisulphide of arsenic may be obtained by melting arsenious acid with sulphur. It occurs as a mineral, *Realgar*, and is a red-colored substance.

The tersulphide is deposited when a stream of sulphureted hydrogen is passed through a solution of arsenious acid. It is a yellow body, and is used in dyeing; it is also known under the name of *Orpiment.*

Arseniureted Hydrogen is prepared by acting on an alloy of zinc and arsenic with dilute sulphuric acid. It is a colorless gas, burns with a blue flame, exhales an odor like garlic. Its specific gravity is 2.695. It is decomposed by chlorine and iodine, and the arsenic is separated by heat and by the rays of the sun.

ANTIMONY. *Sb*=129.

This metal occurs commonly as a sulphide in nature, from which it may be obtained by heating with iron filings, a sulphide of iron forming, and metallic antimony subsiding to the bottom of the crucible. It may also be obtained by fusing the sulphide with black flux, which produces a sulphide of potassium and metallic antimony.

Antimony is a blue-white metal, of a very crystalline structure, and so brittle that it may be pulverized. It melts at 840°. Its specific gravity is 6.7. It possesses, at high temperatures, an intense affinity for oxygen; a fragment of it the size of a pea being ignited on a piece of charcoal before the blowpipe, and then suddenly thrown on the table, takes fire, breaking into a multitude of globules, and filling the air with fumes of the white teroxide. Antimony yields the following compounds:

Where may arsenic exist in the body? What is Realgar? What is Orpiment? How may arseniureted hydrogen be made? How is metallic antimony made? What are its properties?

Teroxide of Antimony	SbO_3	=153
Antimonious Acid	SbO_4	=161
Antimonic Acid	SbO_5	=169
Terchloride of Antimony	$SbCl_3$	=235.5
Perchloride "	$SbCl_5$	=306.5
Tersulphide "	SbS_3	=177
Persulphide "	SbS_5	=209
Oxysulphide "	$2Sb, S_3+SbO_3$	=507

The *Teroxide of Antimony* may be made by adding to an acid boiling solution of chloride of antimony carbonate of soda. It is a gray powder, and is the base of a class of salts, among which tartar emetic may be mentioned. These salts give an orange-colored precipitate with sulphureted hydrogen.

Antimonious Acid is produced by heating the oxide of antimony, or antimonic acid. It is a white powder, and unites with bases, forming antimonites.

Antimonic Acid may be prepared by acting on metallic antimony with nitric acid.

Terchloride of Antimony is made by dissolving one part of sulphide of antimony in five of hydrochloric acid, and distilling. As soon as the matter which passes over becomes solid, the receiver is to be changed, and, continuing the heat, the terchloride is collected. It was formerly known as Butter of Antimony. The perchloride may be made by burning antimony in chlorine gas. The oxychloride is produced when the terchloride is placed in contact with water. It was formerly known as Powder of Algaroth.

The *tersulphide* occurs abundantly as a mineral, as has been said. It is also formed by the action of sulphureted hydrogen on the salts of the oxide of antimony. In this case it is of an orange color, in the former it has a metallic aspect. The *persulphide* is procured when the tersulphide and sulphur are boiled in a solution of potassa, the liquor filtered, and an acid added, a yellow precipitate going down. It was known formerly as the *Golden Sulphuret of Antimony*. The oxysulphide occurs native as the red ore of antimony, and may also be made by boiling the tersulphide with

What color does sesquioxide of antimony yield with sulphureted hydrogen? How is antimonious acid made? How is terchloride of antimony made? What is Powder of Algaroth? What is the golden sulphuret of antimony?

a solution of potassa. On cooling, precipitation of it takes place. It is stated, however, by Berzelius, that this is not a true compound, but merely a mechanical mixture of the oxide and sulphide in irregular proportions. This precipitate is also known under the name of *Kermes Mineral.* From the liquor, after the kermes is separated, an acid throws down the golden sulphuret of antimony.

Antimoniureted Hydrogen.—When hydrogen is evolved from a solution containing tartar emetic (tartrate of antimony and potassa), this substance is produced. It is a gas, having a superficial resemblance to arseniureted hydrogen, and, when used as in Marsh's apparatus, gives a stain on glass resembling that of arsenic. From arsenic it may be distinguished by not being volatile.

The following method of detecting antimony, when dissolved in any organic liquid, is based upon the principle by which copper and other metals may be detected under similar circumstances: Acidulate a portion of the suspected liquid with hydrochloric acid, and place it in a shallow platinum capsule. Touch the platinum through the acid liquid with a piece of pure zinc. Wherever the metals come in contact, antimony in the state of a black powder is deposited on the surface of the platinum. Hydrosulphate of ammonia dissolves the deposit by the aid of heat, giving the orange-red sulphide, which is soluble in hydrochloric acid.

In Marsh's test zinc causes the deposit of the antimony as a black powder, and after a short time barely a trace will escape with the hydrogen from the jet.

Antimony furnishes some valuable alloys; printers' type metal, for example, is an alloy of this substance with lead. It expands in the act of solidifying, and therefore takes accurate impressions of the interior of a mould.

TELLURIUM. *Te*=64.

TELLURIUM is a rare metal, of a white color, very fusible and volatile, having several analogies with selenium,

---

What is kermes mineral? How is antimoniureted hydrogen made? How may antimony be detected in organic liquids? What alloys of antimony are there? What are the properties of tellurium?

and uniting with hydrogen to form tellureted hydrogen, which, with water, yields a claret-colored solution.

Copper. $Cu=32$.

Copper is often found native, and in certain parts of the United States in masses of very great magnitude. It also occurs as a carbonate and sulphide. In the latter combination it is found with the sulphide of iron, as yellow copper ore. This being roasted, the sulphide of iron changes into oxide, the copper sulphide remaining unchanged. The mass is then heated with sand, which yields a silicate of iron, the sulphide of copper separating. This process is repeated until all the iron is parted; and now the sulphide of copper begins to change into the oxide, which is finally decomposed by carbon at a high temperature.

Copper is a red metal, requiring a high temperature for fusion, 1996°. Its specific gravity is 8.8. It has great tenacity, and is ductile and malleable. A polished plate of it, heated, exhibits rainbow colors, and is finally coated with the black oxide. It is one of the best conductors of heat and electricity. Among its compounds the following may be mentioned:

Protoxide of Copper	$CuO=40$
Suboxide "	$Cu_2O=72$
Chloride "	$CuCl=67.5$
Subchloride "	$Cu_2Cl=99.5$
Subsulphide "	$Cu_2S=80$

Protoxide of Copper may be made either by igniting metallic copper in contact with air, or by calcining the nitrate. It is a black substance, not decomposable by heat, but yielding oxygen with facility to carbon and hydrogen, and hence extensively used in organic analysis. It is a base, yielding salts of a blue or green color. The suboxide, called also red oxide, occurs native as ruby copper. It is a feeble base. The subsulphide also occurs native, as copper pyrites.

Copper is easily detected. Caustic potassa gives, with its protosalt, a pale blue hydrate, which turns black on boiling. Ammonia, in excess, yields a beau-

Under what forms does copper occur? How is it reduced? What are its properties? Which oxide is used in organic analysis? How is copper detected?

tiful purple solution; ferrocyanide of potassium, a chocolate-brown precipitate; sulphureted hydrogen, a black; and metallic iron, as the blade of a knife, precipitates metallic copper.

Salts of the Protoxide of Copper.

Carbonate of Copper.—There are several varieties of carbonates. One, which passes under the name of *Mineral Green*, is formed by precipitating with an alkaline carbonate. It occurs naturally in the form of *Malachite*. *Blue copper ore* is another carbonate; the paint called *Green Verditer* has a similar composition.

Sulphate of Copper (*Blue Vitriol*) is prepared for commerce by the oxidation of the sulphide of copper. It crystallizes in rhomboids of blue color, with five atoms of water. It is soluble in four times its weight of cold, and twice its weight of hot water. It is an escharotic, an astringent, and has an acid reaction. With ammonia it forms a compound of a splendid blue color, which may be obtained in crystals; with potassa, also, it forms a double salt. There are also subsulphates of copper.

Nitrate of Copper, formed by the action of nitric acid on metallic copper. It crystallizes in prisms or in plates. It acts with very great energy on metallic tin. There is a subnitrate of copper.

Arsenite of Copper (*Scheele's Green*), produced by adding solution of arsenious acid to the solution of ammonia sulphate of copper.

Copper yields several valuable alloys. Brass is an alloy of copper and zinc; gun metal, bell metal, and speculum metal, of copper and tin. The gold and silver of currency contain portions of this metal; it communicates to them the requisite degree of hardness.

Under what forms does the carbonate occur? What are the method of preparation and properties of the sulphate? What is Scheele's Green? What alloys of copper are there?

LECTURE LXVI.

LEAD.—*Reduction of Galena.—Relations of Lead to Water.—The Oxides of Lead.—Detection of Lead.*—BISMUTH.—SILVER.—*Amalgamation.—Crystallization.—Cupellation.—Properties of Silver.—Salts of Silver.*

LEAD. $Pb=104$.

LEAD occurs under various mineral forms, but the most valuable one is galena, a sulphide. From this it is readily obtained. The galena, by roasting in a reverberatory furnace, becomes partly converted into sulphate of lead; the contents of the furnace are then mixed, the temperature raised, and the sulphate and sulphide produce sulphurous acid and metallic lead, the action being

$$PbO, SO_3+PbS \ldots = \ldots 2SO_2+Pb_2.$$

Lead is a soft metal, of a bluish-white color. Its specific gravity is 11.381. It melts at 612°, and on the surface of the molten mass an oxide (dross) rapidly forms. At common temperatures it soon tarnishes. In the act of solidifying it contracts, and hence is not fit for castings. It possesses, at common temperatures, the welding property; two bullets will cohere if fresh-cut surfaces upon them are brought in contact. Under the conjoint influence of air and water lead is corroded, a white crust of carbonate forming; but when there are contained in the water small quantities of salts, such as sulphates, these form with the lead insoluble bodies, which, coating its surface over, protect it from farther destruction. For this reason lead pipe can be used for distributing water in cities without danger. Lead is one of the least tenacious of the metals. The tartrate of lead calcined in a tube yields one of the best pyrophori; on bringing it into the air at common temperatures, it spontaneously ignites.

Under what forms does lead chiefly occur? How is galena reduced? Why can not lead be used for castings? What is the action of pure water, and water containing salts, upon it?

Of the compounds of lead, the following are some of the more important:

Protoxide of Lead	PbO	$=112$
Sesquioxide "	Pb_2O_3	$=232$
Peroxide "	PbO_2	$=120$
Red oxide "	Pb_3O_4	$=344$
Chloride "	$PbCl$	$=139.5$
Iodide "	PbI	$=230$
Sulphide "	PbS	$=120$

The protoxide is made by heating lead in the air; it is a yellow body, which fuses at a bright red heat. In the first state it is called massicot; in the latter, litharge. It yields a class of salts, being a base. It is slightly soluble in water. The peroxide is made from red lead by digesting it with nitric acid, which dissolves out the protoxide, and leaves the substance as a puce-colored powder. The red oxide, or red lead, is made by calcining lead in a current of air at 600° or 700°. It is used in the manufacture of flint glass. The chloride is made by the action of hot hydrochloric acid on protoxide of lead: on cooling, it is deposited in crystals. The iodide is formed when any soluble iodide is added to a protosalt of lead. It is a beautiful yellow precipitate, soluble in boiling water, forming a colorless solution, which, on cooling, deposits golden crystals. The sulphide is galena; it crystallizes in cubes, and has a high metallic lustre.

Lead is easily detected by sulphureted hydrogen, which throws it down from its solutions as a deep brown or black precipitate, and by the iodide of potassium or chromate of potassa, which give with it a yellow precipitate. Sulphuric acid yields with its salts a white insoluble sulphate of lead.

SALTS OF THE PROTOXIDE OF LEAD.

Carbonate of Lead—White Lead—Ceruse.—This salt forms as a white precipitate when an alkaline carbonate is added to a solution of a salt of lead. Large quantities of it are consumed in the arts as white paint. For commerce it is procured by mixing litharge with water containing a small proportion of acetate of lead;

What are massicot and litharge? What is red lead? How is lead detected? How may white lead be made?

carbonic acid gas is then sent over it, and the carbonate rapidly forms. It is also made by exposing metallic lead in plates to the action of the vapor of vinegar, air, and moisture, the metal becoming oxidized and carbonated.

Nitrate of Lead may be formed by dissolving litharge in dilute nitric acid. It crystallizes in opaque white octahedrons, which dissolve in seven or eight times their weight of cold water. They contain no water of crystallization, and are decomposed at a red heat, as stated in the description of nitrous acid. By the action of ammonia three other nitrates of lead may be obtained.

Among the alloys of lead are the soft solders. Two parts of lead and one of tin constitute plumber's solder; one of lead and two of tin, fine solder.

BISMUTH. $Bi = 213$.

BISMUTH is found both native and as a sulphide. It is of a reddish color, melts at 507°, and may be obtained in beautiful cubic crystals by cooling a quantity of it until solidification commences, then breaking the surface crust and pouring out the fluid portion.

When bismuth is dissolved in nitric acid, and the solution poured into water, the white subnitrate, once used as a cosmetic, is deposited; when this is washed, and subsequently heated, the protoxide is left. There is also a peroxide.

Fusible metal is an alloy of eight parts of bismuth, five of lead, and three of tin. It melts below the boiling point of water, and may be obtained in crystals.

SILVER. $Ag = 108$.

SILVER is found native, and as a sulphide and a chloride, occurring also with a variety of other metals, and in small proportion with galena. When disseminated as a metal through ores, it may be collected from them by amalgamation with quicksilver; on distilling, the quicksilver is driven off.

When it is obtained from the sulphide, that ore is roasted with common salt, which changes it into a

What are the properties of the nitrate? What are solders? What are the properties of bismuth? What is fusible metal? Under what forms does silver occur? How is it obtained from the sulphide?

chloride. This, with the impurities with which it may be associated, is put into barrels, which revolve on an axis, along with water, pieces of iron, and metallic mercury; the iron reduces the chloride to the metallic state, and the silver amalgamates with the mercury. This is washed from the impurities, strained through a bag to separate the excess of mercury, and the residue is driven off by distillation.

The extraction of silver, when it occurs in small quantity with lead, is accomplished by the process of crystallization. It depends upon the fact that an alloy of lead and silver is more fusible than lead. A large quantity of argentiferous lead is melted and allowed to cool. As the setting goes on, the first portions which solidify are pure lead; they may be removed by iron colanders, and by continuing the process there is finally left a portion containing all the silver. This is exposed to a red heat, and a stream of air directed over it; oxidation of the lead takes place, and the litharge is removed by the blast, the process being finally completed by cupellation.

A cupel is a shallow dish made of bone ashes, and is very porous. In this, if an alloy of lead and silver be heated with access of air, the lead oxidizes, and, melting into a glass, soaks into the cupel, or may be driven from the surface by a blast of air directed from a bellows. At the same time, any copper or other base metal oxidizes and is removed along with the lead. The completion of the process is indicated by the silver assuming a certain brilliancy, or flashing, as the workmen term it.

Silver is a white metal, capable of receiving a brilliant polish. It is malleable and ductile, an excellent conductor of heat and electricity. Its specific gravity is 10.5. It melts at 1873°, and when melted absorbs a large quantity of oxygen, giving it out again as soon as it solidifies, and assuming a frosted or porous appearance. The presence of a minute quantity of copper prevents this effect. Silver is so soft that, for making plate or coins, it requires to be alloyed with a portion of cop-

---

What is the process of amalgamation? What is the process of crystallization? What is cupellation? What are the properties of silver? When must it be alloyed?

per; from this it may be purified by dissolving it in nitric acid, and precipitating the silver as chloride by a solution of common salt. Silver shows little disposition to unite with oxygen, though it tarnishes readily by the action of sulphureted hydrogen. It yields three oxides, but of its compounds the following are the most important:

Protoxide of Silver	....	AgO = 116
Chloride "	....	$AgCl$ = 143.5
Iodide "	....	AgI = 234
Sulphide "	....	AgS = 124

The protoxide may be made by the action of caustic potassa on a solution of nitrate of silver, or by boiling recently-prepared chloride in potassa. It is a dark powder, which may be reduced by heat alone. The chloride is sometimes found native, as horn-silver, and may be made by precipitation from the nitrate by hydrochloric acid or a soluble chloride. The sulphide is produced whenever sulphureted hydrogen acts on oxide of silver, or even metallic silver; it is a black compound.

Silver is easily detected by precipitation as a chloride: a curdy, white precipitate, insoluble in water, but soluble in ammonia. It turns dark on exposure to the sun, and is used in photography. The iodide and bromide of silver are very valuable in being sensitive to light. They are formed in that application by dipping a film of collodion containing some soluble iodide and bromide into a solution of nitrate of silver. They also exist native in Mexico.

Salts of the Protoxide of Silver.

Nitrate of Silver (*Lunar Caustic*), procured by dissolving silver in nitric acid diluted with twice its weight of water. It crystallizes in tables, which are not deliquescent, and contain no water of crystallization. It enters into fusion at 426°, but at higher temperatures undergoes decomposition. It is frequently cast into small sticks, and used by surgeons as a cautery. It is soluble in its own weight of cold and half its weight of hot wa-

How may the protoxide be made? What special properties have the chloride, iodide, and bromide? How may silver be detected? How is lunar caustic made? What properties has it?

ter, and, when in contact with organic matter, turns black in the rays of the sun.

Ammoniuret of Silver (*Berthollet's Fulminating Silver*) is formed by digesting precipitated oxide of silver in ammonia. It explodes with the utmost violence under the feeblest friction, with the evolution of nitrogen and the vapor of water.

Hyposulphite of Silver is formed when a compound of silver is dissolved in any of the hyposulphites. The solution of chloride of silver has a sweet taste, as may be observed when removing the excess of chloride from paper that has been used in the positive printing process.

Ammonia Nitrate of Silver, formed by adding ammonia to nitrate of silver till the precipitate at first formed is redissolved, is employed in photographic printing on paper.

Silvering a glass is performed by various processes, the best being that in which a solution of nitrate of silver and ammonia is decomposed by Rochelle salt. The film deposited on glass is so hard as to bear polishing with buckskin, though only $\frac{1}{200,000}$ of an inch thick. This process has been largely used by the author in making the 15½-inch silvered glass reflectors for his telescope at Hastings-on-Hudson, New York. It is fully described in the Contributions to Science of the Smithsonian Institution for 1864.

Brass may be silvered by the aid of chloride of silver, chalk, and carbonate of potassa. In electro-plating a solution of the chloride in cyanide of potassium is used.

What is ammoniuret of silver? What are the properties of the hyposulphite? Of what use is the ammonia nitrate? How is silvering on glass performed? How may brass be silvered?

LECTURE LXVII.

MERCURY.—*Process of Reduction.—The Liquid State of.—Its Oxides.—Calomel and Corrosive Sublimate.—Detection of Mercury.—Its Salts.—Amalgams.*—GOLD.—*Chloride of.—Purple of Cassius.*—PALLADIUM.—PLATINUM.—*Its Catalytic Effects.—Platinum Black.*—IRIDIUM.—RHODIUM.—RUTHENIUM.—OSMIUM.

MERCURY. $Hg = 100$.

MERCURY may be obtained from the sulphide (cinnabar) by distillation with iron filings. It is also, to a certain extent, found native.

The striking characteristic of mercury is its liquid condition. Its melting point is the lowest of that of any of the metals, being −39°. Its specific gravity at 47° is 13.545. It boils at 662°. Kept at that temperature in the air for a length of time, it produces red oxide, but at common temperatures it is not acted on by the air. It may be freed from impurities for the purposes of the laboratory by being kept in contact with dilute nitric acid. It gives the following compounds of interest:

Suboxide of Mercury	Hg_2O	=208
Protoxide "	HgO	=108
Subchloride "	Hg_2Cl	=235.5
Chloride "	$HgCl$	=135.5
Subsulphide "	Hg_2S	=216
Sulphide "	HgS	=116

The suboxide may be made by triturating calomel with potassa water in a mortar. It is a black powder, which is decomposed by light or any of the reducing agents. The protoxide may be formed, as stated above, by the action of air on hot mercury, but more easily by dissolving mercury in nitric acid, and evaporating and heating the salt until no more fumes of nitrous acid are evolved. It is a red powder, and when warmed be-

Under what forms does mercury occur? What are the most striking properties of this metal? How may it be purified? What are the properties of the protoxide?

comes almost black, the color returning as the temperature descends. Like the former, it is a base, and yields a class of salts.

The Subchloride, or *Calomel*, may be made by adding hydrochloric acid to the subnitrate of mercury, or by subliming a mixture of chloride of mercury and mercury. It is a white powder, insoluble in water, and darkens slowly by exposure to sunshine. The *chloride* (or *Corrosive Sublimate*) is formed when mercury burns in chlorine gas, but more economically by sublimation from a mixture of sulphate of mercury and common salt. It is a heavy, white, crystalline body, soluble in water, has a metallic taste, and is poisonous. The antidote for it is albumen (the white of egg).

Of the sulphides of mercury, the subsulphide is black and the sulphide commonly red; in this case it passes in commerce under the name of vermilion, and is used as a paint. It can be obtained, however, quite black; a similar double color is observed in the case of the oxide, and still more strikingly in the iodide, which may be sublimed in beautiful yellow crystals, becoming of a splendid scarlet color by merely being touched.

Mercury may be detected by being precipitated from its soluble combinations by metallic copper as a metal. Its salts, either alone or with carbonate of soda, heated in a tube, yield metallic mercury, which volatilizes.

Salts of the Oxides of Mercury.

Nitrates of the Oxides of Mercury.—When cold dilute nitric acid acts on mercury it gives rise to neutral or basic subsalts, as the acid or mercury is in excess; if the acid be hot, a pernitrate forms; these salts are decomposed by an excess of water, giving rise to basic compounds.

Sulphate of Mercury is formed by boiling sulphuric acid and mercury, and evaporating to dryness. It occurs in the form of a white granular mass, and is decomposed by water, giving a yellow precipitate, a subsulphate called *Turpeth Mineral.*

How may calomel be made? When is corrosive sublimate formed? What is its antidote? What is vermilion? What is the peculiarity of the iodide? How is mercury detected? What is Turpeth Mineral?

The alloys of mercury are called *amalgams;* the amalgam of tin is used for silvering looking-glasses, and that of zinc for exciting electrical machines.

Gold. *Au*=197.

Gold is found native, and may be obtained by washing or by amalgamation with mercury. It may be purified from silver by quartation; that is, fusing it with three times its weight of silver, and then acting on the mass with nitric acid. The gold is left as a dark powder.

From other metals gold is distinguished by its yellow color. Its specific gravity is 19.3. It melts at 2016°. It is the most malleable of all the metals, as is proved by gold leaf, which may be obtained $\frac{1}{200,000}$ inch in thickness; is not acted upon by the air or oxygen. Objects of art covered with it have retained their brilliancy for thousands of years. No acid alone dissolves it, but it is soluble in aqua regia, and also in chlorine.

It can, however, be made to yield two oxides, a protoxide and a teroxide; and two chlorides having the same constitution; the terchloride is formed by the action of nitromuriatic acid (aqua regia) on gold. When evaporated, it yields red, deliquescent crystals. Deoxidizing agents, such as protosulphate of iron, reduce it to the metallic state; this is probably due to their decomposing water and presenting hydrogen to the chloride. Hydrogen gas decomposes the terchloride, and, by heating, it first changes into the protochloride and then into metallic gold. With a solution of tin it forms the *Purple of Cassius.* This and the action of protosulphate of iron serve as a test for it.

Palladium. *Pd*=54.

Palladium is found associated with platinum, and is best obtained from the cyanide of palladium by ignition. It is a white metal, requiring a high temperature for fusion; specific gravity 11.5. It does not tarnish in the air, is dissolved by nitric acid and aqua regia, is one of

---

What are amalgams? How does gold occur? What is quartation? What are the properties of gold? How many oxides has it? What are the tests for it? What is the Purple of Cassius? With what metal is palladium found? What are its properties?

the welding metals, and, when heated, acquires a purple oxidation like watch-spring. It is used to some extent by dentists. Its compounds are not of importance, with the exception of the protochloride, which I have shown to be useful in increasing the opacity of under-exposed negatives. It increases their intensity 16 times, without any liability to staining or injury.

Platinum. *Pt*=99.

Platinum is found native, but always associated with other metals. It may be obtained by first forming a chloride of platinum and ammonium; this, when ignited, leaves pure spongy platinum, which being exposed to powerful pressure, and then alternately made white-hot and hammered, becomes a solid mass.

Platinum is a white metal. Its specific gravity is very high, being 21.15. Malleable platinum is manufactured by Deville's process as follows, 230 pounds having been fused in one mass. The platinum ore is fused with its weight of sulphide of lead and half its weight of metallic lead. Some of the impurities are thus separated in combination with sulphur, while the platinum forms an alloy with the lead, which is freed from the scoriæ, and subjected to the joint action of heat and air until all but about five per cent. of the lead is oxidized. It is then subjected to the intense heat of an oxyhydrogen flame in a furnace of chalk-lime, where the rest of the lead, together with any gold, copper, and osmium, is driven off in fumes. Rhodium and iridium are left in combination.

Platinum is a welding metal, and on this fact the first-mentioned method of preparation depends. It is very malleable and ductile, is not acted upon by oxygen, air, or any acid alone, but dissolves in aqua regia. It possesses the extraordinary property of causing hydrogen and oxygen to unite at common temperatures, an effect which takes place with remarkable energy when the metal is in a spongy state. A jet of hydrogen falling upon spongy platinum in the air makes it red-hot, and

---

Of what value is the protochloride? How is platinum obtained from its ores? What is its specific gravity? What is Deville's process? What fluid dissolves platinum? What relation does it bear to hydrogen?

presently after the gas takes fire. It also brings about the rapid transformation of alcohol into acetic acid, and various other chemical changes.

Fig. 291.

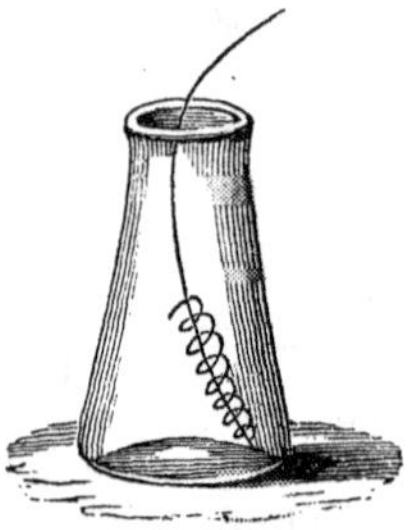

If a quantity of ether be poured into a glass jar, *Fig.* 291, and a coil of platinum wire, recently ignited, be introduced, the metal continues to glow so long as any ether is present.

Platinum is invaluable to the chemist. It furnishes a variety of implements of great value, and is met with under the forms of crucibles, tubes, wire, foil, etc.

Platinum Black is prepared by slowly heating to 212° a solution of chloride of platinum, to which an excess of carbonate of soda and some sugar have been added. It is a dark powder, and possesses the property of determining a variety of chemical changes with much more energy than platinum in mass.

Platinum can be caused to yield two oxides, which are not of any importance; and two analogous chlorides, of which the bichloride, which is the common platinum salt, is made by dissolving the metal in nitrohydrochloric acid, and evaporating to a sirup. It is soluble in water and alcohol, and is used for detecting the salts of potassium, rubidium, and cæsium, which give compounds insoluble in alcohol and almost so in water.

IRIDIUM. $Ir=99$.

Iridium is associated with platinum. It is said to have been found of specific gravity 26. Dr. Hare obtained it 21.8; it is therefore the heaviest of the metals. Its name is derived from the different colors (iris) of its compounds.

RHODIUM. $R=52$.

Like the former metal, rhodium is associated with the platinum ores. It is a hard white metal; its specific gravity is 11, and is sometimes used to form tips to metallic pens.

Describe the flameless lamp. What is platinum black? Of what use is bichloride of platinum? What are the properties of iridium? of rhodium?

Ruthenium. $Ru=52$.

This is one of the metals remaining in that portion of the ore of platinum which resists the action of aqua regia. It is hard, brittle, infusible in the oxyhydrogen flame, but readily oxidized by fusion with nitre, furnishing four oxides.

Osmium. $Os=100$.

Obtained by precipitation, it is in the form of a black powder, which acquires a metallic lustre by friction. The specific gravity slightly exceeds that of platinum. Burned in the air, it oxidizes, exhaling poisonous fumes; hence its name (*osme*, odor). It forms five oxides.

There are some peculiarities belonging to the six preceding metals—namely, platinum and its associates—which deserve notice in reference to their atomic weights and specific gravities, and which have led to their division into two groups of three each:

	Sp. Gr.	At. W't.		Sp. Gr.	At. W't.
Platinum	21.15	99	Palladium	11.8	54
Iridium	21.80	99	Rhodium	12.0	52
Osmium	21.40	100	Ruthenium	11.3	52

It will be observed that the specific gravities and atomic weights of the first group are almost identical; so also are those of the second group; and in the latter case they are almost exactly one half of those in the former.

What are the properties of ruthenium? of osmium? What peculiarities belong to the six preceding metals?

PART IV.

ORGANIC CHEMISTRY.

LECTURE LXVIII.

Composition of Organic Bodies.—Their Proneness to Decomposition.—Formulas of Organic Substances.—The Compound Radical Theory.—Theory of Types and Law of Equivalent Substitution.—Examples of Substitutions.—Homologous Compounds.—Action of Heat.—Eremacausis.—Putrefaction.—Difficulties in the Nomenclature of Organic Bodies.

ORGANIC CHEMISTRY, which treats of the substances derived from the processes of life and the compounds that arise from them, has been defined as the chemistry of the compounds of carbon, because all such bodies, with a few exceptions, as ammonia, contain that element, and are more or less combustible. Three other elements, hydrogen, nitrogen, and oxygen, also enter largely into their constitution, while potassa, soda, lime, magnesia, iron, arsenic, chlorine, fluorine, sulphur, phosphorus, silica, etc., are found to a limited extent, or may be made artificially to become components of them. The variety in the nature of substances which distinguishes inorganic chemistry is here replaced by a variety resulting from the varied groupings of a few elements.

The atomic constitution of organic substances is much more complex than that of inorganic; fibrin, for example, is regarded as having the formula

$$C_{216}H_{169}O_{68}N_{27}S_{2},$$

four hundred and eighty-two atoms entering into the composition of one of its atoms; while soda and potassa consist of only two atoms, carbonic acid of three,

---

What is the definition of organic chemistry? What elements enter into organic bodies? What is the constitution of fibrin?

sulphuric acid of four, etc. As a consequence of this complexity, organic substances exhibit a great proneness to decomposition, very slight causes sufficing to break down their constitution by destroying the balance of affinities, and give origin to simpler and more stable compounds. In the bodies of animals the products of oxidation and metamorphosis of nitrogenized substances have continually to be removed by the process of secretion for a healthy condition to be maintained. As soon as death ensues a general decomposition sets in, and in a short time the tissues are resolved into a few simple substances—water, ammonia, carbonic acid, etc. At the same time, the force which was locked up in them is liberated generally in the form of heat. Before these products can again become organized, a fresh supply of force must be furnished from the sun; the carbon of carbonic acid can not again enter as a constituent into gum, or albumen, or fibrin, unless under the influence of the yellow ray of light. In this respect the vegetable and animal kingdoms stand in a position of antagonism or counterpoise to one another, plants organizing food for animals, and they, in turn, oxidizing organized products so as to be suitable for the sustenance of plants.

Many organic substances, which yield, on analysis, precisely the same percentage amounts of their ingredients, have yet properties altogether distinct. In order to account for such distinctions, chemists have been forced to the conclusion that the nature of a body does not depend alone on its constituent elements nor their relative amounts, but on the varied manner in which a large number of atoms may arrange themselves. This is termed *Grouping*. The different allotropic states that the combining substances may on different occasions present also influence the result (Lecture XXXVII.).

When the formula of an organic substance merely expresses the number of atoms of each element present in it, it is called *empirical;* if, on the contrary, the formu-

---

What occurs to organized products in the bodies of animals? What takes place after death? Whence does the force stored up in vegetable products come? Are substances having the same percentage amount of ingredients identical? What is meant by grouping? What is the difference between an empirical and a rational formula?

la professes to show the actual arrangement of the atoms, it is called *rational*. Thus alcohol has the empirical formula $C_4H_6O_2$; its rational formula is $(C_4H_5)O+HO$, if it be regarded as the hydrated oxide of ethyle on the compound radical theory.

The Compound Radical Theory.—The compound radical theory assumes that in organic chemistry certain groups of atoms play the part that elements do in inorganic chemistry. A compound radical is therefore a body which combines like an element with elementary bodies or with other compound radicals. Some of these radicals have been isolated, as, for example, cyanogen (C_2N), but most of them have only a hypothetical existence. This view of the constitution of organic compounds has many advantages, and greatly facilitates classification. The following table indicates the principal compound radicals:

Table of Compound Radicals.

Amide.	Salicyle.	Propionyle.
Carbonic Oxide.	Cinnamyle.	Butyle.
Cyanogen.	Guiacyle.	Butyryle.
Ferrocyanogen.	Ethyle.	Valeryle.
Ferridcyanogen.	Acetyle.	Caprotyle.
Cobaltocyanogen.	Kakodyle.	Œnanthyle.
Chromocyanogen.	Methyle.	Octyle.
Platinocyanogen.	Formyle.	Caproyle.
Iridiocyanogen.	Cetyle.	Nonyle.
Sulphocyanogen.	Amyle.	Pelargyle.
Mèllone.	Glyceryle.	Rutyle.
Uryle.	Propyle.	Palmityle.
Benzoyle.		

Some of these discharge the duties of electro-negative, some of electro-positive, and some of indifferent bodies.

The Theory of Types.—In the theory of types and law of equivalent substitution substances are divided into classes, and from each class one member is selected as the type. From its formula those of all the rest are derived by substituting for one or more of its atoms, atoms of other elementary substances, or else groups of atoms.

What is the compound radical theory? Are these radicals hypothetical? Name the principal compound radicals. What is the theory of types?

As an example of such a type, we may take water arising from the union of one atom of hydrogen with one of oxygen. Its atom of oxygen may be replaced by one of chlorine or one of iodine, or by a compound radical such as cyanogen, and thus successive classes of the original type arise. Types submitted to substitution yield, therefore, different classes or forms.

These substitutions are not necessarily restricted to one of the constituents; thus, in the chloride of sodium, $Na\,Cl$, we may replace the metal by potassium, calcium, etc.; or we may replace the chlorine by iodine, bromine, or cyanogen, the type in both instances being still preserved.

Instead of the type of water, we may examine the type of ammonia, which is of very frequent occurrence in organic chemistry. It is to be remarked that in the substitution of one element for another, it is not necessary that they should be of the same electro-chemical character; thus electro-positive hydrogen may be substituted by electro-negative chlorine or oxygen.

The following table exhibits instances of substitution in the ammonia type, the compounds being ammonias, that is, bases bearing an analogy to ammonia; the salts are strictly analogous.

$$\text{Type, } \left.\begin{matrix} H \\ H \\ H \end{matrix}\right\} N = \text{Ammonia.}$$

Ethylamine.	Diethylamine.	Triethylamine.
$\left.\begin{matrix} C_4H_5 \\ H \\ H \end{matrix}\right\} N$	$\left.\begin{matrix} C_4H_5 \\ C_4H_5 \\ H \end{matrix}\right\} N$	$\left.\begin{matrix} C_4H_5 \\ C_4H_5 \\ C_4H_5 \end{matrix}\right\} N$

Substitutions are distinguished as partial and complete. Of the former, the substitution of hydrogen by chlorine in Dutch liquid is an example:

$$\text{Dutch liquid, } C_4H_4\,Cl_2 = (C_4H_4)\,Cl_2$$

$$\text{1st substitution, } C_4H_3\,Cl_3 = \left(C_4 \left\{\begin{matrix} H_3 \\ Cl \end{matrix}\right.\right) Cl_2$$

Give an illustration from the type of water; from the chloride of sodium. What must be the electro-chemical character of the substituted element? Give an example of substitution in the ammonia type. Give an explanation of the substitutions in Dutch liquid.

2d substitution, $C_4H_2Cl_4 = \left(C_4 \left\{ \begin{matrix} H_2 \\ Cl_2 \end{matrix} \right.\right) Cl_2$

3d " $C_4HCl_5 = \left(C_4 \left\{ \begin{matrix} H \\ Cl_3 \end{matrix} \right.\right) Cl_2$

4th " $C_4Cl_6 = (C_4Cl_4)Cl_2$

In complete substitution, the displacement, as the term indicates, is immediate and total.

The doctrine of compound radicals and that of substitution have been regarded as being inconsistent with each other. There can be no doubt that the latter facilitates the study of organic chemistry very much; but there can also be no doubt of the actual existence of many compound radicals, since they have been isolated or obtained in a separate state.

The Law of Homologous Series.—A series of compounds is homologous when each member differs from the others by a definite number of equivalents of carbon and hydrogen, or by some multiple of it; and when the properties of these compounds, though they may be similar, differ in degree from each other. The boiling points, as well as the specific gravities, may rise gradually, or the compounds may gradually turn from the liquid into the solid condition. The following example is an illustration of homology:

Formic Acid	$C_2H_2O_4$
Acetic "	$C_4H_4O_4$
Propionic "	$C_6H_6O_4$
Butyric "	$C_8H_8O_4$
Valerianic "	$C_{10}H_{10}O_4$
Palmitic "	$C_{32}H_{32}O_4$
Stearic "	$C_{36}H_{36}O_4$

There is an anology between homologous groups of organic compounds and certain groups of elementary bodies. For instance, chlorine, bromine, and iodine differ from one another precisely as any three continuous homologous compounds might do. Chlorine is an easily-condensible gas, bromine a volatile liquid, iodine a volatile solid. In affinity, bromine is intermediate between the other two, as it is likewise in atomic weight.

Are the compound radical theory and that of types inconsistent? When is a series homologous? Give an example of homology. What analogy is there between homologous groups and certain elements?

The conclusion has been drawn, therefore, that bromine is made up of half an atom of each of the other two, and is therefore a compound body. It is curious that such considerations connected with the obscurer facts of chemistry bring us back to the ancient doctrine of transmutation spoken of in Lecture I. If an atom of chlorine and one of iodine may be considered as capable of originating two atoms of bromine, the discovery of a metal homologous with gold, as sodium is with potassium and lithium, would lead to the expectation that that metal might be transmuted into gold.

From this point it would appear that such substances as chlorine, bromine, and iodine are not to be regarded as elementary bodies, but as homologous compounds, having a common difference between them, just as is the case between formic, acetic, and propionic acids in the last table.

Combination of organic compounds it was supposed can only be produced by the agency of the vital principle, as manifested in animals or plants; but instances are now accumulating which demonstrate that that opinion can no longer be maintained. Thus urea may be made artificially by warming a solution of cyanate of ammonia; formic acid may be prepared from carbonic oxide; and from the formiates so resulting, marsh gas, olefiant gas, and propylene may be obtained; propylene may be converted into glycerine, the proximate principle of fats, and from glycerine a variety of sugar may be produced.

Organic compounds, by reason of their complex constitution, are, as has been said, prone to break up into subordinate groups, and eventually into binary bodies, carbonic acid, water, and ammonia. A slight elevation of temperature is often sufficient to establish these changes both in the absence and in the presence of air. Thus the decay of wood and the turning sour of wine occur. Where the temperature is higher with the copious access of air, the change promptly goes on to its last result, the carbon finally passing into the condition of car-

---

What bearing have these facts on transmutation? What view may be taken of iodine, bromine, and chlorine? How may organic compounds be artificially produced? Give examples. What effect has temperature on organic compounds?

bonic acid, the hydrogen into water, and the nitrogen escaping as free gas. To the slower change the title of *eremacausis* has been given, to the more active *combustion*. No organic substance can withstand a red heat, even in the absence of the air, without being totally destroyed.

When an organic substance is undergoing slow changes, and is brought in contact with another capable of being similarly affected, this last may become involved in the decomposition. Thus, when yeast, a changing nitrogenized body, is diffused through a solution of sugar, the sugar atom is divided into carbonic acid and alcohol, *fermentation*, as it is termed, taking place. In like manner, putrescent animal material will rapidly bring on the putrefaction of fresh animal substance.

Both acids and bases are prone to produce change in organic compounds. The preparation of carbonic oxide by the action of sulphuric acid on oxalic, described in Lecture LIV., is an example of the former, and that of baryta on the acetate of potassa in the production of marsh gas an instance of the latter.

From what has been said respecting the complex constitution of organic bodies, it will be inferred that their classification and nomenclature are attended with very great difficulties. An example of the attempts to indicate the constitution of these substances, not only so far as their grouping is concerned, but also in translating their formulæ into language, will satisfy the reader of the difficulty, if not impossibility, of rendering such attempts available for use.

Thus, $C_{30}H_{13}N_5$ is called dicyanomelaniline; and $C_{28}H_{24}N, O, HO$, that is, $NMeAeAylPh\,O, HO$ is called methylethylamylophenylium.

The arrangement of organic bodies that I shall follow is therefore employed rather from its usefulness than from its scientific propriety.

What are eremacausis and combustion? What is fermentation? How may putrefaction be produced? Give instances of the action of acids and bases on organic compounds. What difficulties are there with the nomenclature of organic bodies? Give examples.

LECTURE LXIX.

ANALYSIS OF ORGANIC SUBSTANCES.—*Proximate and Ultimate Analysis.—Qualitative and Quantitative Analysis.—Processes of Quantitative Analysis.—Description of Instruments.—Dialysis.—Crystalloids and Colloids.*

ORGANIC analysis may be either proximate or ultimate. Blood, for example, analyzed proximately, consists of water, fibrin, albumen, disks, etc., while its ultimate ingredients are carbon, hydrogen, nitrogen, etc.

Ultimate organic analysis may be qualitative or quantitative. In the former, where the nature of the ingredients alone is required, a few simple processes only are necessary. The presence of carbon is ascertained by the charring or blackening produced by heat or sulphuric acid; that of nitrogen by the smell resembling burning hair when raised to a high temperature. Less quantities of nitrogen are detected by the formation of ammonia when the substance is boiled in a solution of caustic potassa. Compounds containing sulphur are oxidized by carbonate of soda and nitrate of potassa at a melting heat, and the sulphuric acid produced precipitated as sulphate of baryta. The same treatment is used for phosphorus, the phosphoric acid being tested for with perchloride of iron and acetate of soda, or with molybdate of ammonia. Inorganic substances are first procured as ash by ignition on platinum, and then tested as usual in inorganic chemistry.

Quantitative Organic Analysis is theoretically simple, but, on account of the many precautions necessary to avoid loss, and the accuracy required to detect the minute differences in composition, is practically difficult.

In the determination of a compound which contains carbon, hydrogen, and oxygen, or only the first two, the object is to oxidize them completely, and, weighing the

---

What is meant by a proximate and what by an ultimate analysis? How is the presence of carbon, nitrogen, sulphur, phosphorus detected? Why is quantitative analysis difficult? How is the analysis of a compound containing carbon and hydrogen conducted?

carbon as carbonic aicd and the hydrogen as water, to estimate the oxygen by the loss, if there be any. This oxidation is accomplished by mixing the finely-powdered body with oxide of copper or chromate of lead, substances readily yielding up their oxygen, and subjecting the whole to heat in a tube closed at one end and communicating by the other with appropriate reagents. The steps of the process will be most easily understood by an example. The analysis of sugar is conducted as follows:

A crystallized variety of sugar being selected and finely powdered, is dried at 212° by the aid of a water-bath, which consists of a cubical chamber surrounded on five sides by boiling water, and with a current of air passing continually through it.

The combustion-tube in which the oxidation is effected is made of hard glass, shaped as in *Fig.* 292, the

Fig. 292.

pointed end or beak being closed. It is a foot or eighteen inches long, and less than half an inch in diameter. A sufficient supply of oxide of copper to fill it is raised to a red heat in a crucible to expel moisture and then allowed to cool. The tube from *c* to the beak is filled with oxide, from *b* to *c* with the oxide ground in a warm mortar with a weighed quantity (about five grains) of the sugar, and from *a* to *b* with oxide. On shaking the tube while in a horizontal position, the contents settle sufficiently to leave a free passage for the evolved gases from one end to the other.

The contents of the tube are freed from any moisture that may have accumulated in them by the apparatus *Fig.* 293. *D* is a wooden trough, *C* the combustion-tube, *B* a tube containing chloride of calcium. The trough is filled with hot sand, as seen below, and the air exhausted from *b* and *C* by the syringe. A fresh supply of air is admitted by the stopcock *a*. The air

In the examination of sugar what is the first step? Describe the combustion-tube. How is it filled? What is the use of the apparatus *Fig.* 293?

Fig. 293.

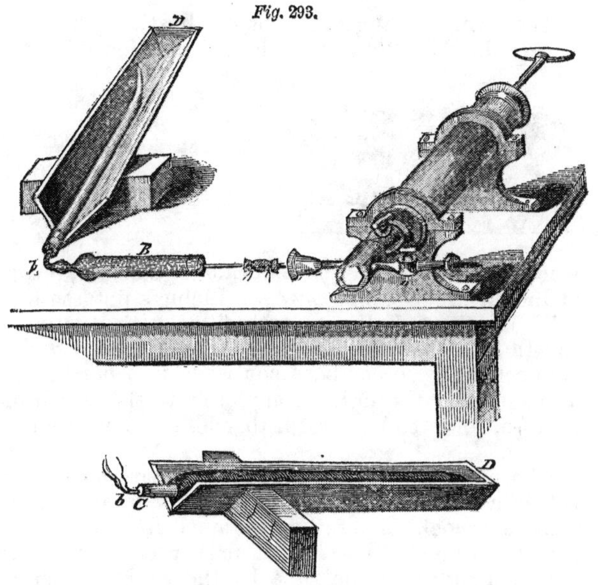

is dried by the chloride of calcium before reaching *C*. This, in its turn, is pumped out, and the process is repeated ten or twelve times.

The next operation is the combustion. This is conducted by the aid of charcoal in a sheet-iron case called a Liebig's combustion-furnace, *Fig.* 294. The bottom

Fig. 294.

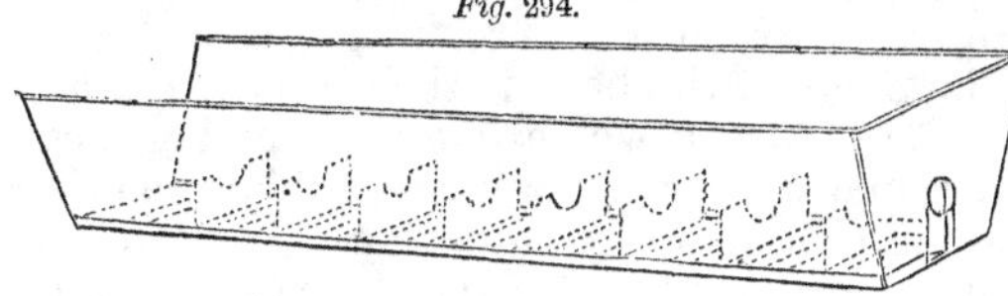

of the case is perforated to admit a draught of air, and from it rise a number of supports of iron to sustain the combustion-tube. As the whole length of the tube is not to be heated at once, a movable screen of sheet-iron has also to be provided.

How is the combustion-furnace constructed?

The entire arrangement for combustion put together is seen in *Fig.* 295: *a* is the combustion-tube, *b B* a

Fig. 295.

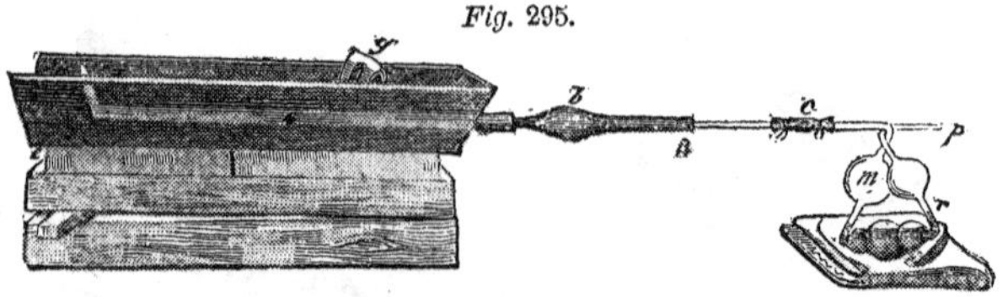

weighed tube containing chloride of calcium, *c* a piece of India-rubber tubing, *m r p* a Liebig's potassa bulbs.

The Liebig's bulbs is partly filled with a solution of caustic potassa of a specific gravity 1.25 and weighed. Its peculiar form has been contrived in order to subject a gas passing through it completely to the action of the potassa, so as to be certain that all carbonic acid is absorbed.

The part of the combustion-tube nearest to the chloride-of-calcium tube is first heated by fragments of ignited charcoal. These are obtained from a small subsidiary furnace. The heat is prevented from affecting the remainder of the tube by the movable screen *g*, which has gradually to be shifted an inch at a time down the length of the combustion-tube, until the whole has been subjected to a red heat. The bubbles of gas should not be made to pass through the caustic potassa faster than two in a second. At the end of the operation the charcoal should be fanned to raise the temperature to such a point that all the sugar may be with certainty burned.

The point of the beak is then broken off, and a small quantity of air drawn through by affixing a cork to the tube *p* and applying the mouth. This removes the remaining products of combustion. It is only necessary, in order to finish the analysis, to weigh again the chloride-of-calcium tube and Liebig's bulbs, and to calculate the results. The increase of weight in the chloride of calcium indicates how much water it has gained, and from that the amount of hydrogen is easily found; the

Describe *Fig.* 295. What is the Liebig's bulbs for, and why is it so shaped? Describe the operation of combustion. What are the last steps of the analysis?

increase in the potassa represents carbonic acid, produced from the oxidation of the carbon.

Where the substance analyzed is not as combustible as sugar, it is necessary to add chlorate of potassa to the oxide of copper, or else to use chromate of lead. The latter salt, not being hygrometric, may also be used for substances that would be decomposed by the warm oxide of copper when ground in a mortar. After it has been heated, it may be allowed to become perfectly cool before being put into the combustion-tube.

Volatile fluids are weighed out in bulbs shaped as in *Fig.* 296, the bulbs being filled by the acid of a spirit-lamp and the neck then sealed. After weighing, but before being put with the oxide of copper in the combustion-tube, the neck is broken off.

Fig. 296.

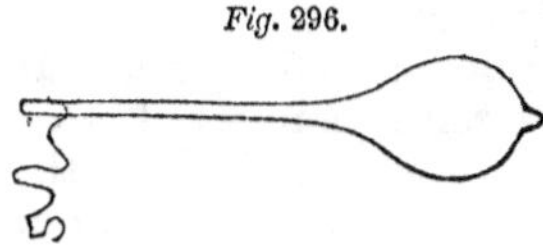

The nitrogen in organic compounds may be determined either as free gas or as ammonia. In the former case, the apparatus *Fig.* 297 is used. The combustion-tube is two feet long, closed at one end like a test-tube. At the closed end, n, dry bicarbonate of soda is placed for six inches, then an inch and a half of oxide of copper. These are followed by a weighed quantity of the dried organic substance that has been ground with warm oxide of copper, a little pure oxide, and, finally, a layer of copper turnings five inches long.

Fig. 297.

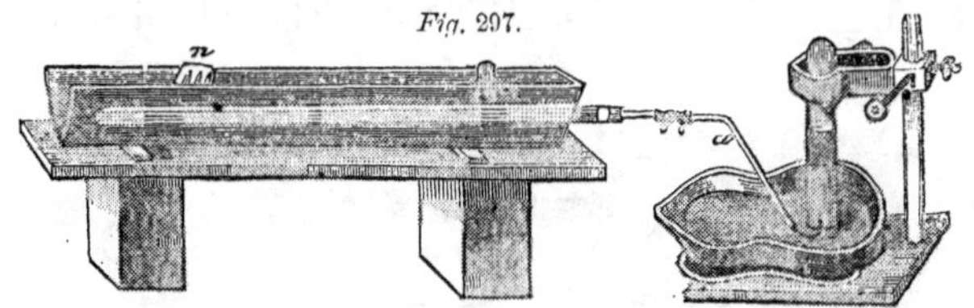

The copper turnings are for the purpose of setting free any nitrogen that may have united with oxygen to

When the substance is not combustible, what must be done? How are volatile fluids treated? How may nitrogen be determined? Describe the determination as free gas. What are the copper turnings for?

form nitrous acid. The heating is conducted as in the former case, except that a part of the bicarbonate of soda must first be decomposed to drift out the air in the apparatus, and that the copper turnings must be kept at a red heat while the actual combustion is in progress. Instead of the analysis being terminated by drawing air through the apparatus, the remaining part of the bicarbonate is decomposed, so that the residue of nitrogen may be expelled by carbonic acid.

The nitrogen is collected by the aid of the tube *a*, *Fig.* 297, in a graduated cylinder over mercury, the cylinder containing enough solution of caustic potassa to absorb the carbonic acid that comes over.

In determining nitrogen as ammonia by the method of Varrentrapp and Will, the nitrogen is converted into ammonia by igniting the compound containing it with soda-lime, a mixture of caustic soda, 1, and caustic lime 2 parts. The soda-lime furnishes hydrogen from its water to the nitrogen, forming ammonia; the oxygen of the water unites with the carbon of the organic body. The ammonia is passed through a Varrentrapp and Will's bulbs, partly filled with hydrochloric acid of a specific gravity of 1.13, *a*, *Fig.* 298. The tube *d* serves

Fig. 298.

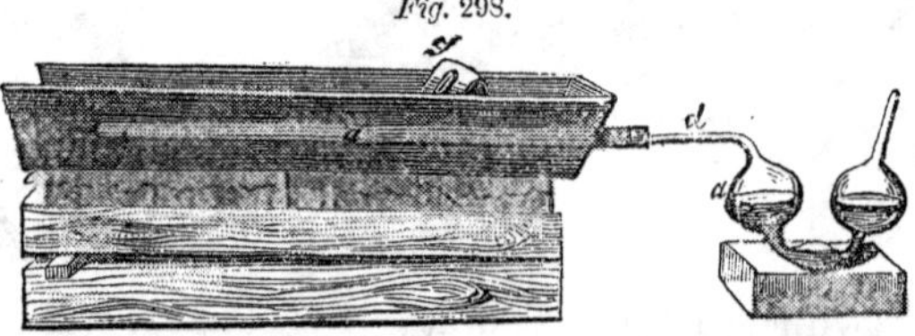

to connect it with the combustion-tube *a* in the furnace *g*. By treating the contents of the bulb apparatus with bichloride of platinum the double chloride of platinum and ammonia is formed, and may be, after evaporation to dryness and washing with ether and alcohol, collected on a filter and weighed.

Sulphur compounds are oxidized as stated in the beginning of this lecture, and weighed as a sulphate of

How is the nitrogen collected? How may the nitrogen be converted into ammonia? Describe the determination of nitrogen as free ammonia. How are sulphur and chlorine determined?

baryta. Chlorine is determined as chloride of silver, the substance being ignited with soda-lime, the mass dissolved in dilute nitric acid, and precipitated with nitrate of silver.

A process of analysis has been invented by Graham called *Dialysis*, in which the principles of liquid diffusion are brought into use. He divides all substances into two classes, *crystalloids* and *colloids;* the former being capable of crystallization, and possessing a tendency to diffusion through porous septa; the latter being of a viscid, glutinous nature, like solutions of gum. The following table shows the time of diffusion, hydrochloric acid, the most diffusible of known bodies, being taken as unity.

Table of Times of Equal Diffusion.

Hydrochloric Acid	1	Sulphate of Magnesia	7
Chloride of Sodium	2.33	Albumen	49
Cane Sugar	7	Caramel	98

The process of dialysis is best understood from an instance. Make a shallow tray by stretching a piece of parchment paper—that is, paper modified by sulphuric acid—over a hoop of gutta percha. Having placed the mixture to be analyzed in the tray, float it in a dish of water. In a day or two the crystalloids will have diffused through the paper into the water, from which they may be obtained by evaporation. Arsenious acid may in this manner be separated from the contents and tissues of a stomach.

Some colloids are found among inorganic bodies, as in the case of gelatinous silicic acid; but they are mostly organic, and of complex atomic constitution. The crystalloid is the stable, while the colloid is the changeable condition of matter. The latter is perpetually tending to turn into the former.

What is dialysis? What classes are substances divided into, and what are the properties of each? Give the rate of diffusion of some bodies. Describe the process of dialysis. What is the difference between the crystalloid and colloid states of matter?

LECTURE LXX.

THE NON-NITROGENIZED BODIES.—*The Starch Group.* —*Starch.*—*Properties.*—*Tests for various Forms of Starch.*—*Production of Dextrine.*—*Action of Diastase.*—*British Gum.*—*Cane Sugar.*—*Action of Sugar on Polarized Light.*—*Grape Sugar.*—*Milk Sugar.*—*Gum.*—*Lignine.*

THE non-nitrogenized bodies which we shall first consider are characterized by the peculiarity that they form a group, each member containing twelve atoms of carbon united with hydrogen and oxygen in the proportions to form water. They are, for the most part, indifferent bodies.

The Starch Group.

Starch	$C_{12}H_{10}O_{10}$
Cane Sugar (crystallized)	$C_{12}H_{11}O_{11}$
Grape Sugar	$C_{12}H_{14}O_{14}$
Fruit Sugar	$C_{12}H_{12}O_{12}$
Milk Sugar	$C_{12}H_{12}O_{12}$
Gum	$C_{12}H_{11}O_{11}$
Lignine	$C_{12}H_{8}O_{8}$
Etc.	Etc.

Starch, Fecula, Amylum ($C_{12}H_{10}O_{10}$), is found abundantly in the vegetable kingdom, and may be obtained from potatoes by rasping, and washing the mass upon a sieve, the starch being carried off by the water. It may also be obtained from flour by making the flour into a paste with water and then washing it. The starch separates, and gluten is left behind.

It is a white substance, commonly met with in irregular prismatic masses, which shape it assumes while drying. It is almost insoluble in cold water, and entirely so in alcohol and ether. It consists of granules of different sizes, as in *Fig.* 299, those of the potato being about the two hundred and fiftieth of an inch in diameter.

What is the peculiarity of the starch group? Name the members of the starch group, and give their composition. Whence is starch obtained? What are its properties?

Fig. 299.

When starch is heated in water, the covering membrane of each granule bursts open, and the interior matter dissolves out. If the proportion of starch be considerable, the whole forms a jelly-like mass, which may be dried into a yellowish body having the same constitution as starch itself. Gelatinous starch passes under the name of *Amidine.*

With free iodine starch strikes a deep blue color. When water containing this compound is heated to 212° the color totally disappears, and is not restored on cooling; but if the source of heat be removed as soon as the color disappears, and when the temperature is not much above 160°, the color returns. Starch and iodine constitute an exceedingly delicate test for each other. Paper impregnated with starch and iodide of potassium is blued by chlorine and bromine, and is useful as a test for ozone.

In commerce, starch is found under various modifications, such as *Arrow-root*, *Tapioca*, *Cassava*, *Sago*, *Wheat starch*, *Potato starch*, *Rice starch*, *Tous les Mois*, etc. It forms an important constituent of respiratory or heat-making food, and is said to exist in the ventricles of the brain. *Inuline*, which is derived from the dahlia and other plants, is a substance approaching starch in many respects.

When starch is boiled in water with a small quantity of sulphuric, hydrochloric, or nitric acid, it changes into *Dextrine*, a substance of the same composition ; the sulphuric acid can be subsequently removed by carbonate of lime and filtration, dextrine being procured on evaporation as a gummy mass. But if the ebullition be continued for a longer time, the dextrine disappears and grape sugar comes in its stead. Starch may also be converted into grape sugar by the action of a peculiar ferment, *Diastase*, contained in an infusion of malt.

What effect has hot water on it? What is amidine? What are the reactions of starch with iodine? What are the commercial forms of starch? How is dextrine produced? How may grape sugar arise? What are the properties of diastase?

Gelatinous starch may in the course of a few minutes, at 160°, be converted into dextrine by this substance, and soon after into sugar. In either of these cases the presence of atmospheric air is not required, the final action being that the starch simply assumes four atoms of water, and becomes converted into grape sugar.

When baked at a temperature of about 400°, starch becomes soluble in water, and passes in commerce under the name of *British Gum*, or *Leiocome.*

Starch is used for stiffening various fabrics, and for making thin calicoes appear of greater substance than they really are. It is also largely employed in confectionery, cheap sugar-plums being composed of refuse starch, chalk, gypsum, etc.

Cane Sugar, Sucrose ($C_{12}H_9O_9+2HO$), is found in the juices of many plants, as the sugar-cane, beet-root, sweet maple, Indian corn, and date-tree. It is chiefly extracted from the sugar-cane, which, after being crushed between rollers, yields a juice that is mixed with lime and boiled. A coagulum having been removed from it, it is rapidly evaporated at as low a temperature as possible, and then crystallized. In this state, after a brownish sirup, molasses, has drained from it, it passes in commerce under the name of *Muscovado*, or brown sugar. This is purified by boiling in water with albumen, which, coagulating, separates many of the impurities. The solution is then decolorized by animal charcoal, evaporated in a vacuum-pan at a temperature of 150°, solidified in conical vessels, and, being washed with a little clean sirup, is thrown into commerce as loaf sugar. Maple sugar is manufactured to a large extent in the United States, and beet-root sugar in France.

From a strong solution sugar crystallizes in rhombic prisms, which are colorless; they pass under the name of *Sugar Candy*. It is soluble in one third its weight of cold water. It has a sweet and proverbially characteristic taste. When heated it melts, and gives rise to a yellowish, transparent body called *Barley Sugar*; but if kept at a temperature of 400°, it turns of a reddish-

Why is air unnecessary? How is British gum formed? What are the uses of starch? What are the sources of cane sugar? How is it manufactured? What is sugar candy? What is barley sugar?

brown color, constituting *Caramel.* Sugar unites with various bodies, such as lime and oxide of lead, and with common salt yields a crystallized product. By casein it is transformed into lactic acid.

A solution of sugar candy produces circular polarization in a beam of light transmitted through it. This property, which is also exhibited by quartz, oil of lemon, oil of turpentine, and some other substances, has been made use of to determine the purity of sirups. Crystallizable cane sugar, for example, produces a right-hand rotation; molasses, or uncrystallizable sirup, a left-hand rotation. The apparatus necessary for these operations is seen in *Fig.* 300. A glass tube, *o o*, full

Fig. 300.

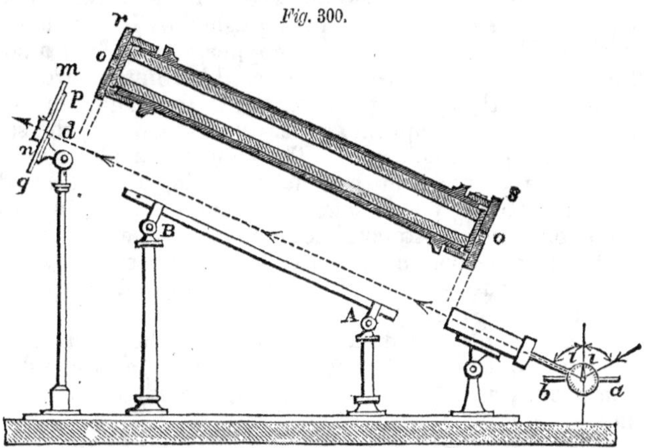

of the solution, and closed at the ends by plates of glass, is placed in a metallic case, *r s*. A beam of red light is polarized by reflection from the mirror *a b*. At *n* is a Nicol's prism capable of rotation around the line *d i;* its angular movement is measured by a graduated circle, *p q*, and vernier, *m*. The eye-piece, *n*, being so adjusted that the polarized beam is no longer visible, the tube full of solution, *o o*, is placed on the supports *A B*. A certain amount of light then passes, and the eye-piece must be rotated to the right or left, as the case may be,

What is caramel? What effect has sirup on light? Describe the apparatus *Fig.* 300. Of what use is this property?

to cut it off. The amount of rotation necessary expresses the rotatory power of the liquid.

Grape Sugar — Glucose — Starch Sugar — Diabetic Sugar ($C_{12}H_{14}O_{14}$)—is the substance just described as arising from the transmutation of starch under the influence of acids, a process largely carried on in France as a commercial manufacture. It occurs naturally in many vegetable juices and in honey.

Compared with cane sugar, it is much less soluble in water and less disposed to crystallize. It requires 1½ parts of water for solution. It may be distinguished by its action with caustic alkalies and sulphuric acid, the former turning it brown and the latter dissolving it without blackening, while cane sugar is little acted on in the former instance and blackened in the latter. The two varieties may also be distinguished by being mixed with a solution of sulphate of copper, to which, if a solution of caustic potassa be added, blue liquids are obtained, and these being heated, the grape sugar throws down a green precipitate, which turns deep red, the solution being left colorless. The cane sugar alters very slowly, a red precipitate gradually forming and the liquid remaining blue. Grape sugar, like cane sugar, gives with common salt a crystallized compound. When heated to 212° it loses two atoms of water and becomes $C_{12}H_{12}O_{12}$. At 284° it passes into caramel, $C_{12}H_9O_9$, and at a higher temperature is decomposed.

Grape sugar and milk sugar possess the interesting property of causing the reduction of silver, as a metallic film, from ammoniacal solutions of the nitrate of that metal.

Milk Sugar—Lactine ($C_{12}H_{12}O_{12}$)—may be obtained by evaporating whey to a sirup, and the crystals which then form are to be purified by animal charcoal. It is sparingly soluble, requiring five or six times its weight of water. The crystals are gritty between the teeth. It is through the alcoholic fermentation of this body that the Tartars produce intoxicating milk—koumiss.

What are the sources of grape sugar? What are its properties compared with cane sugar? Describe the respective reactions with sulphate of copper and caustic potassa. What substances can reduce silver on glass? How is milk sugar prepared? What are its properties?

Besides the foregoing, there are several subordinate varieties of sugar, among which may be cited

Ergot Sugar.............................. $C_{12}H_{13}O_{13}$,
Eucalyptus Sugar......................... $C_{12}H_{14}O_{14}$,

and others, as honey, liquorice sugar, and mushroom sugar, or mannite.

GUM.—*Gum Arabic* is obtained from several species of the mimosa or acacia, from the bark of which it exudes. It is in white or yellowish tears of a vitreous aspect. It dissolves in cold water, forming *mucilage*, from which it may be precipitated pure as *Arabine* ($C_{12}H_{11}O_{11}$) by alcohol. Arabine produces arabinates with several metallic oxides, as those of lead and iron.

Bassorine is the principle of *Gum Tragacanth.* It does not dissolve in water, but merely forms a jelly-like mass. With this substance should be classed *Pectine* ($C_{64}H_{48}O_{64}$), the jelly obtained from currants and other fruits. It furnishes *Pectic Acid* ($C_{32}H_{20}O_{28}$, $2HO$) by the action of bases.

Gelose is the gelatinizing principle of Algæ, Fuci, and Lichens. It is known as Japan Isinglass, and differs from animal gelatin or isinglass in not putrefying nor precipitating with tannin. It contains no nitrogen, being ($C_{24}H_{21}O_{24}$). The birds' nests used in China for soup are constructed by a species of swallow from this substance.

LIGNINE.—This substance, with *Cellulose* and other bodies, forms the woody fibre or ligneous tissue of plants. It occurs in a state of purity in the fibres of fine linen and cotton, and is, as is well known, of perfect whiteness, insoluble in water and alcohol, and tasteless. Its specific gravity is about 1.5, and its durability turns on the association of resins, tannin, etc. Those kinds of wood which would otherwise decay rapidly from the absence of such preservatives, may be kept by artificially introducing into their pores solutions of corrosive sublimate, arsenic, chloride of zinc, tar, etc.

Strong and cold sulphuric acid converts lignine into

What other varieties of sugar are there? Whence is gum arabic procured? What are its properties? What is the source of bassorine and pectine? What is gelose? Where is lignine found? How may wood be preserved? What is the action of sulphuric acid on lignine?

dextrine, as may be shown by adding to that liquid pieces of linen, taking care that the temperature does not rise so as to blacken the mixture, which is to be well stirred and suffered to stand for a time. On dissolving it then in water, and neutralizing by the addition of chalk, dextrine is obtained; or if, before neutralizing, the solution be well boiled, grape sugar is produced.

Parchment paper, a valuable substitute for parchment, is made by steeping unsized paper in a mixture of equal parts of sulphuric acid and water at a temperature of 60°.

LECTURE LXXI.

ACTION OF AGENTS ON THE STARCH GROUP.—*Action of Sulphuric Acid on Sugar.—Action of Lime on Sugar.—Production of Oxalic Acid.—Properties of Oxalic Acid.—Its Constitution.—Its Salts.—Saccharic Acid.—Mucic Acid.—Pyroxyline, its Preparation and Properties.*

IN the preceding Lecture we have explained the change of starch into sugar, and of lignine into dextrine, under the influence of sulphuric acid. In the vegetable world there can be no doubt that these and other similar modifications arise from the action of many causes. On inspecting the constitution of this group, it will be seen that in theory this is to be done by the addition or abstraction of water.

When melted grape sugar is mixed with strong sulphuric acid, and the diluted solution neutralized with carbonate of baryta, the sulphosaccharate of baryta is found in the solution. *Sulphosaccharic Acid* is a sweetish liquid, readily decomposing into sugar and sulphuric acid.

When, in the process of converting cane sugar into grape sugar by boiling with sulphuric acid, the action is long continued, a dark-colored substance is formed, consisting of two different bodies, *Ulmine* and *Ulmic*

How is parchment paper made? How are changes in the starch group theoretically effected? How is sulphosaccharic acid made?

Acid, or, as they are termed by Liebig, *Sacchulmine* and *Sacchulmic Acid.* The latter is converted into the former by continued boiling in water.

When a solution of grape sugar containing lime is kept for some time, the alkaline reaction of the lime finally disappears through the formation of *Glucic Acid*, the constitution of which is ($C_{12}H_5O_5+3HO$). Under the influence of heat it becomes *Apoglucic Acid* ($C_{18}H_9O_8+2HO$). Glucic acid is soluble, deliquescent, of a sour taste, and yields, for the most part, soluble salts. If grape sugar be boiled with potassa water, it becomes dark from the formation of glucic acid: this is *Moore's Test* for grape sugar. A dark substance is precipitated by an acid from this solution—*Melasinic Acid* ($C_{12}H_6O_5$).

These are some of the less important results of the action of acid and alkaline bodies on the starch group; there are others of far more interest.

Oxalic Acid ($C_2O_3, HO+2Aq$).—Oxalic acid, found in many plants, as sorrel (Oxalis acetosella), is formed artificially by the action of nitric acid on starch or sugar, or any other of the starch group except gum and sugar of milk. One part of sugar is to be mixed with four of nitric acid and two of water; carbonic oxide and carbonic acid are evolved. The nitric acid is to be distilled off until the residue will deposit crystals on cooling. These, being collected, are to be purified by redissolving and crystallizing. Oxalic acid may be also manufactured very economically from sawdust, by mixing it with hydrate of potassa and soda in solution. The mixture becomes soluble in water, and must then be raised to 400° for some hours. Subsequently the heat is increased, but not to the point of destructive distillation. The result is a mixture of the oxalates of potassa and soda. Two pounds of sawdust yield one of oxalic acid. Thus prepared, it is consumed by the ton for calico printing, dyeing, and bleaching.

The crystals of oxalic acid are oblique rhombic prisms, more soluble in hot than cold water, of an intensely acid

When do sacchulmine and sacchulmic acid arise? How are glucic and apoglucic acids made? How is oxalic acid artificially formed? Describe the production of oxalic acid from sawdust. What are the properties of crystallized oxalic acid?

taste, and poisonous to animals, death being produced in a few minutes. Chalk or magnesia is the antidote. The crystals contain one equivalent of saline water and two of water of crystallization. The latter may be removed by exposure to a low heat, the crystals then becoming a white powder and subliming without difficulty. Any attempt to remove the saline water and isolate the oxalic acid (as $C_2 O_3$) leads to its decomposition. Thus, when the acid is heated with oil of vitriol total decomposition results; equal volumes of carbonic oxide and carbonic acid are set free, for the constitution of oxalic acid is such that we may regard it as composed of an atom of each of these bodies.

$$C_2 O_3 = C O_2 + C O.$$

Upon this fact is founded one of the methods of preparing carbonic oxide. The gaseous mixture which results from the action of the oil of vitriol in the flask *a*, *Fig.* 301, is passed through a bottle, *b*, containing potassa water, which absorbs the carbonic acid, and the carbonic oxide may be collected at the water-trough.

Fig. 301.

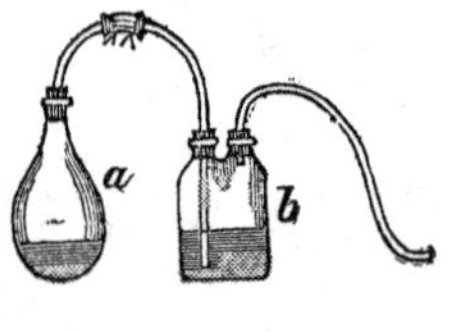

The production of oxalic acid from sugar by nitric acid is due to the replacement of hydrogen by an equivalent quantity of oxygen.

$$C_{12} H_9 O_9 + O_{18} = C_{12} O_{18} + H_9 O_9;$$

that is, one atom of dry sugar, with eighteen of oxygen, yield six of oxalic acid and nine of water.

Salts of Oxalic Acid.

There are three potassa salts: 1st. *Neutral Oxalate of Potassa*, made by neutralizing oxalic acid with carbonate of potassa, crystallizes in rhombic prisms, soluble in three times their weight of water. 2d. *Binoxalate of Potassa*, made by dividing a solution of oxalic acid into two parts, neutralizing one with carbonate of potassa and then adding the other. It crystallizes in rhombic prisms, has a sour taste, dissolves in forty parts of water, and is found naturally in many plants, as sorrel and

---

What change takes place when the water of crystallization is driven off? Describe the production of carbonic oxide. Name the potassa salts of oxalic acid. How are they made?

rhubarb. 3d. *Quadroxalate of Potassa*, made by dividing a solution of oxalic acid into four parts, neutralizing one and adding the rest. It crystallizes in octahedra, and is less soluble than either of the foregoing. These salts are sometimes used for the removal of ink stains from linen.

Oxalate of Ammonia, prepared by neutrâlizing a hot solution of oxalic acid with carbonate of ammonia. It crystallizes in rhombic prisms which are efflorescent. Its solution is used, as has been already stated, as a test and precipitant of lime. When exposed to heat in a retort, it is for the most part decomposed into water, ammonia, carbonic acid, cyanogen, and other compounds; but a flocculent substance called *Oxamide* also sublimes, the constitution of which is

$$C_2O_2, NH_2 = NH_2 + 2(CO),$$

that is, containing the constituents of one atom of amidogen and two of carbonic oxide. This remarkable substance, when boiled with potassa, yields, through the decomposition of water, oxalate of potassa and ammonia. *Oxamic Acid*, $C_4H_3O_6N$, is one of the results of the destructive distillation of binoxalate of ammonia at 450°. It is a yellowish powder, which, boiled in water, is reconverted into binoxalate of ammonia.

Oxalate of Lime occurs naturally, forming the skeleton of many lichens, and also as *Raphides*, crystalline bodies found in the cells of plants. Mulberry calculi are composed of it. It may be obtained by precipitating a lime-salt, as has just been said. It is soluble in nitric acid, and, ignited in a covered crucible, is converted into carbonate of lime, and finally into quicklime. When dried, this salt stands at the head of substances which become positive by friction.

SACCHARIC ACID ($C_{12}H_5O_{11} + 5HO$), *Oxalhydric Acid*, is made by the action of dilute nitric acid in sugar. It is a pentabasic acid.

RHODIZONIC ACID ($C_7O_7 + 3HO$) is obtained by the action of potassium on carbonic oxide at a red heat. When boiled it changes into *Croconic Acid*, a yellow body having the constitution $C_5O_4 + HO$.

Describe the properties of oxalate of ammonia. What is oxamide? What is oxamic acid? How does oxalate of lime occur? How are saccharic and rhodizonic acids made?

Mucic Acid ($C_{12}H_8O_4+2HO$), obtained by the action of dilute nitric acid on gum or sugar of milk, as in the preparation of oxalic acid by other members of the starch group. It requires sixty times its weight of water for solution. Decomposed by heat, it yields pyromucic acid.

Xyloidine ($C_6H_4O_4$, NO_5), made by the action of nitric acid, specific gravity 1.5 on starch, which is converted into a gelatinous body, and yields this substance as a white precipitate when acted on by water. Its origin is apparent from a comparison of its formula with that of starch. Xyloidine is insoluble in boiling water, but by the continued action of nitric acid changes into oxalic acid; 100 parts of starch yield 128 of xyloidine.

Gun-Cotton—*Pyroxyline.* A remarkable compound, proposed in 1846 as a substitute for gunpowder by Schonbein. It may be prepared by the action of monohydrated nitric acid on cotton, paper, or sawdust, and still more conveniently by a mixture of nitric acid, specific gravity 1.5, three parts, and sulphuric acid five parts, on those substances.

It may also be made by soaking cotton for a few minutes in a mixture of pulverized nitrate of potassa and oil of vitriol, washing the result in hot water to free the cotton from the potassa salt, and finishing the washing by a weak solution of ammonia; 100 parts of cotton yield 170 of gun-cotton. Gun-cotton appears white like ordinary cotton, the fibre being little changed; it is harsh to the touch when dry, highly electric, and explodes when heated to 400°, or when struck by a hammer. Its mechanical force much exceeds that of gunpowder, but the suddenness of its explosion has hitherto rendered it difficult to replace powder by this substance. Baron Leuk, under the auspices of the Austrian government, has carried on a series of experiments for this purpose.

A special variety of gun-cotton, possessing the explosive property in a minor degree, but completely dissolving in ether and alcohol, and forming a solution called *Collodion*, is extensively used as a photographic

How is mucic acid made? How is xyloidine made, and what are its properties? How is pyroxiline made? Give another process. How does it compare with gunpowder? What is collodion?

agent. Photographic pyroxyline requires many precautions in its manufacture, and attention must be particularly directed to the specific gravity of the acids, their relative quantities, the temperature of the mixture, and the time of immersion. Ten ounces, by measure, of sulphuric acid (1.84), and five ounces of nitric acid (1.37), with two ounces of water, are to be mixed. When the temperature has fallen to 130°, five drachms of cotton are to be added tuft by tuft, and kept in ten minutes. The cotton must, on removal from the vessel, be well pressed and washed. The qualities of the collodion as to tenacity, transparency, etc., depend principally on the gun-cotton. For its use in photography, *see* Lecture XXIV.

Pyroxyline is prone to spontaneous decomposition, with the evolution of nitrous acid. It is to be regarded as the nitrite of an organic base, having the formula ($C_{24}[H_{16}(NO_4)_4O_{20}]$). It is a substitution compound, in which four atoms of hydrogen of cellulose, $C_{24}H_{20}O_{20}$, are replaced by four of nitrous acid.

LECTURE LXXII.

METAMORPHOSES OF THE STARCH GROUP BY NITROGENIZED FERMENTS.—*Action of Leaven.—Bread.—Fermentation of Sugar.—Yeast.—Making of Alcoholic Preparations.—Ferments.—Effect of Temperature on Fermentation.—Nature of Ferments.—Making of Wine.—The Bouquet.*

IN the preceding Lecture we have traced the action of the more powerful inorganic agents on the amyles, and seen how a variety of bodies of different characters arise, some of which, as oxalic acid, are of very considerable importance.

But there is another system of changes which can be impressed on this group of bodies far more curious in its nature, and leading to far more important results.

When flour, made into a paste with water, is brought

Describe the manufacture of photographic pyroxyline. What is the chemical nature of pyroxyline?

in contact with *leaven*, that is to say, a similar dough, undergoing an incipient putrefactive fermentation, at a temperature of 60° or 70° bubbles of gas are disengaged, the paste swells up, and, when baked, forms leavened bread. This ancient process, which is now in use all over the world, depends on the action of the changing leaven being propagated to the sugar which the flour contains. The sugar is resolved into alcohol and carbonic acid, the former of which may be obtained by distilling the dough, and the bubbles of the latter, entrapped in the yielding mass, give to the bread the lightness for which it is prized.

But the process may be better traced by observing the phenomena of alcoholic fermentation in the case of pure sugar. If we take a solution of sugar in water, it may be kept for a length of time without undergoing any change; but if nitrogenized matters, such as blood, albumen, leaven, etc., in a state of putrescent decay, be mixed with it at a temperature of 70°, the sugar rapidly disappears, carbonic acid is given off, and alcohol is found in the solution. The change is obvious.

$$C_{12}H_{12}O_{12}=2(C_4H_6O_2)+4(CO_2);$$

that is, one atom of dry sugar yields two of alcohol and four of carbonic acid. The final action, therefore, of the ferment is to split the sugar atom into carbonic acid and alcohol.

Of all ferments, *Yeast*, for these purposes, is the most powerful. It is a substance which arises during the fermentation of beer. The active part of yeast is composed of minute cells, which germinate to produce a microscopic fungus, the Torula cerevisiæ. It is probable that, in the various sugars, the first action is to bring them into the condition of grape sugar, and then the metamorphosis ensues.

By an analogous transformation of the sugar contained in fruits, different wines and intoxicating liquids are formed. Thus, if we take the expressed juice of grapes which has not been exposed to the contact of the air, it may be kept for a length of time without change; but

---

What is leaven? What effect has it on dough? What is the cause of the fermentation? Describe the alcoholic fermentation of sugar. Give the formulas. What is yeast? Describe the change in grape-juice.

if a small quantity of oxygen be admitted to it, fermentation at once sets in, the grape sugar diminishes, and alcohol comes in its stead, carbonic acid being disengaged, and the nitrogenized ferment being deposited. If a solution of pure sugar be added, it is involved in the change, and portion after portion will disappear; but, finally, the ferment itself is exhausted, and then any excess of sugar remains unacted on. By the aid of a jar upon the mercurial trough these changes may be made manifest.

It is obvious that the primary action is a change in the ferment, and the moment its particles are set in motion, the motion is propagated to the adjacent body, the particles of which submit in succession, and therefore the fermentation is not a sudden action, but one requiring time. Moreover, it is plain that the action is limited; a given quantity of ferment will transmute only a definite quantity of sugar.

The ferments, or bodies which possess this singular quality, are nitrogenized bodies; and inasmuch as non-nitrogenized bodies never spontaneously ferment while oxidizing, we impute the qualities in question to the nitrogen.

Temperature has a remarkable control over ferment action. The juice of carrots or beets, fermenting at 50°, will yield alcohol, carbonic acid, and yeast; but the same juices fermenting at 120° produce lactic acid, gum, and mannite. Under these circumstances, therefore, alcohol is the product of fermentation at low, and lactic acid at high temperatures.

But when milk ferments at 50° lactic acid is the chief product, while at 80° the casein acts like a yeast ferment, the milk sugar becoming transformed into grape sugar, and then resolving itself into alcohol and carbonic acid. In this instance the action is the reverse of the former, lactic acid being the product of a low, and alcohol of a high temperature.

A very remarkable decomposition takes place when casein ferment acts on sugar at 80° in presence of carbonate of lime. Under these circumstances, carbonic

What is the eventual result? What is the action of the ferment? What element do all ferments contain? What effect has temperature? What difference is there in its effect on beet-juice and milk?

acid gas and hydrogen are evolved, and *Butyric Acid* appears. On comparing the constitution of butyric acid with alcohol, it will be seen that the latter contains the elements of the former, with an excess of hydrogen, so that during this fermentation the alcohol atom is divided.

In the *acetous fermentation* of alcohol the alcohol absorbs four equivalents of oxygen, and is resolved into one of acetic acid and two of water,

$$C_4H_6O_2 + O_4 = C_4H_4O_4 + 2HO.$$

This change only takes place in the presence of decaying nitrogenized matter.

All ferments possess certain properties in common, but each has its specific powers, and products which are evolved differ in different cases. Most commonly the activity of these bodies is excited by an incipient oxidation, the results of which would be to bring the ferment itself to a simpler constitution. In this respect, therefore, the first stage of fermentation is a combustion at common temperatures, or an eremacausis of the ferment itself; but this action is speedily propagated to the surrounding mass, which becomes involved in the change. Whatever, therefore, prevents the incipient oxidation of the ferment puts a stop to the whole process. By raising their temperature to 212°, and then cutting off the access of air, substances which would otherwise undergo a very rapid change may be kept for any length of time without alteration. On this principle, meats, milk, and other viands may be preserved, as is seen in the case of canned fruits, vegetables, etc.

We have now pointed out the peculiarities of fermentation, showing that two successive stages may be traced in the process, the first arising in the oxidation of the ferment, by which its molecules are decomposed; and the second, which consists in the propagation of this movement to the surrounding particles, upon which changes are impressed, the nature of which differs with the temperature and the specific action of the ferment itself. In fermentations the actual contact of the fer-

---

Under what circumstances does butyric acid arise? Describe the change during the acetous fermentation. What is the nature of a ferment? What effect has high temperature on ferments? What stages are there in the action of ferments? Is contact of the ferment necessary?

ment itself is essential; if it be separated by a diaphragm of bibulous paper, the action is not propagated.

Wine is made from the expressed juice of grapes, which, containing a nitrogenized body (albumen), when exposed to the air undergoes spontaneous fermentation; the course of the action being, 1st, the oxidation of the vegetable albumen; 2d, the propagation of its action to the grape sugar. If the sugar be in excess, the wine remains sweet; if the albumen be in excess, the wine is dry. The wine, as soon as the first action is over, is removed into casks. The preservation of a certain amount of carbonic acid in solution gives rise to the effervescence seen in Champagne. During these changes, the bitartrate of potassa, which exists naturally in grape-juice, and which, though sparingly soluble in water, is much less so in alcohol, is deposited. It goes under the name of Argol. Most other fruit-juices contain free acid, such as malic or citric, and hence good wine can not be made from them, because, if all the sugar be removed, they possess a sharp taste; and if, as is commonly the case, a portion be left to correct the acidity, it is liable to run into a second fermentation.

The *bouquet* of wine, a substance having the characters of an essential oil, is partly natural to the grape and partly a result of fermentation. It has been represented as a true ether, a combination of oxide of ethyle and œnanthic acid, $C_{18}H_{18}O_3$.

Inferior liquors, such as cider, perry, etc., are made from other vegetable juices, as those of apples and pears. Beer, porter, and ale are made from an infusion of malt, which is barley, a portion of the starch having been transmuted into sugar by partial germination. The principles of the fermentation are in all these cases the same.

What is the ferment in the case of wine? When is the wine sweet and when dry? What is argol? What is the bouquet? What are beer, porter, and ale made from?

LECTURE LXXIII.

THE DERIVATIVES OF FERMENTATIVE PROCESSES.—*Alcohol.—Its Properties.—Its Existence in Wines.—Lactic Acid.—Production and Properties.—Sulphuric Ether.—Preparation and Properties.—The Ethyle Group.—Chloride, Iodide, etc.—Œnanthic Ether.*

ALCOHOL (*Hydrated Oxide of Ethyle*), $C_4H_6O_2$.

BY the distillation of wine or any other fermented saccharine juice, spirits of wine may be obtained. As first prepared, it contains a large quantity of water, which comes over with it. This product, being rectified, and the first portion preserved, yields a spirit containing twelve or fifteen per cent. of water. By putting this into a retort with half its weight of quick-lime, keeping the mixture a few days, and then distilling at a low temperature, absolute or anhydrous alcohol is obtained.

Anhydrous alcohol is a colorless liquid, of a burning taste, and pleasant odor. Its specific gravity at 60° is 0.794. It boils at 173°, and at a still lower point if slightly diluted with water, though the boiling point rises if the water be in greater proportion. It has not yet been frozen, though it thickens at −160°. The specific gravity also varies with the amount of water present, and hence the purity of spirits of wine may be determined by ascertaining its density. Alcohol is very inflammable, and its vapor forms an explosive mixture with oxygen gas. It burns with a pale blue flame, with the production of carbonic acid and water, and is much used in chemical investigations as furnishing a lamp-flame free from smoke, and as possessing an extensive range of solvent powers acting upon sulphur, resins, oils, and other bodies which are not acted on by water.

Alcohol mixes with water in every proportion, heat being extricated at the moment of mixture, and the volume of the combined liquids not equaling the sum of

What is the chemical name for alcohol? How is spirit of wine made? How is absolute alcohol produced? Describe the properties of anhydrous alcohol. What are its uses?

the original quantities. *Rectified Spirit* has a specific gravity 0.825, containing 89 per cent. of absolute alcohol. *Proof Spirit*, so named from the old gunpowder test, has a specific gravity 0.920, containing 49 per cent. of absolute alcohol. In that test the alcohol was mixed with gunpowder and fired. If the gunpowder took fire the spirit was over proof; if it did not, owing to excess of water in the spirit rendering it damp, it was under proof. Alcohol combines with many saline bodies, being apparently substituted for the water of crystallization. Such compounds are called *Alcoholates*.

The strong wines, such as port and sherry, contain from 19 to 25 per cent. of absolute alcohol, the light wines from 12 per cent. upward, beer, porter, etc., from 5 to 10 per cent. The amount of alcohol may be determined by carefully distilling a specimen of the wine, rendered alkaline, until one half its bulk has passed over, then filling the distilled portion up with water until the original bulk is reached, and ascertaining the specific gravity. By reference to a table the strength may be known. Or by decolorizing the wine and mixing it with dry carbonate of potassa, the water may be abstracted, and the alcohol left floating on top of the aqueous solution of carbonate. The specific gravity of the supernatant alcohol is 0.825; it therefore contains 89 per cent. of absolute alcohol. Brandy, rum, gin, and whisky contain nearly half their weight of alcohol, and resemble proof spirit.

The value of alcoholic liquids turns mainly on the flavoring principle, which is much modified by age. Varieties in wine depend partly on the grape used and partly on the methods of manipulation. The flavor of gin is derived from juniper berries; that of whisky from the malt, particularly if it has been dried over peat; that of rum from molasses; Curaçoa from orange berries, etc.

Many alcohols are enumerated—the Propylic, C_6H_7O, HO; the Butylic, C_8H_9O, HO; the Caproic, $C_{12}H_{14}O_2$;

What is rectified spirit? Why is proof spirit so named? What are alcoholates? What percentage of alcohol do port, sherry, etc., contain? How may the amount of alcohol be determined? Describe the process with carbonate of potassa. What are the flavoring principles of gin, whisky, etc.? Name some of the alcohols.

the Caprylic, $C_4H_{18}O_2$, etc. The Amylic and Methylic alcohols will be hereafter described.

Lactic Acid Fermentation.

We have already seen that vegetable juices, as well as milk, will, under certain circumstances of temperature, yield, during fermentation, lactic acid instead of alcohol. This acid may therefore be made by dissolving a quantity of sugar of milk, putting it in a warm place, and allowing it to turn sour spontaneously. A part of the casein of the milk here acts as the ferment, and as lactic acid is set free, it coagulates the rest, and makes it insoluble. By the addition of carbonate of soda to neutralize the acid, this is prevented, and the ferment, retaining its activity, produces more lactic acid. When, by this process, all the sugar is exhausted, the liquid is boiled, filtered, evaporated to dryness, and the lactate of soda dissolved out by hot alcohol. From this alcoholic solution the acid may be obtained by precipitating the soda by sulphuric acid.

Lactic Acid ($C_6H_5O_5+HO$) is obtained as a sirupy solution by concentration in a vacuum over oil of vitriol. It is colorless, has a specific gravity of 1.215, is very sour, soluble in water and alcohol, dissolves phosphates of lime, and coagulates albumen. It yields a complete series of salts, most of which are soluble. Among these salts, the most interesting are those of lime and zinc.

ETHER. — *Sulphuric Ether.* — *Oxide of Ethyle*, C_4H_5O. — Ether is prepared by distilling equal weights of alcohol and oil of vitriol, receiving the resulting vapor in a Liebig's condenser, *a d h c*, *Fig.* 302, the condenser being cooled by water from the reservoir *i* flowing into the funnel *c*, the waste passing into the vessel *b*, and the ether distilling into the bottle *e*. The process is to be stopped as soon as the mixture begins to blacken. The first product may be rectified by redistillation from caustic potassa.

Ether is a colorless and limpid liquid, of a peculiar odor and hot taste. It boils at 96°, and has not been

How may lactic acid arise? Why is carbonate of soda added? What are the properties of lactic acid? How is ether made? Describe the apparatus. What are the properties of ether?

Fig. 302.

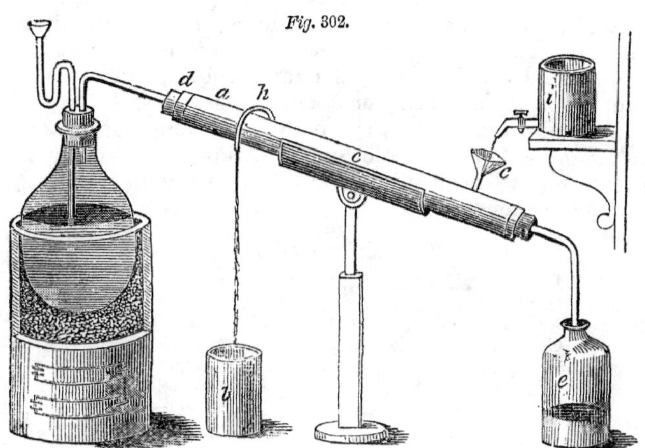

frozen; its specific gravity at 60° is .720. It volatilizes with rapidity, and therefore produces cold; a drop of water covered by ether, upon which a current of air is directed, may be frozen. It is combustible, and burns with the evolution of much more light than alcohol. The specific gravity of the vapor is 2.586. With oxygen or atmospheric air it forms an explosive mixture, and, kept in contact with air, becomes acid from the production of acetic acid. It dissolves in alcohol in all proportions, but nine parts of water are required to dissolve one of it; it also dissolves many fatty substances, and hence is of considerable use in organic chemistry.

But its most important applications are, first, as a solvent for pyroxyline in the making of photographic collodion; and, second, as an *anæsthetic* agent. The vapor of ether, when respired, produces at first an exhilarating effect, but a prolonged use eventually causes a complete stupefaction, and permits the most painful operations to be performed without the knowledge of the patient. It can scarcely be regarded as dangerous to life if administered with care in the recumbent posture and with an unembarrassed chest. The clothes

---

In what fluids is ether soluble? What are its most important applications? What precautions are necessary in its use as an anæsthetic?

should always be loosened, corsets removed, etc., and an examination of the heart and lungs made in the first instance. The anæsthetic powers seen in ether are also possessed by chloroform, the former being an American, the latter a subsequent Scotch discovery.

Ether is regarded as the oxide of a compound radical, *Ethyle*, C_4H_5, a colorless gas, liquefying under a pressure of 2¼ atmospheres at 37°, and giving rise to a series of other bodies.

The Ethyle Group.

Ethyle, C_4H_5	$=Ae$
Oxide of Ethyle..........................	$=AeO$
Hydrated Oxide of Ethyle..........	$=AeO+HO$
Chloride of Ethyle....................	$=AeCl$
Bromide "	$=AeB$
Nitrate "	$=AeO+NO_5$
Hyponitrite "	$=AeO+NO_3$
Etc.	Etc.

The oxide of ethyle, as has just been stated, is ether; the hydrated oxide, alcohol.

Numerous ethers are produced by the action of a variety of acids upon alcohol; they are called *compound Ethers.*

Chloride of Ethyle (*Hydrochloric Ether*) may be made by saturating rectified spirits of wine with dry hydrochloric acid gas, and distilling the result at a low temperature, conducting the vapor through a bottle of warm water, and then condensing in a receiver surrounded by a freezing mixture. It is a colorless, volatile liquid, of a peculiar aromatic smell; specific gravity .874. It boils at 60°, and at −10° crystallizes in cubes; is soluble in 50 parts of water, and in all proportions in alcohol and ether.

Iodide of Ethyle (*Hydriodic Ether*) is interesting as being the liquid from which ethyle was isolated by the action of zinc. It is obtained by the distillation of alcohol, iodine, and phosphorus.

Bromide, *Sulphide*, *Cyanide*, *Sulphocyanide*, and *Fluoride* of Ethyle are not of importance. *Hydrosul-*

What is the constitution of ether? Describe ethyle. Give the ethyle group. What are the compound ethers? How is chloride of ethyle made? What are its properties? Why is the iodide of ethyle interesting? What other ethyle compounds are there?

phuric Ether, or *Mercaptan* (C_4H_5S+HS), is procured by distilling hydrosulphate of sulphide of barium with sulphovinate of baryta. It is a colorless liquid, smelling like garlic; specific gravity .832; boils at 97°, and has a powerful affinity for mercury; hence its name, *mercurium captans*.

Nitrate of Ethyle (*Nitric Ether*) may be made on a small scale by distilling equal weights of alcohol and nitric acid with a small quantity of nitrate of urea. The latter substance is used to prevent the nitric acid deoxidizing and giving rise to the production of nitrous ether. It is insoluble in water, has a density of 1.112, boils at 185°, and has a sweet taste. Its vapor explodes when heated.

Hyponitrite of Ethyle (*Nitrous Ether*) may be made by passing the hyponitrous acid, disengaged from one part of starch and ten of nitric acid, through alcohol diluted with half its weight of water and kept cold. It is a yellowish, aromatic liquid, having the odor of apples; boils at 70°; specific gravity .947. The sweet spirits of nitre is a solution of this ether with aldehyde and other substances in alcohol.

Carbonate of Ethyle—Carbonic Ether ($Ae\,O$, Co_2), made by the action of potassium on oxalic ether, and distillation of the product with water. It floats on the surface of the distilled liquid, is an aromatic fluid, and boils at 259°.

Oxalate of Ethyle (*Oxalic Ether*), prepared by distilling four parts of binoxalate of potassa, five of sulphuric acid, and four of alcohol into a warm receiver. The product is washed with water to separate any alcohol or acid, and redistilled. It is an oily liquid, of an aromatic odor, boiling at 353°, and slightly heavier than water. With an excess of ammonia it yields *Oxamide* and alcohol; with a smaller proportion of ammonia it yields *Oxamethane*, $C_6H_7NO_6$.

Many other ethers are formed by the union of oxide of ethyle with anhydrous acids, as perchloric, silicic, boracic, arsenic, cyanic, hydrocyanic, formic, acetic, benzoic, succinic, and citric.

What is mercaptan? How is the nitrate of ethyle made? How is nitrous ether procured? How are the carbonate and oxalate made? What other ethers are there?

Œnanthic Ether ($Ae\,O, C_{14}H_{13}O_2$) is prepared from an oily liquid which passes over during the distillation of certain wines. It may be obtained by agitating the oil derived from brandy with carbonate of soda, which neutralizes the *Œnanthic acid*, and then distilling from chloride of calcium. It has a powerful vinous odor, is a colorless liquid, specific gravity .862; boils at 440°; soluble in alcohol and ether, but not in water. It gives a peculiar aroma to the wines in which it is found. Œnanthic acid is prepared from it by the successive action of potassa and sulphuric acid. It is an oily body, becoming a soft solid at 55°.

The compound ethers are found ready formed in many plants, and often give origin to their special odors and flavors; hence many of these can be imitated. *Pine-apple Oil*, for example, is butyric ether (C_4H_5O, $C_8H_7O_3$), and may be made from butter or glycerine. *Pear Oil* is an alcoholic solution of the acetate of amyle, and *Apple Oil* is the valeriate of the same radical.

LECTURE LXXIV.

Derivative Bodies of Alcohol.—*Sulphovinic and Phosphovinic Acids.—Products of Sulphovinic Acid at different Boiling Points.—The continuous Ether Process.—The continuous Olefiant Gas Process.—Dutch Liquid.—Successive Substitutions of Chlorine in it.—Heavy and Light Oil of Wine.—Sulphate of Carbyle and its derivative Acids.*

Sulphovinic Acid—*Bisulphate of Ether* (C_4H_5O, $2SO_3+2HO$). A mixture of sulphuric acid with an equal weight of alcohol is to be heated to the boiling point, and then allowed to cool. It is diluted with water, and neutralized with carbonate of baryta, the sulphate of baryta subsiding. After filtering and evaporating, the solution is allowed to cool, and the sulphovinate of baryta crystallizes. From this the sulphovinic acid may be obtained by precipitating the baryta with

How is œnanthic ether prepared, and what are its properties? Give the composition of pine-apple oil, etc. How is sulphovinic acid made?

dilute sulphuric acid, and evaporating the resulting solution in vacuo. It is a sirupy liquid, of a sour taste, giving rise to a series of soluble salts, which decompose at the boiling point, as will be presently seen.

PHOSPHOVINIC ACID (C_4H_5O, PO_5+2HO) is made on the same principles as the foregoing, phosphoric acid being substituted for sulphuric, and the resulting baryta salt being decomposed in the same way. It is a sirupy liquid, of a sour taste, and dissolves in water, alcohol, and ether readily. It is decomposed by heat.

If sulphovinic acid be diluted so as to bring its boiling point below 260°, it is resolved at that temperature chiefly into sulphuric acid and alcohol. If the boiling point be from 260° to 310°, the distillation results chiefly in the production of hydrated sulphuric acid and ether. If, by the addition of sulphuric acid, the boiling point be carried above 320°, the action is more complex, but the chief product which passes over is olefiant gas.

The ordinary method of preparing ether is therefore very disadvantageous, because it is only within a particular range of temperature that that body is evolved. At first the low temperature yields alcohol; and, as the heat rises, the mixture begins to blacken, and olefiant gas to be evolved.

To obviate these difficulties, a very beautiful process, the *continuous ether* process, has been introduced. It consists in taking a mixture of eight parts by weight of sulphuric acid and five of alcohol specific gravity .834, the boiling point of which is about 300°. This is brought to that temperature in a flask by a spirit-lamp, as seen in *Fig.* 303; and alcohol of the same density is allowed slowly to flow into the flask from a bottle provided with a stopcock, the temperature being steadily kept at 300°, and the mixture kept in a state of violent ebullition. Water and ether distill over together, and may be passed through a Liebig's condenser; they collect in the receiver in separate strata; or, if this does not take place at first, the addition of a little water in the receiver insures it.

How is phosphovinic acid made? What effects result from variations in the boiling point of sulphovinic acid? Why is the ordinary method of making ether disadvantageous? Describe the continuous ether process.

Fig. 303.

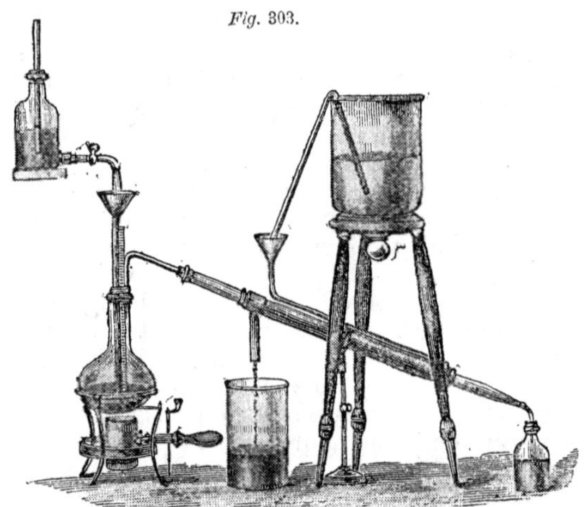

In this manner a very large quantity of alcohol may be converted into ether and water by the action of a limited amount of sulphuric acid; and in a similar manner, by adjusting the boiling point so as to be between 320° and 330°, olefiant gas may be continuously obtained. All therefore that is required is to convey the alcoholic vapor through a mixture of oil of vitriol with half its weight of water which has the required boiling point. In this process the acid does not blacken, and it is therefore much more advantageous than that described for the preparation of olefiant gas heretofore.

Chloride of Olefiant Gas—Dutch Liquid ($C_4H_4Cl_2$) —is prepared by mixing equal volumes of chlorine and olefiant gas in a large glass globe. It is a colorless and fragrant liquid, soluble in alcohol and ether, but less so in water. It boils at 180°, and when acted on by a solution of caustic potassa in alcohol it yields chloride of potassium, and a substance, C_4H_3Cl, which, on being cooled by a freezing mixture, condenses into a liquid. This liquid, brought in contact with chlorine, absorbs that substance, and yields a compound, $C_4H_3Cl_3$, which

What are the advantages of the continuous process? At what temperature does olefiant gas arise? How is Dutch liquid prepared? What are its properties?

may be decomposed by an alcoholic solution of potassa into chloride of potassium water and a new volatile body, $C_4H_2Cl_2$.

There are an iodide and bromide of olefiant gas which possess a constitution analogous to the chloride.

When chlorine gas is made to act on Dutch liquid, four different substances may be successively formed by the gradual abstraction of hydrogen, and its equivalent substitution by chlorine. These substances are as follows:

Dutch liquid	$C_4H_4Cl_2$
(1.)	$C_4H_3Cl_3$
(2.)	$C_4H_2Cl_4$
(3.)	$C_4H\ Cl_5$
(4.)	$C_4\ \ Cl_6$

The first and second of these products are volatile liquids, the fourth is the perchloride of carbon, in which it appears that all the four atoms of hydrogen in the Dutch liquid have been removed, and their places occupied by four atoms of chlorine. This *perchloride of carbon* is a white crystalline body, soluble in alcohol and ether; its melting point is 320°. By passing its vapor through a red-hot porcelain tube it is decomposed, yielding C_4Cl_4 and free chlorine, and this again gives rise to *subchloride of carbon*, C_4Cl_2, by being passed through a white-hot porcelain tube. The former of these bodies is a colorless liquid, the latter a silky solid.

Heavy Oil of Wine (C_4H_5O, SO_3) may be procured by the destructive distillation of sulphovinate of lime, or by distilling 2½ parts of oil of vitriol and one of spirit of wine. It has a yellow color, a penetrating aromatic odor, and a specific gravity of 1.133. It can not be distilled without decomposition; at 270° it is converted into alcohol, sulphurous acid, and olefiant gas. When boiled in water it yields sulphovinic acid, alcohol, and *Light Oil of Wine*, which, after standing a few days, deposits white inodorous crystals of *Etherine*, C_4H_4. The residue, which still remains liquid, is *Etherole*, C_4H_4. It is a yellow liquid, lighter than water, and soluble in alcohol and ether.

Describe the action of chlorine on Dutch liquid. Describe the products that arise. How is heavy oil of wine procured? What changes occur on heating it?

Sulphate of Carbyle ($C_4H_4, 4SO_3$) arises when the vapor of anhydrous sulphuric acid is absorbed by pure alcohol. It is a white crystalline body. When dissolved in alcohol and water added, the solution neutralized by carbonate of baryta, filtered, concentrated, and then mixed with alcohol, the *Ethionate of Baryta* precipitates. This, when decomposed by dilute sulphuric acid, yields *Hydrated Ethionic Acid* ($C_4H_5O, 4SO_3+2HO$). Ethionic acid yields a series of salts, many of which can be obtained in crystals. On being boiled, a solution of ethionic acid yields sulphurous acid and *Isethionic Acid*, the peculiarity of which is that it is isomeric with sulphovinic acid, both containing $C_4H_5O, 2SO_3+HO$. *Methionic Acid* arises from the action of hot sulphuric acid on ether. *Althionic Acid* is made by acting on alcohol with great excess of sulphuric acid. The last two acids are by some regarded as compounds of sulphovinic with isethionic acid.

LECTURE LXXV.

OXIDATION OF ALCOHOL.—*The Acetyle Series.—Aldehyde.—Its Preparation and Properties.—Aldehydic Acid.—The Flameless Lamp.—Acetal produced by Platinum Black.—Acetic Acid, Production of.—Nature of the Change from Alcohol to Acetic Acid.—Salts of Acetic Acid.*

IT has already been stated that when alcohol is burned in contact with oxygen gas or atmospheric air, the sole products of the combustion are carbonic acid and water; but when the oxidation is partial, the hydrogen is removed by preference, and a new series of bodies is the result, designated as

The Acetyle Series.

Acetyle, C_4H_3	$=Ac$
Oxide of Acetyle	$=AcO$
Hydrated Oxide of Acetyle (Aldehyde)	$=AcO+HO$
Acetylous Acid (Aldehydic Acid)	$=AcO_2+HO$
Acetic Acid	$=AcO_3+HO$

Describe the sulphate of carbyle. How is ethionic acid produced? How are isethionic, methionic, and althionic acids made? What is the difference between the partial and total oxidation of alcohol? Give the acetyle series.

Acetyle differs from ethyle by containing only three atoms of hydrogen instead of five.

Hydrated Oxide of Acetyle (*Aldehyde*) may be obtained by distilling a mixture of 4 parts of alcohol, 6 of oil of vitriol, 4 of water, and 6 of binoxide of manganese, into a receiver cooled by ice. The product is redistilled from chloride of calcium. It then consists of aldehyde, acetal, ether, and alcohol. It is next mixed with twice its volume of ether and saturated with dry ammonia. Two parts of the crystalline compound of aldehyde and ammonia, dissolved in two of water, with a mixture of three of oil of vitriol and four of water, are distilled, and the distillate redistilled from chloride of calcium at a temperature of 87°. It is a colorless liquid; of a suffocating odor; specific gravity .790, boiling point 68°. It is soluble in water, alcohol, and ether. It slowly oxidizes in the air, and more rapidly under the influence of the black powder of platinum, producing acetic acid. Heated with caustic potassa it yields aldehyde resin, a brown body of a resinous aspect. Aldehyde is so called because it contains the elements of alcohol minus two atoms of hydrogen (*Alcohol Dehydrogenatus*).

When pure aldehyde is kept for a length of time at 32° in a close vessel it yields *Elaldehyde*, a substance isomeric with itself, but possessing different properties, the specific gravity of its vapor, for example, being three times that of the vapor of aldehyde. From it there is also produced, at common temperatures, a second isomeric body, *Metaldehyde.*

Fig. 304.

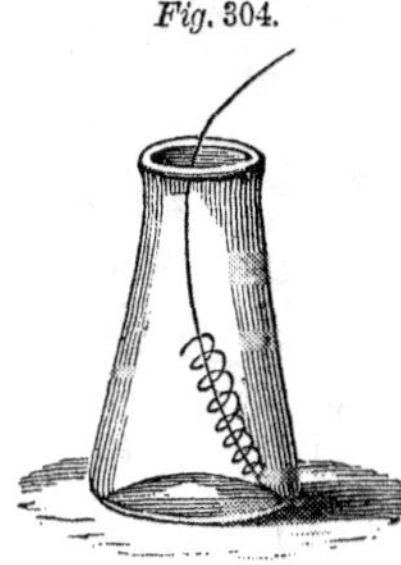

Aldehydic Acid may be obtained by digesting oxide of silver with aldehyde, and precipitating the metal with sulphureted hydrogen. It contains one atom of oxygen less than acetic acid, and is one of the products of the slow combustion of ether in Davy's flameless lamp, which may be made by putting a small quantity of ether in a jar, *Fig.* 304, and suspending in the vapor, as it mixes with atmospheric air, a coil of platinum

How is aldehyde made? What are its properties? What is the origin of its name? What is elaldehyde? How is aldehydic acid made? What is Davy's flameless lamp?

wire which has been recently ignited. The wire remains incandescent as long as any ether is present. The same result may be obtained by putting a spiral of platinum wire, *Fig.* 305, or a ball of spongy platinum over the wick of a spirit-lamp. The lamp being lighted for a short time and then blown out, the platinum continues incandescent, evolving a peculiarly acrid vapor.

Fig. 305.

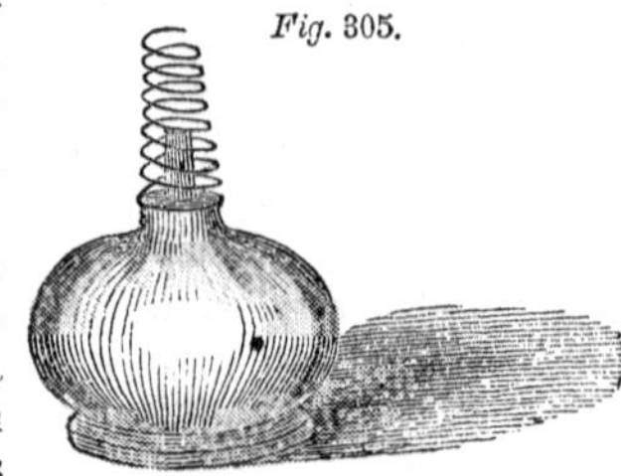

Acetal ($C_8H_9O_3$), containing the elements of ether and aldehyde, is produced by the oxidation of the vapor of alcohol by black powder of platinum, the alcohol being placed in a jar with moistened platinum black in a capsule above it. In the course of several days the alcohol will be found to have become sour; it is then to be neutralized with chalk and distilled. Chloride of calcium separates an oily liquid from the distilled product. This, on being distilled at a temperature of 200°, yields acetal. It is a colorless, aromatic, mobile liquid, specific gravity .825, and boiling at 203°. It produces, under the influence of an alcoholic solution of caustic potassa, by absorbing oxygen from the air, resin of aldehyde.

Acetic Acid—Pyroligneous Acid—Vinegar ($C_4H_3O_3+HO$). When dilute alcohol is dropped on platinum black, oxidation takes place, and the vapors of acetic acid are formed. On the large scale it is also made by allowing a mixture of alcohol, water, and a small quantity of yeast, *b*, *Fig.* 306, to flow over wood shavings which have been steeped in vinegar, contained in a barrel, through which atmospheric air is allowed to circulate by the apertures *c c c*. The temperature rises and the acetification goes on with rapidity, the product being collected in the receiver *a*. Vinegar also is formed by

Fig. 306.

Describe *Fig.* 305. Describe the preparation and properties of acetal. How is acetic acid made? Describe *Fig.* 306. How does vinegar arise?

the spontaneous souring of wines or beer containing ferment, and kept in a cask to which atmospheric air has access. During the destructive distillation of dry wood, acetic acid, hence called pyroligneous acid, in an impure state is found among the products.

Anhydrous acetic acid is obtained by distilling eight parts of dry acetate of potassa with three of oxychloride of phosphorus. It is a colorless liquid, specific gravity 1.07, boiling point 280°, and dissolves in water, falling at first to the bottom like a heavy oil.

Very strong acetic acid may be made by distilling powdered anhydrous acetate of soda with three times its weight of oil of vitriol. The product is then redistilled and exposed to a low temperature, when crystals of hydrated acetic acid form; the fluid portion is poured off and the crystals suffered to melt. It is a colorless liquid, crystallizing below 60° in plates or tufts; has a very pungent odor, and, placed on the skin, blisters it; it boils at 243°, the vapor being inflammable. The specific gravity of the liquefied crystallized acid is 1.0635, and the density increases on dilution until the acid contains one equivalent of anhydrous acid to three of water, when it is 1.073; on farther dilution the density diminishes. It also dissolves in alcohol and ether. In an impure state as vinegar, its taste, odor, and applications are well known. Acetic acid is largely used in photographic operations to retard the action of reducing agents upon nitrate of silver, constituting an important ingredient of *developers*, as they are called.

If acetic acid be compared with alcohol,

Alcohol.................................... $C_4H_6O_2$,
Acetic Acid................................ $C_4H_4O_4$,

it is seen to differ in the circumstance that two hydrogen atoms have been removed from the alcohol and their places taken by two oxygen atoms; hence the various processes for its production are easily explained. Acetic acid gives rise to several important salts.

Acetate of Potassa ($KO, C_4H_3O_3$) is obtained by neu-

How is anhydrous acetic acid made? How may strong acetic acid be made? What are its properties? Of what use is acetic acid? What is the difference between acetic acid and alcohol? Name some of the salts of acetic acid, and give their methods of preparation.

tralizing acetic acid with carbonate of potassa, evaporating to dryness, and fusing. This salt is very deliquescent, has an alkaline reaction, is soluble in its weight of water and twice its weight of alcohol.

Acetate of Soda is made on the large scale by saturating the impure pyroligneous acid formed in the destructive distillation of wood with lime, and then decomposing the acetate of lime with sulphate of soda. The sulphate of lime precipitates, the solution being crystallized, and the crystals subsequently purified by draining, fusion, solution, and recrystallization. The crystals effloresce in the air, and are soluble in water and alcohol.

Acetate of Ammonia (*Spirit of Mindererus*).—The solution is made by saturating acetic acid with carbonate of ammonia, and the solid by distilling acetate of lime and chloride of ammonium. The acetate of ammonia passes over, and chloride of calcium is left.

Acetate of Alumina is made by the decomposition of a solution of alum by acetate of lead. It is much used by dyers as a mordant.

Acetates of Lead.—1st. *Neutral Acetate* (*Sugar of Lead*) may be made by dissolving litharge in acetic acid. It occurs in colorless prismatic crystals, and also in crystalline masses. It has a sweetish, astringent taste, from which its commercial name is derived. It is soluble in less than two parts of water and in eight of alcohol. The crystals effloresce. 2d. *Subacetate of Lead* (*Sesquibasic Acetate*) is formed by partially decomposing the neutral acetate by heat. Its solution is known as *Goulard's Water.* Two other subacetates may be made by the action of ammonia on the neutral salt. Their solutions have an alkaline reaction, absorb carbonic acid from the air, giving rise to a precipitate of the basic carbonate, and constituting a delicate test for that gas.

Acetates of Copper.—1st. *Neutral Acetate* (*Crystallized Verdigris*), made by dissolving verdigris in hot acetic acid. On cooling, it yields green crystals, soluble in five parts of boiling water, and also in alcohol. It is used as a paint. 2d. *Bibasic Acetate of Copper* (*Ver-*

How many acetates of lead are there? Give their properties. How is verdigris made?

digris) may be made by the action of vinegar and air conjointly on metallic copper. Plates of copper are alternated with pieces of cloth steeped in acetic acid. As they become covered with verdigris, it is removed as a blue-green powder, and the operation continued until the copper is exhausted. Verdigris is a mixture of several acetates, one of which may be obtained by digesting it in warm water; a second arises on boiling this; the insoluble residue contains a third.

LECTURE LXXVI.

Derivatives of Acetyle.—The Kakodyle Group.—*Chloracetic Acid.—Acetone.—Chloral.—Hydrochloric Ether.—Action of Chlorine on the Ethers.——Xanthic Acid.—The Kakodyle Group.—Oxide.—Chloride.—Kakodylic Acid.*

Chloracetic Acid ($HO, C_4H_2O_3Cl$) is produced when chlorine is passed through a mixture of two parts of glacial acetic acid and one of water, the sunlight being excluded. It is said to produce definite salts.

When glacial acetic acid is exposed to sunshine in a jar of chlorine, white flocculi are formed; they are *trichloracetic acid* ($HO, C_4Cl_3O_3$). The hydrogen of the acetic acid is replaced by chlorine. Chlorocarbonic, carbonic, and oxalic acids are at the same time formed. The salts of this acid are soluble in water.

Sulphacetic Acid arises when acetic acid is acted on by anhydrous sulphuric acid. Its composition is ($C_4H_2O_2, 2SO_3, 2HO$). It forms deliquescent crystals and bibasic salts.

Acetone—Pyroacetic Spirit ($C_6H_6O_2$)—is obtained when acetate of lime is distilled with excess of quicklime, or when two parts of acetate of lead and one of lime are heated in an iron retort. The distillate is redistilled from a water-bath. It is a limpid, colorless, and volatile liquid, specific gravity .792, boiling at

---

What is the composition of verdigris? How is chloracetic acid produced? What results on exposing acetic acid to chlorine? How is sulphacetic acid formed? What are the properties of pyroacetic spirit?

132°; burns with a bright flame, and is soluble in water, ether, and alcohol. By the action of chlorine, three substitution products are obtained, in which two, three, and four atoms of hydrogen are replaced by chlorine. Nordhausen oil of vitriol, distilled with acetone, yields an oily body, the constitution of which is C_3H_2; it is lighter than water, and has an odor of garlic.

Sir R. Kane considers acetone to be the hydrated oxide of a radical, *Mesityle*, C_6H_5, and has produced the oxide and chloride of mesityle. Zeise also discovered a compound consisting of the oxide of mesityle and bichloride of platinum.

CHLORAL ($C_4HCl_3O_2$).—When dry chlorine is passed into anhydrous alcohol, and the action finished by the aid of heat, hydrochloric acid is produced; and on its ceasing to appear, if the product be agitated with three times its volume of oil of vitriol, and the mixture warmed, an oily liquid floats on the acid; this is chloral. It may be purified by successive distillation from oil of vitriol and quick-lime. It is an oily, colorless liquid, which causes a flow of tears, leaves a transient greasy stain on paper, has a density of 1.502, boils at 206°, is soluble in water and alcohol, and gives no precipitate with nitrate of silver. When kept for a length of time in a sealed tube, it spontaneously becomes white, solid chloral. In this condition it is little soluble in water, and reverts to its other state by being warmed.

If chlorine acts on alcohol containing water, *heavy Hydrochloric Ether* is formed. It is a colorless and volatile liquid.

The action of chlorine upon common ether, and also upon the compound ethers, is very interesting. It consists in the gradual removal of hydrogen, chlorine being substituted for it. This, in many instances in which the aid of the sunlight is resorted to, terminates in the entire removal of the hydrogen. In the compound ethers it is the basic hydrogen which is removed, while that of the acid escapes, as in the case of *chloruretted acetic* and *chloruretted formic* ethers. When the vapor of light hydrochloric ether is acted upon by chlorine gas,

What is the chemical composition of acetone? How is chloral made? What changes may occur in it? What is the action of chlorine on the ethers?

a complete series of compounds may be obtained, the hydrogen eventually disappearing:

Hydrochloric Ether	C_4H_5Cl
Monochlorureted Hydrochloric Ether	$C_4H_4Cl_2$
Bichlorureted " "	$C_4H_3Cl_3$
Trichlorureted " "	$C_4H_2Cl_4$
Quadrichlorureted " "	$C_4H\ Cl_5$
Perchloride of Carbon	$C_4\ \ Cl_6$

furnishing, therefore, a very striking instance of the doctrine of substitution.

Xanthic Acid ($C_6H_5S_4O+HO$).—Hydrate of potassa is to be dissolved in twelve parts of alcohol, specific gravity .8, and bisulphide of carbon dropped into the solution until it ceases to have an alkaline reaction. On cooling to zero, the xanthate of potassa crystallizes; it is to be dried in vacuo. It is soluble in water and alcohol, but not in ether, and from it xanthic acid may be procured by the action of dilute hydrochloric acid. Xanthic acid is an oily liquid, heavier than water, which first reddens and then bleaches litmus paper. At 75° it is decomposed into alcohol and bisulphide of carbon. It is also decomposed by the action of the air.

KAKODYLE ($C_4H_6As=Kd$) is a compound radical which gives rise to an extensive group of bodies, in which it acts the part of a metal.

The Kakodyle Group.

Kakodyle, C_4H_6As	$=Kd$
Oxide of Kakodyle	$=KdO$
Chloride "	$=KdCl$
Iodide "	$=KdI$
Sulphide "	$=KdS$
Etc.	Etc.

Kakodyle may be obtained by decomposing the chloride of kakodyle with metallic zinc in an apparatus filled with carbonic acid, and may be purified by redistillation from zinc, similar precautions being taken to exclude atmospheric air. It is a colorless liquid, of a highly offensive odor, whence its name, taking fire spontaneously in the air, oxygen, or chlorine, boils at 338°, crystallizes at 20° in square prisms, and is decomposed

Give the compounds of hydrochloric ether. How is xanthic acid produced? What is kakodyle? How is it obtained? What are its properties?

at a red heat into olefiant gas, light carbureted hydrogen, and arsenic. It is very poisonous.

Oxide of Kakodyle—Alkarsine—Cadet's Fuming Liquor—is prepared by the distillation of acetate of potassa and arsenious acid, receiving the products in an ice-cold vessel, the temperature being finally carried to a red heat. The oxide comes over in an impure state, sinking to the bottom of the other products. It is to be decanted, washed with water, boiled, and then distilled in a vessel full of hydrogen from hydrate of potassa. It is a colorless liquid, spec. gr. 1.64; boils at 300°, and solidifies at 9°; is insoluble in water, but dissolves in alcohol and ether; is excessively poisonous, possessing a smell like concentrated garlic. Heated in the air, it burns, producing carbonic acid, water, and arsenious acid.

Chloride of Kakodyle may be procured by the action of a dilute solution of corrosive sublimate on a dilute alcoholic solution of oxide of kakodyle. A white precipitate falls, which, distilled with strong hydrochloric acid, yields corrosive sublimate; water and the chloride of kakodyle passes over. When purified by chloride of calcium and distilled in an atmosphere of carbonic acid, it is a colorless liquid, of a very offensive odor, heavier than water and insoluble therein, but soluble in alcohol. It is very poisonous. It boils at about 212°, the vapor taking fire in the air.

Kakodylic Acid—Alcargen ($Kd\,O_3$)—may be made by the action of oxide of mercury upon oxide of kakodyle under the surface of water at a low temperature. Kakodylic acid forms crystals which deliquesce in the air, are soluble in water and alcohol, but not in ether. It is not acted upon by oxidizing agents, such as nitric acid, but is reduced to oxide of kakodyle by several deoxidizing bodies. It is not poisonous, though it contains 56 per cent. of arsenic: seven grains of it produced no effect on a rabbit.

Kakodyle furnishes a complete series of bodies—the iodide, sulphide, cyanide, and a substance isomeric with the oxide, which has the name of parakakodylic oxide.

How is the oxide of kakodyle made? What are its properties? How is the chloride formed? What are its properties? How is kakodylic acid made? How much arsenic does it contain? What properties have the kakodyle compounds in common?

The preparation of these compounds is very dangerous, from their explosibility and poisonousness. The cyanide of kakodyle, diffused in the smallest quantity through the atmosphere, produces a sudden cessation of muscular power in the hands and feet, giddiness, and insensibility.

LECTURE LXXVII.

THE WOOD-SPIRIT GROUP.—*Methyle.—Its Oxide and Hydrated Oxide.—Salts of Methyle.—Formic Acid, Natural and Artificial Production of.—Chloroform.—Its Anæsthetic Properties.—Action of Chlorine on the Oxide of Methyle.—Substitutions in Chloride of Methyle.*

IN the destructive distillation of wood in the preparation of pyroligneous acid, there passes over about one per cent. of a body to which the name of wood-spirit has been given. This is the hydrated oxide or alcohol of a compound radical, *Methyle.*

Methyle, C_2H_3	$=Me$
Oxide of Methyle	$=MeO$
Hydrated Oxide of Methyle	$=MeO+HO$
Chloride of Methyle	$=MeCl$
Etc.	Etc.

Methyle is a gas, specific gravity 1.036, produced by decomposing iodide of methyle by zinc.

Oxide of Methyle—Methylic Ether—Wood Ether (C_2H_3O).—This substance is made from the hydrated oxide, on the same principle that ether is made from alcohol. One part of wood-spirit and four of oil of vitriol being heated in a flask, the vapor is passed through a small quantity of caustic potassa solution and received at the mercurial trough. It is a permanently elastic gas, colorless, and has a specific gravity of 1.59; burns with a pale blue flame, is very soluble in water, which takes up thirty-three times its volume of it, and yields it unchanged when heated.

When does wood-spirit arise? What is the composition of methyle? How is methyle produced? How is the oxide of methyle made?

Hydrated Oxide of Methyle— Wood-Spirit—Pyroxylic Spirit—may be separated from crude wood vinegar by distillation. It passes over with the first portions along with a little acid, which, being neutralized with hydrate of lime, the wood-spirit may be separated from the oil which floats on its surface and redistilled. The product thus obtained may be rectified in the same manner as common alcohol, and rendered anhydrous by quick-lime. It is then a colorless liquid, of a hot taste and peculiar smell, like peppermint. It boils at 150°, and has a specific gravity of .7398 at 60°; the density of its vapor is 1.125. It is soluble in water, alcohol, and ether in all proportions, and may be burned like spirit of wine, exhaling a peculiar odor. Its solvent powers resemble those of alcohol; it may be used as a substitute for that fluid in the manufacture of fulminating silver. It dissolves resins and oils. It is a powerful antiseptic. *Methylated Spirit* contains ten per cent. of methylic alcohol and 90 per cent. of alcohol.

Chloride of Methyle ($Me\,Cl$) may be made from the reaction of sulphuric acid upon common salt and wood-spirit. It is a colorless gas, which may be collected over water, and has a density of 1.731. It has a peculiar odor, is inflammable, and may be decomposed by passing through a red-hot tube. Chloride of lime, acting upon pyroxylic spirit, gives rise to methylic chloroform.

Sulphate of Oxide of Methyle ($Me\,O, S\,O_3$) may be prepared by distilling one part of wood-spirit with eight or ten of oil of vitriol; the product is to be washed with water and redistilled from caustic baryta. It is an oily, neutral liquid, smelling like garlic; specific gravity 1.324; boils at 370°. It is not soluble in water, but is decomposed by that liquid, especially at the boiling point, into sulphomethylic acid and hydrated oxide of methyle. It is to be observed that in the series of wine alcohol there is no compound corresponding to this.

Nitrate of Oxide of Methyle ($Me\,O, N\,O_5$) is obtained

How is pyroxylic spirit made and what are its properties? What is methylated spirit? What are the properties of chloride of methyle? Describe the preparation and properties of sulphate of oxide of methyle. State the properties of nitrate of oxide of methyle.

by the action of a mixture of wood-spirit and oil of vitriol upon nitrate of potassa. It is a colorless liquid, heavier than water; boils at 150°; burns with a yellow flame; its vapor explodes when heated. In a solution of caustic potassa, it decomposes into nitrate of potassa and wood-spirit.

Oxalate of Oxide of Methyle ($Me\,O\,C_2\,O_3$) is made by distilling oxalic acid, wood-spirit, and oil of vitriol. The liquid which is collected is allowed to evaporate; it yields crystals of the oxalate. When pure, it is colorless, melts at 124°, and boils at 322°. It is decomposed by hot water into oxalic acid and wood-spirit; by solution of ammonia into oxamide and wood-spirit.

Sulphomethylic Acid ($Me\,O,\ 2S\,O_3+HO$) is the compound corresponding to sulphovinic acid, and is prepared in the same way by substituting wood-spirit for alcohol. It is thus procured as a sirup, or in small crystals, soluble in water and alcohol. It is an instable body, and possesses many analogies with sulphovinic acid.

Formic Acid (C_2HO_3+HO).—This acid, in the wood-spirit series, is the analogue of acetic acid in the alcohol series. It may be procured on principles similar to those involved in the preparation of acetic acid, as by the gradual oxidation of the vapor of wood-spirit in the air under the influence of platinum black. It may be prepared by the distillation of 10 parts of starch, 37 of peroxide of manganese, 30 of water, and 30 of sulphuric acid; or by distilling 10 parts of tartaric acid, 3 of peroxide of manganese, 3 of sulphuric acid, and 3 of water. The dilute acid distillate is saturated with carbonate of lead, and the formate of lead purified by crystallization. It is then decomposed by an equivalent of sulphuric acid or by sulphureted hydrogen. It has been prepared by distilling oxalic acid and glycerine at 212°. The oxalic acid is resolved into formic and carbonic acids, the glycerine remaining unchanged,

$$2(HO,\ C_2O_3)=(HO,\ C_2HO_3)+2(CO_2).$$

Formic acid occurs naturally in the bodies of red ants, and hence obtained its name, having been first derived from the distillation of those animals. Anhy-

What body does sulphomethylic acid resemble? What acid is formic acid the analogue of? How is formic acid made? Describe its production from oxalic acid. What is the origin of the name?

drous formic acid (C_2HO_3) obviously contains the elements of two atoms of carbonic oxide and one of water. It yields two hydrates, respectively containing one and two atoms of water. The first, for which the formula has already been given, is a very acrid fuming liquid, specific gravity 1.22, crystallizable below 32°, boiling at 220°, and yielding an inflammable vapor whose density is 2.125. Formic acid is represented as the teroxide of *Formyle* ($C_2H + O_3$); it yields a complete series of salts, and has been used as a reducing agent in photographic developers.

Chloroform (C_2HCl_3) is made by distilling wood-spirit with a solution of chloride of lime, or chloral with lime and water, or a mixture of 1 part of alcohol, 24 parts of water, and 6 parts of chloride of lime. It is a colorless, transparent liquid; specific gravity 1.5; boils at 140°; the density of its vapor is 4.2. It burns with a green flame; is readily vaporizable; almost insoluble in water, but soluble in alcohol and ether; dissolves resins, bromine, iodine, the alkaloids, and many other substances.

When the vapor of chloroform is respired, diluted with air, it speedily produces insensibility, and is therefore useful as an *anæsthetic* in surgical operations, and on occasions when pain is to be avoided. It should never be respired by persons having affections of the lungs or heart, and must always be taken in the recumbent position, with the clothing loosened. Notwithstanding these precautions, death occasionally results from its use.

The relation between formic acid and chloroform is obvious, consisting in the substitution of three atoms of chlorine for three of oxygen. It is regarded as the terchloride of formyle. There are two analogous compounds:

Bromoform	C_2HBr_3
Iodoform	C_2HI_3

Formomethylal ($C_3H_4O_2$) is prepared by distilling wood-spirit, oxide of manganese, and dilute sulphuric

---

What compounds does formic acid yield? How is chloroform made? Of what use is chloroform? What precautions should be taken in respiring it? What are the formulas for chloroform, bromoform, and iodoform? How is formomethylal prepared?

acid. On saturating the product with potassa, formo-methylal separates as a colorless oily liquid, specific gravity .855, boiling at 107°, and soluble in water.

Methyle-mercaptan.—Formed like the common mercaptan by substituting sulphomethylate of potassa for sulphovinate of lime. It is analogous to common mercaptan.

When chlorine is made to act on the oxide of methyle at common temperatures, it removes one of the hydrogen atoms, and, by continuing the action, a second may be taken away, and the process of substitution, as shown in the following series, may be carried so far as to end in the entire removal of the hydrogen and oxygen, and the production of chloride of carbon:

Oxide of Methyle		C_2H_3O
1st substitution		C_2H_2O, Cl
2d "		$C_2H\ O, Cl_2$
3d "		$C_2\ \ O, Cl_3$
4th "	Chloride of Carbon	$C_2\ \ \ Cl_4$

Other methylic compounds furnish similar series, thus:

Chloride of Methyle		C_2H_3Cl
1st substitution		$C_2H_2Cl_2$
2d "	Chloroform	$C_2H\ Cl_3$
3d "	Chloride of Carbon	$C_2\ \ Cl_4$

LECTURE LXXVIII.

THE POTATO-OIL GROUP.—*Amyle.*—*Fusel Oil.*—*Chloride of Amyle.*—*Sulphamylic Acid.*—*Amylene.*—*Valerianic Acid.*

THE BENZOYLE GROUP.—*Oil of Bitter Almonds.*—*Benzoic Acid.*—*Sulphobenzoic Acid.*—*Chloride of Benzoyle.*—*Benzamide.*—*Hydrobenzamide.*

IN the distillation of brandy from potatoes a volatile oil passes over; it is the hydrated oxide of a compound radical, *Amyle*, $C_{10}H_{11}$.

What is the action of chlorine on oxide of methylal? Give the series resulting from the action of chlorine. What is the composition of amyle?

The Potato-Oil Group.

Amyle, $C_{10}H_{11}$	$=Ayl$
Amylic Ether	$=AylO$
Amylic Alcohol (Potato Oil)	$=AylO+HO$
Chloride of Amyle	$=AylCl$
Etc.	Etc.
Amylene	$=C_{10}H_{10}$
Valerianic Acid	$=C_{10}H_9O_3$

Amyle is isolated by acting with zinc amalgam on iodide of amyle under pressure. It is a colorless, pellucid liquid, of an ethereal odor and burning taste. Cooled to 18° it thickens, but does not solidify; specific gravity at 52°, .7704; boils at 310°. It does not ignite at ordinary temperatures, but burns when heated; is insoluble in water, but is soluble in all proportions in alcohol and ether. It forms a hydride.

Amylic Ether acts like the ethers of ethyle and methyle.

Hydrated Oxide of Amyle—Amylic Alcohol—Potato Oil—Fusel Oil—passes over toward the end of the first distillation of potato spirit or corn spirit, and communicates to it a milky tint. On standing, it floats to the surface, and may be purified by washing with water, drying with chloride of calcium, and redistillation at 268°. It is a colorless liquid, of a peculiar, nauseating, suffocating, and persistent odor, and acrid taste; specific gravity .812; boils at 270°, crystallizes at 4°. It is sparingly soluble in water, but dissolves in alcohol, ether, and fixed and volatile oils. When acted on by oxidizing agents it yields *Valeric* acid.

Chloride of Amyle is made by distilling equal weights of potato oil and perchloride of phosphorus, washing with potassa water, and redistilling from chloride of calcium. It is an aromatic liquid, boils at 215°, and burns with a green flame. Under the influence of sunshine eight of its hydrogen atoms may be removed, eight chlorine atoms being substituted for them, $C_{10}H_{11}Cl$ yielding $C_{10}H_3Cl_9$, forming chlorureted chloride of amyle.

Mention the members of the potato-oil group. What are the properties of amyle? Describe fusel oil. How is chloride of amyle made? What effect has chlorine on it under the influence of sunshine?

The *Iodide* and *Bromide of Amyle* are compounds analogous to the chloride.

Acetate of the Oxide of Amyle is obtained by distilling acetate of potassa, potato oil, and sulphuric acid. It is a colorless liquid, which boils at 257°.

Sulphamylic Acid ($Ayl\,O, 2SO_3H+HO$) is generated when sulphuric acid is made to act on an equal weight of potato oil. From this, by the successive action of carbonate of baryta and sulphuric acid, it may be procured by operating on the same principles as for sulphovinic acid, to which, both in constitution and properties, it is analogous. It is a sirupy or crystalline body, and is decomposed by ebullition into potato oil and sulphuric acid.

Amylene is obtained by the action of anhydrous phosphoric acid on potato oil. It is an oily liquid, lighter than water, boiling at 102°. It is a hydrocarbon, isomeric with olefiant gas and etherine. The density of its vapor is 5.06, five times that of olefiant gas; each volume of it therefore contains ten volumes of hydrogen and ten atoms of carbon. It has been used as an anæsthetic, but has caused fatal results several times. It occupies the same position that olefiant gas does in the wine-alcohol series.

Valerianic Acid ($C_{10}H_9O_3$) bears the same relation to the amyle group that acetic acid does to the wine-alcohol group, or formic acid to the wood-spirit group. It is formed when warm potato oil is dropped on platinum black in contact with the air. It occurs naturally in the root of the *Valeriana Officinalis* and other plants, and in train oil, but is best made by heating potato oil in a flask with a mixture of quick-lime and hydrate of potassa for several hours at a temperature of 400°. The white residue is immersed in cold water and distilled with a slight excess of sulphuric acid, so as to drive off hydrated valerianic acid and water. It is an oily liquid, of an acid taste, specific gravity .944, combustible, and boiling at 270°. The anhydrous acid is formed by the action of oxychloride of phosphorus on

---

How is sulphamylic acid generated? What are the properties of amylene? What are its relations to olefiant gas? What is the analogue of valerianic acid? How is it made? How is anhydrous valerianic acid made?

valerate of potassa. It is a limpid oil, with a smell of apples, but on contact with water it is hydrated, and acquires an intense odor of valerian.

When acted on by chlorine in the dark, and the action aided by heat, it gives rise to *Chlorovalerisic Acid* ($C_{10}H_6 Cl_3 O_3+HO$), in which there has been a removal of three hydrogen atoms and a substitution of three of chlorine. Under the influence of the sunshine, by the same process, another hydrogen atom is removed, and *Chlorovalerosic Acid* forms ($C_{10}H_5 Cl_4 O_3+HO$).

The Benzoyle Group.

Benzoyle $C_{14}H_5O_2$	....	$=Bz$
Hydride of Benzoyle	....	$=BzH$
Oxide of "	(Benzoic Acid)............	$=BzO$
Chloride of "	....	$=BzCl$
Etc.		Etc.

Benzoyle has been obtained as an oil by the dry distillation of benzoate of copper. It discharges the functions of a metallic body.

Hydride of Benzoyle (*Oil of Bitter Almonds*) is obtained by the distillation of bitter almonds, from which the fixed oil has been expressed, with water, and arises from the action of the water upon *Amygdaline* contained in the seed. It may be purified from the hydrocyanic acid it contains; it must be agitated with milk of lime and protochloride of iron and redistilled. It is a colorless liquid, of an agreeable odor and pungent flavor, not poisonous, as it is used in cookery, slightly soluble in water, but very soluble in alcohol and ether. It boils at 350°, is inflammable, and oxidizes in the air into benzoic acid.

Oxide of Benzoyle—Benzoic Acid—is obtained by sublimation from gum benzoin, a resinous exudation of the *Styrax Benzoin*, a tree growing in Sumatra, Borneo, and Java. It is found in the pods of vanilla. The benzoin is placed in a shallow vessel, over the top of which a covering of filtering paper is pasted, and this covered by a taller cylinder of stouter paper. On heating, the

What effect has chlorine on it in the dark and in the sunshine? How is the hydride of benzoyle made? What are the properties of oil of bitter almonds? What is the source of benzoic acid? Describe the method of preparation.

vapor passes through the filtering paper, and, condensing in feathery crystals in the space above, falls down upon the paper, and is retained by it. A better method is to boil a mixture of the gum with hydrate of lime, filter, concentrate the solution, add hydrochloric acid, and the benzoic acid crystallizes in thin plates on cooling. It may be subsequently sublimed.

When pure, benzoic acid has no odor, but is generally scented with a trace of volatile oil; it has a sour and acrid taste. It melts at 250°, boils at 460°, the vapor having a specific gravity of 4.26, and exciting coughing. It is six times more soluble in hot than cold water, and is dissolved by alcohol, ether, and the fixed and volatile oils. The crystals contain an atom of water, the equivalent being 122. It forms a series of salts, and is sometimes used for the separation of iron from other metals.

Sulphobenzoic Acid ($C_{14}H_4S_2O_8+2HO$) a bibasic acid, formed by the action of anhydrous sulphuric acid upon benzoic acid, the mass being dissolved in water and neutralized by carbonate of baryta. By filtering, and adding hydrochloric acid to the hot solution, on cooling, the sulphobenzoate of baryta crystallizes; it may be decomposed by dilute sulphuric acid. It is a white crystalline mass.

Chloride of Benzoyle ($Bz{\cdot}Cl$).—When chlorine gas is passed through oil of bitter almonds, hydrochloric acid is formed, and, after expelling the excess of chlorine by heat, chloride of benzoyle remains. It is a colorless liquid, of a disagreeable odor, heavier than water, combustible, and decomposed by boiling water into benzoic and hydrochloric acids.

Benzamide ($C_{14}H_7NO_2$) is formed by the action of chloride of benzoyle on dry ammonia, the chloride of ammonium being removed from the resulting white mass by cold water. From a solution in boiling water the benzamide crystallizes. It melts at 239°, and corresponds in its chemical relations to oxamide.

Hydrobenzamide ($C_{14}H_{18}N_2$).—Made by the action of pure oil of bitter almonds on a solution of ammonia, the

---

What are the properties of benzoic acid? Describe sulphobenzoic acid. What is the reaction of chlorine on oil of bitter almonds? How are benzamide and hydrobenzamide formed?

product being washed with ether, and from its alcoholic solution this substance crystallizes; but when impure almond oil is employed, three other compounds may be obtained — benzhydramide, azobenzoyle, and nitrobenzoyle.

LECTURE LXXIX.

THE SALICYLE AND CINNAMYLE GROUPS.—*Benzoine.— Benzone.—Benzole.— Sulphobenzide.—Nitrobenzide. — Chlorobenzide.— Hippuric Acid.— Oil of Spiræa. —Salicylic Acid.— Compounds of Cinnamyle.*

BENZOINE ($C_{28}H_{12}O_4$) is a body isomeric with bitter almond oil, but inodorous and tasteless. It is found in the residue after purifying that oil from hydrocyanic acid by distillation from lime and oxide of iron, and may be obtained by dissolving out those bodies by hydrochloric acid. It crystallizes from an alcoholic solution, on cooling, in colorless crystals, which melt at 248°. It dissolves in an alcoholic solution of caustic potassa, which, by boiling until the violet color has disappeared, furnishes benzilate of potassa, a salt from which benzilic acid may be obtained by hydrochloric acid. The constitution of *Benzilic acid* is ($C_{28}H_{11}O_5+HO$).

Benzone ($C_{26}H_{10}O_2$) is obtained by the distillation of dry benzoate of lime with a little lime at a high temperature. A red liquid passes over; it is redistilled, and the liquid that comes over between 600° and 620° is benzone, or *Benzophenone.* It crystallizes in transparent rhombic prisms, insoluble in water, soluble in alcohol and ether. The formation from benzoate of lime is thus explained:

$$2(CaO,\ C_{14}H_5O_3)=2(CaO,\ CO_2)+C_{26}H_{10}O_2.$$

Benzine — Benzole — Phene ($C_{12}H_6$) — occurs in the volatile liquids condensed from coal naphtha, but is best obtained by distilling one part of benzoic acid with three of slacked lime. It is a limpid, colorless liquid, specific gravity .85, boils at 170°, solidifies at 40°, insol-

---

How is benzoine procured? What are its properties? What is the process for obtaining benzone? Where does benzole occur? What are its properties?

uble in water, soluble in alcohol and ether. In its formation, one equivalent of benzoic acid yields two equivalents of carbonic acid and one of benzine:

$$C_{14}H_5O_3+HO=C_{12}H_6+2CO_2.$$

Sulphobenzide—Sulphobenzole ($C_{12}H_5SO_2$)—is made by taking the substance which arises from the union of benzine with anhydrous sulphuric acid, and acting upon it with an excess of water. The sulphobenzide, which is insoluble in that liquid, may be obtained in crystals from its ethereal solution. It melts at 212°. From the acid liquid from which it has been separated *hyposulphobenzidic* acid may be obtained. The constitution is $C_{12}H_5S_2O_5+HO$.

Nitrobenzide—Nitrobenzole ($C_{12}H_5NO_4$)—is produced by adding benzine to fuming nitric acid gently heated. It is an oily liquid, of a sweet taste, heavier than water, boils at 415°, and is used by perfumers and confectioners under the name of *Essence of Mirbane*, or bitter almonds. From it *Azobenzide*, $C_{12}H_5N$, may be obtained by distillation with an alcoholic solution of caustic potassa, in the form of red crystals; the liquid contains *Aniline*. *Binitrobenzole*, $C_{12}H_4O_8N_2$, results from the action of nitric and sulphuric acids upon benzine. In these derivatives of benzole one and two atoms of hydrogen are replaced by one and two atoms of NO_4.

Chlorobenzine—Chlorobenzole ($C_{12}H_6Cl_6$)—is formed by the union of benzine and chlorine in the sun's rays. When distilled, the solid yields hydrochloric acid and a liquid, *Chlorobenzide* ($C_{12}H_3Cl_3$).

Hippuric Acid ($C_{18}H_8O_5N=HO$) is found in the urine of graminivorous animals, and occurs in the urine of persons who have taken benzoic acid. It may be prepared by evaporating the fresh urine of the cow, and acidulating the concentrated liquor with hydrochloric acid; crystals of hippuric acid are deposited, which may be purified by dissolving in boiling alcohol. On cooling, colorless four-sided prisms separate; they are soluble in 500 parts of cold water, but are abundantly dissolved by boiling water and alcohol. By a high temperature and the action of dilute nitric or hydrochloric

How are sulphobenzide and nitrobenzide made? Give some of the other derivations of benzole. Where does hippuric acid naturally occur? What are its properties?

acid, it yields benzoic acid and *gelatin sugar*, or *Glycocol.* It gives rise to a series of salts—the hippurates.

The Salicyle Group.

There is contained in the bark of the willow and other trees a bitter crystalline principle, Salicine ($C_{26}H_{18}O_{14}$). It may be extracted by boiling the bitter bark in water, and digesting the concentrated solution with oxide of lead to decolorize it, removing any dissolved lead by sulphureted hydrogen, and evaporating until the salicine crystallizes. It forms white needles of a bitter taste, much more soluble in hot than cold water. The solutions have a bitter taste, and are lævo-gyrate with regard to polarized light. Salicine is characterized by the deep red color it gives with strong sulphuric acid; when boiled with sulphuric acid, grape sugar is a product; nitric acid converts it into oxalic and carbazotic acids.

The Salicyle Group.

Salicyle, $C_{14}H_5O_4$	$=Sa$
Salicylous Acid	$=SaH$
Iodide of Salicyle	$=SaI$
Chloride of "	$=SaCl$
Etc.	Etc.

Salicylous Acid—Oil of Spiræa Ulmaris, or Meadow Sweet ($C_{14}H_5O_4H$)—is prepared by distilling one part of salicine, one of bichromate of potassa, two and a half of sulphuric acid, and twenty of water, the salicine being dissolved in one part of the water and the acid mixed with the rest. The yellow oil which comes over is rectified from chloride of calcium. It may also be obtained by distilling the flowers of meadow-sweet with water. It is transparent, but turns red in the air; is slightly soluble in water, but very soluble in alcohol; specific gravity 1.173; boils at 385°; contains the same elements as benzoic acid.

Salicylic Acid ($C_{14}H_5O_4+O$) is obtained by the action of hydrate of potassa on the foregoing body by the assistance of heat. After the disengagement of hydrogen is over, the mass is dissolved in water, and salicylic

How is salicine obtained? Give the tests for it. What is the composition of oil of spiræa? How is it made? Describe salicylic acid.

acid separates in crystals on the addition of hydrochloric acid. It is more soluble in hot than cold water, and is charred by hot oil of vitriol. The *salicylates* strike a violet-blue color with a persalt of iron. Certain insects oxidize salicine in their bodies, and, irritated on paper impregnated with a persalt of iron, produce violet spots. *Oil of Wintergreen* is a salicylate of the oxide of methyle.

Chloride of Salicyle is made by the action of chlorine on salicylous acid. Its crystals are insoluble in water, but soluble in solutions of the fixed alkalies, from which it separates on the addition of an acid, resisting decomposition even when boiled with caustic potassa. It unites with caustic potassa.

Bromide and *Iodide of Salicyle* are not of interest.

Chlorosamide ($C_{42}H_{15}N_6O_5Cl_3$).—Ammoniacal gas is absorbed by the chloride of salicyle, producing a yellow body, crystallizing from a boiling ethereal solution. It is insoluble in water. When acted upon by hot acids it yields a salt of ammonia and chloride of salicyle; an alkali forms with it ammonia and chloride of salicyle. There is an analogous bromosamide.

Salicylide of Potassium (KSl) is formed by the action of oil of meadow-sweet on a solution of caustic potassa. It forms in yellow crystals from its alcoholic solution, and has an alkaline reaction.

Melanic Acid ($C_{10}H_4O_5$) is produced when the crystals of salicylide of potassium are exposed in a moist state to the air. They first turn green and then black, and alcohol extracts from them melanic acid.

CINNAMYLE.

The essential oil of cinnamon is supposed to be the hydride of a compound radical, *Cinnamyle.*

The Cinnamyle Group.

Cinnamyle, $C_{18}H_7O_2$	$= Ci$
Hydride of Cinnamyle (Oil of Cinnamon)......	$= CiH$
Oxide of " (Cinnamic Acid).........	$= CiO$
Chloride of "	$= CiCl$
Etc.	Etc.

What is oil of wintergreen? How is chlorosamide produced? Describe the production of salicylide of potassium. How is melanic acid made? Describe the cinnamyle group.

Hydride of Cinnamyle (*Oil of Cinnamon*) is obtained by infusing cinnamon in a solution of salt and then distilling the whole. It is heavier than water, and may be separated from that liquid by contact with chloride of calcium.

Cinnamic Acid is formed when oil of cinnamon is exposed to oxygen, the oil becoming a white crystalline mass—hydrated cinnamic acid. It may also be made by dissolving the oil of Balsam of Tolu in an alcoholic solution of potassa, evaporating to dryness, dissolving in hot water, and adding to the cinnamate of potassa an excess of hydrochloric acid. It melts at 248°, and boils at 560°. It is soluble in boiling water and alcohol; is decomposed by nitric acid into oil of bitter almonds and benzoic acid.

Chlorocinnose ($C_{18}H_4Cl_4O_2$) arises from oil of cinnamon by the substitution of four atoms of chlorine for four of hydrogen, and is made by the action of chlorine on oil of cinnamon by the aid of heat. It crystallizes from its alcoholic solution in colorless needles.

Cinnamole ($C_{16}H_8$), which bears the same relation to cinnamic acid that benzole does to benzoic acid, is formed when cinnamic acid is distilled with baryta. It is isomeric with *Styrole.*

LECTURE LXXX.

THE NITROGENIZED PRINCIPLES. — AMMONIA *and its Salts.*—CYANOGEN.—*Hydrocyanic Acid.—Amygdaline.—Cyanide of Potassium.—Cyanic, Fulminic, and Cyanuric Acids.*

AMMONIA.—I have already described in Lecture LVI. the compounds of hydrogen and nitrogen, under the names of amidogen, ammonia, and ammonium, and have also shown the relation there is between the salts of potassa and soda and those of the oxide of ammonium. This compound metal is a hypothetical body; its exist-

How is oil of cinnamon procured? What is cinnamic acid? State its properties. How are chlorocinnose and cinnamole made? What compounds of nitrogen and hydrogen are there? How may ammonium be produced?

ence may, however, be illustrated by passing a Voltaic current through a globule of mercury in contact with moist chloride of ammonium, or by putting an amalgam of mercury and potassium in a strong solution of that salt. The mercury rapidly increases in volume, retaining its metallic aspect, becomes of the consistency of butter, with a very trivial increase of weight, the resulting substance being *Ammoniacal Amalgam.* All attempts to insulate ammonium from it have failed.

The most important salts of ammonia are the following:

Chloride of Ammonium—Sal Ammoniac—was formerly brought from Egypt, but is now made from the ammoniacal liquors resulting from the destructive distillation of animal matters, coal, etc. It is soluble in an equal weight of boiling water, crystallizes in cubes or octahedrons, and sublimes below a red heat unchanged. It is decomposed by lime and potassa, and is formed when the vapors of ammonia mingle with those of hydrochloric acid. It is much used for tinning metals in soldering.

Nitrate of Ammonia is formed by neutralizing nitric acid with ammonia. It is deliquescent, and therefore very soluble in water. At 22° it fuses, at 356° boils, at 400° is resolved into protoxide of nitrogen and water, at 600° decomposes with slight explosion.

Carbonates of Ammonia.—The *neutral carbonate* only exists in combination. With the carbonate of water it unites, forming *Bicarbonate of Ammonia*, which may be prepared by washing the commercial *Sesquicarbonate* with water or alcohol, which leaves it undissolved. The carbonate of ammonia of commerce, *Salt of Hartshorn*, is prepared by sublimation from a mixture of sal ammoniac and chalk. Its constitution is not uniform, though it is commonly regarded as a sesquicarbonate ($2NH_3$, $3\,CO_2$, $2HO$).

Sulphate of Ammonia may be made by neutralizing sulphuric acid with carbonate of ammonia. It is soluble in twice its weight of cold water, and crystallizes in six-sided prisms.

Can ammonium be isolated? What are the sources of sal ammoniac? What change occurs in nitrate of ammonia at 400°? Describe the carbonates of ammonia.

Hydrosulphate of Ammonia — Sulphide of Ammonium—is made by passing sulphureted hydrogen into water of ammonia until no more is absorbed. Though colorless at first, it absorbs oxygen, and sulphur being liberated, it turns yellow. It is of considerable use as a test for metals.

CYANOGEN—*Bicarburet of Nitrogen* (C_2N).—The mode of preparing this remarkable body, and also its leading properties, have been described in Lecture LVI. It is of great interest in organic chemistry, as being the first distinctly-established compound radical, and the best representative of the electro-negative class of those bodies.

We may call to mind that it is easily made by the decomposition of cyanide of mercury at a low red heat, is a condensible gaseous body, soluble in water, and therefore must be collected over mercury. It is combustible, and burns with a purple flame.

The Cyanogen Group.

Cyanogen, C_2N	Cy
Hydrocyanic Acid	CyH
Cyanic Acid	CyO
Fulminic Acid	Cy_2O_2
Cyanuric Acid	Cy_3O_3
Etc.	Etc.

Paracyanogen (C_4N_2).—When the cyanide of mercury is decomposed in the process for preparing cyanogen, a brownish substance is set free, which is paracyanogen. It is insoluble in water and alcohol, and is polymeric with cyanogen.

Hydrocyanic Acid—Prussic Acid—Cyanide of Hydrogen—may be obtained in a state of purity by passing dry sulphureted hydrogen over dry cyanide of mercury in a tube, and conducting the vapor which is evolved when the tube is warmed into a vial immersed in a freezing mixture. The result of the decomposition is sulphide of mercury and hydrocyanic acid. In a state of aqueous solution, it is best obtained by the action of dilute sulphuric acid on ferrocyanide of potassi-

How is hydrosulphate of ammonia made? What is the composition of cyanogen? Why is it of interest? Name some members of the cyanogen group. What is paracyanogen? How is hydrocyanic acid made? Give another method.

um in a retort, and receiving the vapor in a Liebig's condenser. Having ascertained the strength of the product, it may then be diluted to the proper point. This examination may be conducted by precipitating a known weight of the acid with nitrate of silver in excess, collecting the cyanide of silver on a weighed filter, washing, drying, and reweighing, which gives the weight of the cyanide. This, divided by five, is the weight of the pure hydrocyanic acid nearly.

Anhydrous hydrocyanic acid is a colorless and very volatile liquid, which exhales a strong odor of peach-blooms; specific gravity .696; boils at 80°, congeals at 4°. It mixes with water and alcohol in any proportion. A drop of it held in the air on a glass rod becomes solidified by the rapid evaporation from its surface. In the sunlight it decomposes rapidly, producing a dark-colored substance; the same change goes on, though more slowly, in the dark. It is one of the most insidious and terrible poisons, a few drops producing death in a few seconds; and its vapor, even largely diluted with air, brings on very unpleasant symptoms. Under the action of strong acids it is decomposed into ammonia and formic acid, the change being very simple:

$$C_2N, H+3HO=NH_3+C_2HO_3.$$

Under such circumstances, hydrocyanic acid yields chloride of ammonium and hydrated formic acid. Hydrocyanic acid may to a certain extent be preserved from spontaneous change by the presence of a minute quantity of any mineral acid.

Prussic acid may be detected by its smell, and by yielding a precipitate of Prussian blue when acted upon in solution successively by sulphate of iron, potassa, and an excess of hydrochloric acid. The liquid in which the poison is suspected to exist should be acidulated with sulphuric acid and distilled: the hydrocyanic acid will be found in the first portions which come over.

Amygdaline ($C_{40}H_{27}O_{22}N$), a crystallizable substance found in bitter almonds, the leaves and berries of the cherry laurel, and the kernels of peaches, etc., is of considerable interest in connection with hydrocyanic acid,

---

What are the properties of hydrocyanic acid? What effect has it on animals? How may prussic acid be detected? What are the sources of amygdaline?

inasmuch as these organic bodies yield, when distilled with water, that substance. The change consists in the action of water upon amygdaline by the aid of an azotized ferment called *Synaptase* or *Emulsine*, which constitutes the larger portion of the pulp of almonds; the bitter almond oil at the same time makes its appearance. Amygdaline may be extracted from the paste of bitter almonds when the fixed oil has been expressed by the aid of boiling alcohol. The alcohol being subsequently distilled off, the sugar contained in the sirupy residue is destroyed by fermentation with yeast. The liquid, being evaporated again to a sirup, is mixed with alcohol, which precipitates the amygdaline as a white crystalline powder, purified by being redissolved in alcohol and left to cool. It is soluble in hot and cold water, but sparingly soluble in cold alcohol. A weak solution of it in water, under the influence of a small quantity of an emulsion of sweet almonds, yields at once oil of bitter almonds and hydrocyanic acid; sugar and formic acid are also produced. When amygdaline is boiled with an alkali it gives rise to *Amygdalic Acid*, which forms a salt with the alkali, and ammonia is evolved.

Cyanide of Potassium may be formed by the direct union of cyanogen and potassium, or by the ignition of the ferrocyanide of potassium in a close vessel. For common purposes in the arts it may be formed in a state somewhat impure by mixing eight parts of ferrocyanide of potassium, rendered anhydrous by heat, with three parts of carbonate of potassa, also dry, and fusing the mixture in a crucible, stirring it until the fluid part of the mass is colorless. The sediment of iron, etc., is allowed to settle and the clear liquid poured off: it is the substance in question, mixed with cyanate of potassa.

$$2(K_2FeCy_3+2(KO,\ CO_2)=5(KCy)+KO,\ CyO+Fe_2 +2\,CO_2.$$

The formation of the cyanate may be prevented by adding to the mixture, before fusing, one eighth its weight of powdered charcoal; the fused mass may then be digested in boiling alcohol, from which the cyanide crystallizes on cooling.

Cyanide of potassium is very soluble in water, yields

What ferment is found in almonds? What are the properties of amygdaline? How is cyanide of potassium made?

colorless octahedral crystals which deliquesce in the air, melts without change at a red heat, and exhales the odor of prussic acid. It is very poisonous, and is useful as a reducing agent in mineral analysis, and as a solvent for bromide and iodide of silver in photographic operations.

Cyanide of Mercury may be made by dissolving red oxide of mercury in hydrocyanic acid, or by the action of a solution of ferrocyanide of potassium on persulphate of mercury, the cyanide crystallizing from the filtered hot solution. It forms fine prismatic crystals, soluble in eight parts of water at 60°, and sparingly soluble in alcohol. It is poisonous, and, when decomposed at a low red heat, yields cyanogen gas.

Cyanic Acid ($CyO+HO$) is procured by heating in an air-tight retort cyanuric acid deprived of its water of crystallization. A colorless liquid comes over into the receiver; it is hydrated cyanic acid, and has a strong odor like acetic acid; is intensely corrosive, and produces blisters on the skin. It is decomposed by contact with water, carbonic acid being evolved; carbonate and cyanate of ammonia are formed, and, by evaporation, crystals of urea may be obtained. Cyanic acid is a very unstable body, spontaneously changing in a short time into *Cyamelide*, a body of the same constitution, but a white opaque solid, insoluble in water and alcohol, and decomposed by hot oil of vitriol into sulphate of ammonia, while carbonic acid escapes.

Fulminic Acid (Cy_2O_2) has not yet been isolated, but some of its salts, presently to be described, are characterized by the violence with which they detonate under very trivial disturbances. It is a bibasic acid.

Cyanuric Acid (Cy_3O_3+3HO) may be made by heating urea, which disengages ammonia; the residue is dissolved in hot sulphuric acid, and nitric acid added until the liquid becomes colorless; on mixing it with water and allowing it to cool, the cyanuric acid separates. Its crystals are efflorescent; it is sparingly solu-

---

What substances result from the decomposition? What use is made of cyanide of potassium? How is cyanide of mercury made? Give the process for procuring cyanic acid. What is the characteristic of the fulminates? How is cyanuric acid made? What are its properties?

ble in water, and is a tribasic acid; and, as has already been stated, at a red heat may be distilled, yielding cyanic acid without any other product.

LECTURE LXXXI.

BODIES ALLIED TO CYANOGEN.—*Salts of the Oxycyanogen Acids.*—FERROCYANOGEN.—*Prussiate of Potassa.*—*Prussian Blue.*—FERRICYANOGEN.—*Ferricyanide of Potassium.*—COBALTOCYANOGEN.—SULPHOCYANOGEN.—*Sulphocyanide of Potassium.*—*Melam.*

CYANATE OF POTASSA (KO, $Cy\,O$) may be prepared by oxidizing cyanide of potassium by oxide of lead in an earthen crucible. The result, boiled with alcohol, yields, on cooling, crystals of cyanate of potassa in thin transparent plates, which undergo no change in dry air, but with moisture become converted into bicarbonate of potassa and ammonia.

Cyanate of Ammonia—Urea ($C_2H_4N_2O_2$).—The vapor of hydrated cyanic acid, mixed with ammoniacal gas, yields cyanate of ammonia. The solution in water, when heated, gives off ammonia, and the cyanate changes into *Urea*, from which caustic alkalies can not disengage ammonia. Urea may also be made from the action of sulphate of ammonia on cyanate of potassa.

Fulminate of Silver ($2Ag\,O$, $C_4N_2O_2$) is made by dissolving silver in warm nitric acid, and adding alcohol. It separates from the hot liquid as a white powder, which, being washed in water, is dried in small portions, at a temperature of 100°, on filtering paper. It detonates with wonderful violence when either struck or rubbed. It is sparingly soluble in hot water, and crystallizes from that solution on cooling. It yields, by digestion with water and metals, salts, as those of zinc and copper.

Fulminate of Mercury ($2Hg\,O$, $C_4N_2O_2$) is prepared in the same manner as the foregoing, and, like it, is very explosive. It is used for making percussion caps. The

---

How is cyanate of potassa made? What is the relation between cyanate of ammonia and urea? How is fulminate of silver made? What peculiarities has it? What use is made of the fulminate of mercury?

gases evolved by its explosion are carbonic acid, nitrogen, and the vapor of mercury.

Chloride of Cyanogen ($Cy\,Cl$) is prepared by the action of chlorine on moist cyanide of mercury in the dark. It is a colorless gas, soluble in water, specific gravity 1.1244, congeals at 0°, boils at 11°, condenses into a liquid under a pressure of four atmospheres. A *liquid* compound, $Cy_2\,Cl_2$, is obtained from the same substances under the action of sunlight; it is a heavy, yellow, oily liquid, insoluble in water, but soluble in alcohol. The *solid* chloride is procured by adding hydrocyanic acid to dry chlorine, and exposing to the sunshine. These compounds are deadly poisons.

FERROCYANOGEN.

Ferrocyanogen ($C_6N_3Fe = Cfy$ or Fcy) is a compound radical, in which iron is an important constituent.

Hydroferrocyanic Acid (Cfy, H_2) may be obtained by decomposing the insoluble ferrocyanide of lead by sulphureted hydrogen while suspended in water. The solution, being filtered, is to be evaporated with sulphuric acid in vacuo until the acid is left solid. It may also be prepared by dissolving tartaric acid in alcohol, and pouring it into an aqueous solution of ferrocyanide of potassium, the acid separating on evaporation in small crystals. Another method consists in adding hydrochloric acid to a strong solution of ferrocyanide of potassium, and then mixing it with ether, which precipitates the acid. It is soluble in water, to which it gives a powerful acid reaction. It decomposes the alkaline carbonates with effervescence, forming ferrocyanides of their bases; is inodorous, and not poisonous; is permanent in the dry state, but, moistened and exposed to air, forms Prussian blue.

Ferrocyanide of Potassium — Prussiate of Potassa ($2K, Cfy + 3HO$).—This salt is made on the large scale by igniting carbonate of potassa, iron filings, and animal matter in an iron vessel. The mass is then acted on by hot water, dissolving out a large quantity of cy-

What are the properties of chloride of cyanogen? What is the composition of ferrocyanogen? How may hydroferrocyanic acid be made? What change occurs in it when exposed to air? Give the process for making prussiate of potassa.

anide of potassium, which is converted into the ferrocyanide by the iron. The filtered solution, on cooling, yields lemon-yellow crystals, soluble in four parts of cold water. It is not poisonous. At a red heat it decomposes, cyanide of potassium forming. It is a valuable reagent, forming insoluble precipitates in many metallic solutions: *white* with the salts of manganese, zinc, tin, cadmium, lead, bismuth, antimony, protosalts of iron, mercury, and silver; *yellowish-green* with those of cobalt; *reddish-brown* with those of copper and uranium; *blue* with the persalts of iron; *pea-green* with the salts of nickel.

Common Prussian Blue ($3\,Cfy+4Fe$) is prepared by precipitating a persalt of iron by solution of ferrocyanide of potassium. When dry, it is of a deep blue color, with a lustre of coppery-red. It is insoluble in water; is decomposed by alkaline solutions, which yield alkaline ferrocyanide, and precipitate oxide of iron. It is soluble in solution of oxalic acid, and then constitutes the basis of blue writing inks. It is also much employed as a paint.

Basic Prussian Blue ($3\,Cfy, 4Fe+Fe\,O_3$) is formed when the white precipitate yielded by a protosalt of iron with ferrocyanide of potassium is exposed to the air. As its formula shows, it is common Prussian blue with peroxide of iron. It differs from Prussian blue in the remarkable peculiarity that it is soluble in water.

FERRICYANOGEN.

Ferricyanogen—Ferridcyanogen ($C_{12}N_6Fe_2 = Cfdy$, or $Fdcy$)—is a hypothetical compound radical, which yields some compounds of interest.

Ferricyanide of Potassium ($3K+Cfdy$) may be made by passing chlorine through a dilute solution of ferrocyanide of potassium until it ceases to yield a precipitate with a persalt of iron. The liquid, being concentrated, produces, on cooling, deep red crystals, the solution of which is of a greenish color. It gives no precipitate with peroxide of iron, but with the protosalts a bright blue, lighter than Prussian blue, and known as Turnbull's blue.

What are the properties of prussiate of potassa? What precipitates does it yield? How is Prussian blue made? What is basic Prussian blue? How is ferricyanide of potassium made?

COBALTOCYANOGEN ($Cy_6 Co_2 = Cky$) is analogous to ferricyanogen in composition, and, like it, is tribasic.

SULPHOCYANOGEN ($C_2NS_2 = Csy$) is obtained by saturating a concentrated solution of sulphocyanide of potassium with chlorine, as well as by boiling a soluble metallic sulphocyanide in diluted nitric acid. It falls in the form of a yellow precipitate, which preserves its color when dry; is insoluble in water, alcohol, and ether, but soluble in hyposulphuric acid.

Sulphocyanic Acid—Hydrosulphocyanic Acid ($Cy S_2H$)—may be obtained by decomposing sulphocyanide of lead by dilute sulphuric acid, taking care to leave excess of the salt of lead, which may afterward be removed by sulphureted hydrogen. It is also formed when sulphocyanide of lead or silver is decomposed by sulphureted hydrogen. The hydrated acid is colorless, decomposed by exposure to air or heat, yields with the peroxide of iron a blood-red color, and exists in the saliva of man and the sheep.

Sulphocyanide of Potassium ($KCsy$) may be made by heating powdered ferrocyanide of potassium with half its weight of sulphur and one third of carbonate of potassa, and keeping it melted for a short time. The mass is then boiled with water, which dissolves out the sulphocyanide, and the solution, being concentrated, yields prismatic crystals of the salt. It is soluble in water and alcohol, and deliquesces in the air; melts at a red heat; its solution with peroxide of iron gives a blood-red color.

Melam ($C_{12}H_9N_{11}$) is produced when sulphocyanide of ammonium is distilled at a high temperature, or by heating dry sulphocyanide of potassium with twice its weight of sal ammoniac. It is insoluble in water, but dissolves in strong sulphuric acid. When heated, it yields mellone and ammonia.

Melamine ($C_6H_6N_6$), *Ammelin* ($C_6N_5H_5O_2$), and *Ammelid* ($C_{12}N_9H_9O_6$) are products of the decomposition of melam.

What precipitates does ferricyanide of potassium yield? How is sulphocyanogen made? Describe the production of sulphocyanic acid. What precipitate does sulphocyanide of potassium yield? When does melam arise? What products arise from the decomposition of melam?

LECTURE LXXXII.

MELLONE.—UREA.—*Mellone.—Mellonides of Hydrogen and Potassium.—Natural and Artificial Formation of Urea.—Uric Acid.—Its Derivatives.—Murexide.—Xanthic and Cystic Oxides.*

MELLONE ($C_6N_4=Me$).—If sulphocyanide of potassium be acted upon by chlorine or nitric acid, a yellow powder is deposited. This, when heated, gives off bisulphide of carbon and sulphur, and there is left a yellowish powder, which is mellone. The relation of its constitution with cyanogen is obvious. It resists a moderate heat without change, and combines directly with metals to form *Mellonides.*

Hydromellonic Acid (MeH_3).—By adding hydrochloric acid to a hot solution of mellonide of potassium, this acid separates as a white powder on cooling. It is partly soluble in hot water, and possesses strong acid powers.

Mellonide of Potassium (MeK_3) may be prepared by melting ferrocyanide of potassium with half its weight of sulphur, and adding, when the fusion is complete, five per cent. of dry carbonate of potassa. The resulting mass is acted on by water, and the solution, being filtered, is evaporated until, on cooling, it forms a mass of crystals, from which the sulphocyanide may be removed by alcohol, and the mellonide left. It is soluble in water, and yields, by double decomposition with the salts of baryta, lime, etc., mellonides of these bodies, for the most part sparingly soluble.

Urea ($C_2H_4O_2N_2$) may be obtained from urine by adding to it, when concentrated, a strong solution of oxalic acid. The precipitated oxalate of urea is to be boiled with powdered chalk, and the filtered solution concentrated until the urea crystallizes on cooling. It may also be made artificially by adding to a strong solution of cyanate of potassa an equal weight of dry sul-

What relation does mellone bear to cyanogen? How is hydromellonic acid made? How is mellonide of potassium prepared? How is urea obtained from urine? How may it be made artificially?

phate of ammonia; the solution is evaporated to dryness in a water-bath, and the urea dissolved out by alcohol. It crystallizes in prisms, very soluble in water, but permanent in the air. At a high temperature it gives off ammonia and cyanate of ammonia, cyanuric acid remaining. Urea contains the elements of cyanate of oxide of ammonium, has neither an acid nor alkaline reaction, is decomposed by hot alkaline solutions with evolution of ammonia, and by uniting with two atoms of water yields carbonate of ammonia, a result which takes place during the putrefaction of urine, the change being brought on by a nitrogenized ferment—the mucus of the bladder. Urea unites with acids, and forms, with nitric and oxalic acids, characteristic salts. According to Dr. J. C. Draper, it arises principally from the excess of nitrogenized matter taken into the system.

Uric Acid—Lithic Acid ($C_{10}H_4N_4O_6$)—may be obtained from the solid urine of serpents, which, being boiled in solution of caustic potassa and filtered, yields uric acid by the addition of hydrochloric acid, as a white, inodorous, and sparingly soluble powder, soluble without change in sulphuric acid, from which it is precipitated by water. Uric acid also exists in human urine, and appears to be always a product of the action of the animal economy. It may be precipitated by adding hydrochloric acid, and allowing the urine to stand twenty-four hours. Of its salts the urate of soda is interesting; it is the chief ingredient of gouty concretions in the joints, called chalk-stones. The urate of ammonia occurs as a urinary calculus, and is often deposited from urine as a reddish cloud or powder. *Guano*, the excrement of aquatic birds, contains a large proportion of uric acid.

Allantoin ($C_4H_3N_2O_3$) is prepared by boiling uric acid with peroxide of lead; the filtered solution, being concentrated, deposits prismatic crystals of allantoin on cooling. It is soluble in 160 parts of cold water. By a solution of caustic alkali it is decomposed into ammonia and oxalic acid, assuming during this change the elements of three atoms of water.

What are the properties of urea? What is the source of uric acid? What are chalk-stones? How is allantoin prepared?

Alloxan ($C_8H_4N_2O_{10}$) is made by the action of concentrated nitric acid on uric acid in the cold. The uric acid is to be added in small portions successively until about one third the weight of the nitric acid has been used. An effervescence takes place, nitrogen and carbonic acid are evolved, and there is left a white mass, from which the excess of acid is to be drained. The substance is then to be dissolved in hot water and crystallized. Its solution has an acid reaction and a bitter taste, and stains the skin purple, and with a protosalt of iron and an alkali yields a characteristic blue compound. If the nitric acid be very dilute, *Alloxantin* ($C_8H_5N_2O_{10}$) arises.

Alloxanic Acid (C_4HNO_4+HO) may be prepared by decomposing the alloxanate of baryta by dilute sulphuric acid. The alloxanate itself is obtained by the addition of barytic water to a warm solution of alloxan. It is a strong acid, decomposing carbonates, and even water, by the aid of zinc.

Mesoxalic Acid (C_3O_4+2HO) may be obtained by boiling a solution of alloxan with acetate of lead, the resulting mesoxalate of lead being decomposed by sulphureted hydrogen. It is a strong acid, resists a boiling heat, and is bibasic.

Mykomelinic Acid ($C_8N_4H_5O_5$) is prepared by boiling a solution of alloxan with an excess of ammonia, and then precipitating by an excess of dilute sulphuric acid. It is a light yellow powder.

Parabanic Acid ($C_6H_2O_6N_2$) is formed by the action of strong nitric acid on alloxan or uric acid by the aid of heat. The crystals form on cooling, and may be dried by draining and then recrystallized. It is soluble in water, reddens litmus, and forms beautiful prismatic crystals.

Oxaluric Acid ($C_6H_4O_8N_2$) may be made by decomposing a hot solution of the oxalurate of ammonia by dilute sulphuric acid and cooling rapidly. The ammonia salt is itself procured by boiling a solution of the parabonate of ammonia, when it crystallizes, on cooling, in small needles. Oxaluric acid is a white crystalline powder. It contains the element of one atom of para-

---

Give the process for making alloxan. How is alloxanic acid made? How are mykomelinic, parabanic, and oxaluric acids made?

bonic acid and three of water, and its solution by boiling yields oxalic acid and oxalate of urea.

Thionuric Acid ($C_8H_7N_3S_2O_{14}$), a bibasic acid, prepared by decomposing thionurate of lead with sulphureted hydrogen. It contains the elements of one atom of alloxan, one of ammonia, and two of sulphurous acid.

Uramile ($C_8H_5O_6N_3$).—When an excess of saturated solution of sulphurous acid in water is mixed with a cold solution of alloxan, and an excess of carbonate of ammonia with caustic ammonia added, and the whole boiled, the thionurate of ammonia is deposited on cooling. From this the lead salt used in the preparation of the foregoing acid may be obtained by acetate of lead. The thionurate of ammonia, with a little hydrochloric acid, being boiled in a flask, there separates a white body, which is uramile. It differs from thionuric acid in not containing the elements of two atoms of sulphuric acid. If the thionurate of ammonia is mixed with dilute sulphuric acid and evaporated in a water-bath, *Uramilic Acid* ($C_{16}H_{10}N_5O_{15}$) is deposited.

Murexide—Purpurate of Ammonia—may be made by the action of dilute nitric acid on uric acid, and then adding ammonia; or by boiling equal weights of uramile and red oxide of mercury with eighty times their weight of water rendered alkaline by ammonia. The liquid turns of a deep purple color, and, when filtered, deposits, on cooling, crystals of murexide in square prisms, which by reflected light are of a green metallic lustre, and by transmitted light of a purple. It is sparingly soluble in cold water, but much more so in hot, and is one of the most splendid colors known. It is made largely from guano, to be used as a dye pigment.

Murexan—Purpuric Acid.—Murexide is to be dissolved in a solution of caustic potassa, and dilute sulphuric acid added. It forms a yellow powder, and, dissolved in ammonia, gives rise to the foregoing body.

Xanthic Oxide ($C_4H_2N_2O_2$) occurs as a urinary calculus of a brown color and waxy aspect. The calculus may be dissolved in dilute potassa, and xanthic ox-

---

What is the composition of thionuric acid? How is uramile prepared? How does it differ from thionuric acid? When does murexide arise? What is its composition? Of what value is it? How is xanthic oxide made?

ide precipitates as a white powder by carbonic acid. It is a waxy body.

Cystic Oxide ($C_6H_6NS_2O_4$) occurs also as a urinary calculus. It is remarkable for the large quantity of sulphur it contains.

LECTURE LXXXIII.

THE VEGETABLE ACIDS.—*Tartaric Acid and its Salts.—Citric Acid.—Tannic Acid.—Gallic Acid.—Pyrogallic Acid.—Metagallic Acid.*

OF the vegetable acids, several will be described, with their associated alkalies. The following are treated of in this Lecture:

Tartaric	$C_8H_4O_{10}+2HO$	Maleic	$C_8H_2O_6+2HO$
Paratartaric	$C_8H_4O_{10}+2HO$	Fumaric	C_4HO_3+HO
Pyrotartaric	$C_6H_3O_5+HO$	Tannic	$C_{18}H_5O_9+3HO$
Tartralic	$C_8H_4O_{10}+3HO$	Gallic	C_7HO_3+2HO
Tartrelic	$C_8H_4O_{10}+HO$	Ellagic	$C_7H_2O_4$
Citric	$C_{12}H_5O_{11}+3HO$	Pyrogallic	$C_6H_3O_3$
Aconitic	C_4HO_3+HO	Metagallic	$C_6H_2O_2$
Malic	$C_8H_4O_8+2HO$		

Besides acids such as these, which constitute a very numerous group, there is another class, which pass under the name of *Coupled Acids*, the peculiarity of which is that they consist of an acid affixed or coupled to another body, which, without affecting the neutralizing power of the acid, accompanies it in all its combinations. Thus hyposulphuric acid couples with napthaline to form hyposulphonapthalic acid, which neutralizes just as much of any base as hyposulphuric acid could do, the napthaline not changing its powers.

Tartaric Acid ($C_8H_4O_{10}+2HO$).—A bibasic acid, which occurs, as has already been stated, in the juice of grapes and other fruits as bitartrate of potassa. It may be obtained by dissolving cream of tartar in boiling water and adding powdered chalk, a tartrate of lime precipitating. The rest of the tartaric acid may be obtained from the solution by the addition of chloride of

---

How is cystic oxide made? Name some of the vegetable acids, and give their composition. What are coupled acids? Give an example. Describe tartaric acid.

calcium, which yields another portion of tartrate of lime, and may be decomposed by digesting with an equivalent proportion of dilute sulphuric acid. The concentrated and filtered solution yields crystals acid to the taste, inodorous, and soluble both in water and alcohol. The solution decomposes by keeping. Tartaric acid gives several valuable salts.

Tartrate of Potassa—Soluble Tartar ($2KO, C_8H_4O_{10}$) —may be made by adding carbonate of potassa to cream of tartar. It is very soluble, and is useful as an aperient.

Bitartrate of Potassa—Cream of Tartar ($KO, HO, C_8H_4O_{10}$)—is the salt deposited from the juice of the grape during fermentation as *Argol.* It may be purified from the coloring matter it contains by solution in hot water and the action of animal charcoal. In cold water it is very sparingly soluble. It produces black flux when ignited in a close vessel, the black flux being carbonate of potassa enveloped in carbonaceous matter.

Tartrate of Potassa and Soda—Rochelle Salt—Salts of Seignette ($KO, NaO, C_8H_4O_{10}+8HO$)—may be procured by neutralizing a solution of the foregoing salt with carbonate of soda. On evaporation and cooling, it separates in large prismatic crystals. It possesses the property of causing the deposit of metallic silver on glass from solutions of the ammoniacal nitrate of silver, and has been much used by the author in the making of silvered glass specula for his reflecting telescope at Hastings.

Tartrate of Antimony and Potassa—Tartar Emetic ($KOSbO_3, C_8H_4O_{10}+HO$).—This valuable medicinal agent is made by boiling oxide of antimony in a solution of cream of tartar; on cooling, the crystals are deposited. They are much more soluble in hot than in cold water, and dissolve without decomposition. It has emetic and alterative qualities, and is poisonous in large doses.

Paratartaric Acid—Racemic Acid—has the same constitution as tartaric acid, and resembles it very closely, being found in the grapes of certain parts of Germany and France. Racemic acid, however, differs from

---

What is argol? What is the difference between soluble tartar and cream of tartar? What is the composition of Rochelle salt? What peculiar property has it? Give the formula for tartar emetic.

tartaric in yielding a precipitate with a neutral salt of lime.

Pyrotartaric Acid ($C_6H_3O_5+HO$) is obtained by the destructive distillation of tartaric acid at 400°, as a liquid.

The action of heat on tartaric acid is remarkable. When exposed to a temperature approaching 400° it melts, throws off water, and yields in succession the following group:

Crystallized Tartaric Acid	$C_8H_4O_{10}+2HO$
Metatartaric Acid......................	$C_8H_4O_{10}+2HO$
Isotartaric Acid	$C_8H_5O_{11}+HO$
Tartralic Acid..........................	$2(C_8H_4O_{10})+3HO$
Tartrelic Acid	$C_8H_4O_{10}+HO$
Anhydrous Tartaric Acid	$C_8H_4O_{10}$

All these, by the continued contact of water, pass back into the condition of tartaric acid.

Citric Acid ($C_{12}H_5O_{11}+3HO$), a tribasic acid occurring abundantly in the juice of lemons and other sour fruits, and separated therefrom by the aid of chalk and sulphuric acid. It is clarified by digestion with animal charcoal, and yields prismatic crystals of a pleasant taste, and soluble in both hot and cold water. The crystals are of two different forms, according to the conditions of their formation: those which separate in the cold by spontaneous evaporation contain five atoms of water, three of which are basic; but those which are deposited from a hot solution contain only four.

The citrates form a very numerous family of salts; for, as the acid is tribasic, we have them with three atoms of metallic oxide, or two of oxide and one of water, or one of oxide and two of water, besides subsalts.

Aconitic Acid—Equisetic Acid ($C_{12}H_6O_{12}+HO$)—is formed by fusing citric acid and dissolving the resulting brown product in water, the change being

$$C_{12}H_8O_{14}=C_{12}H_6O_{12}+2HO\,;$$

that is, one atom of hydrated citric acid yields one of aconitic acid and two of water. Aconitic acid occurs naturally in the varieties of aconite and in the equisetums. *Itaconic* and *Citraconic Acids* are produced

---

Describe the action of heat on tartaric acid. What is the source of citric acid? What crystalline forms has it? How may aconitic acid be formed?

by the continued influence of heat; their formula is $C_{10}H_4O_6+2HO$.

Malic Acid ($C_8H_4O_8+2HO$), a bibasic acid occurring in the juice of apples and other fruits. It may also be prepared from the stalks of rhubarb, in which it is found with oxalate of potassa. It is a colorless solid, soluble in water, the solution changing by keeping. When heated in a retort to 400°, it melts and then boils, emitting a volatile acid, *Maleic Acid* ($C_8H_2O_6+2HO$), which condenses with water in the receiver. If the heat be carefully maintained between 270° and 280°, the product is *Fumaric* or *Paramaleic Acid*, which is isomeric with maleic, and also aconitic acid.

Tannic Acid ($C_{54}H_{22}O_{34}$), an astringent principle found in the bark of the oak, nut-galls, and other vegetable productions. It may be separated by placing in a vessel, *b*, *Fig.* 307, powdered galls. On pouring on them sulphuric ether containing ten per cent. of water, a liquid drops through the funnel tube, *c*, into the bottle, *a*, spontaneously separating into two portions. The lower, which is a solution of tannic acid in water, is to be decanted and evaporated in the presence of sulphuric acid in vacuo. It yields tannic acid, or tannin, in the form of an uncrystallized mass. This acid is soluble in water, but much less so in ether, has an astringent taste, and reddens litmus paper. With the persalts of iron, it gives a characteristic and valuable precipitate of a black color, the basis of common writing ink. The following furnishes a good writing ink. Digest three quarters of a pound of bruised galls in a gallon of cold water, then add six ounces of sulphate of iron, with an equal weight of gum arabic, and a few drops of creasote. Let the mixture digest for two or three weeks, with occasional agitation, then decant. Tannic acid forms insoluble compounds with starch, gelatine, and other organic bodies, that with gelatine being of considerable interest: it is the basis of leather. From the

Fig. 307.

b
c
a

Where does malic acid occur? What effect has heat on malic acid? Where is tannic acid found? How is it prepared? Give the process for making ink. What is leather?

characteristic precipitate it gives with iron, it is used as a test for that metal, which must, however, be in the state of peroxide, as the protosalts are unacted upon. The gradual darkening of pale writing inks is due to the slow oxidation of the iron they contain. Tannin is very valuable as a preservative in the dry collodion process; a solution of it, being allowed to dry upon the sensitized collodion, will keep it in a condition impressible to light for many months.

Tannigenic Acid—Catechine ($C_{15}H_6O_6$)—is extracted by hot water from catechu. It forms white, silky crystals, not giving an insoluble precipitate with gelatine, but producing a green color with persalts of iron. By the action of caustic potassa in excess it yields a black insoluble substance, *Japonic Acid* ($C_{12}H_4O_4+HO$). Carbonate of potassa converts it into a red acid, *Rubinic Acid* ($C_{18}H_6O_4$).

In coffee and tea there exist similar acids, the Caffeotannic ($C_{14}H_8O_7$) and Boheic ($C_{14}H_6O_8$).

Gallic Acid ($C_{14}H_6O_{10}$) may be formed by exposing a solution of tannic acid to the air, or by making powdered galls into a paste with water, and keeping it exposed in a warm place to the air for some weeks. The mass is then pressed, and boiled with water. On cooling, the solution precipitates a quantity of gallic acid, which may be purified by recrystallization. Like tannic acid, this substance yields no precipitate with a protosalt of iron, but a deep blue black with a persalt. It does not, however, precipitate gelatine; the crystals are soluble in one hundred parts of cold and three parts of boiling water; the solution has an astringent taste. Gallic acid is used in photography for reducing silver from the nitrate and iodide of silver under the influence of light.

Tannic acid passes into gallic acid by oxidation, carbonic acid and water being evolved. Dilute hydrochloric or sulphuric acid convert it into gallic acid and sugar,

$$C_{54}H_{22}O_{34}+10HO=3(C_{14}H_6O_{10})+C_{12}H_{14}O_{14};$$

What precipitate does it yield with iron? Of what use is it in photography? How does tannigenic acid originate? What acids exist in coffee and tea? How may gallic acid be prepared? Of what use is it? When tannin oxidizes, what products arise?

that is, one atom of tannic acid and ten of water produce three of gallic acid and one of sugar.

Ellagic Acid ($C_{14}H_2O_7+3HO$) is produced with gallic acid when moistened galls are exposed to the air. It is a gray, crystalline powder, and has been found in the intestinal concretions called *Oriental Bezoars.*

Pyrogallic Acid ($C_{12}H_6O_6$) is largely manufactured for photographic uses, being one of the principal *developers* in the collodion process. It is obtained by subliming gallic acid at a temperature between 410° and 420°: pyrogallic and carbonic acids are produced,

$$C_{14}H_6O_{10}=C_{12}H_6O_6+2\,CO_2;$$

that is, one atom of gallic acid yields one of pyrogallic and two of carbonic. It is in the form of white acicular and lamellar crystals, feebly acid, and of an astringent, bitter taste, soluble in water, alcohol, and ether. It is permanent when dry or dissolved in alcohol, but the aqueous solution oxidizes and turns brown. On account of this property, it may be used for the analysis of air, especially when associated with an alkali. It gives a black precipitate with the salts of silver.

Metagallic Acid ($C_{12}H_4O_4$) is formed when gallic acid is suddenly heated in a retort to 500°. It is a black mass, insoluble in water, but soluble in alkalies, from which it is precipitated as a black powder by acids.

LECTURE LXXXIV.

THE VEGETABLE ALKALOIDS. — *General Properties of the Vegetable Alkaloids. — Morphia. — Narcotina. — Codeia. — Meconic Acid. — Quinia. — Cinchonia. — Strychnia. — Brucia. — Artificial Alkaloids. — Aniline, and the Dyes formed from it.*

THE vegetable alkaloids constitute an extensive class of bodies, which are for the most part the active medicinal agents of the plants in which they occur. They are generally sparingly soluble in water, but more solu-

What do Oriental Bezoars contain? How is pyrogallic acid manufactured? What relation does it bear to gallic acid? What are its properties? What is metagallic acid? What general properties have the alkaloids?

ble in boiling alcohol, of a bitter taste, and characterized by containing nitrogen. In their natural state they are united with an acid, and, possessing basic properties in a very marked manner, neutralize acids completely. This quality seems to depend on the nitrogen they contain, and has no reference to their oxygen, for the quantity of this latter element present seems to have no relation to their neutralizing power, and, indeed, in some of them it is not present at all. In many respects they are analogous to ammonia, their salts, unlike those of some of the compound radicals, such as ethyle, undergoing decomposition in the same manner as the salts of ammonia. Thus the chloride of ethyle does not decompose the nitrate of silver, but the analogous compounds of ammonia and the vegetable alkaloids do; and these bodies may therefore be separated from the natural combinations in which they occur precisely as we should separate lime, or potassa, or magnesia in their salts. Most of the vegetable alkaloids are poisonous bodies, and, indeed, among them we meet with some of the most terrific poisons known. There are several artificial substances, such as *Aniline*, and those containing arsenic and platinum, which ought to be classed with these basic bodies.

One general method is applicable to the separation of these bodies. The substance containing them is boiled with dilute hydrochloric acid, the solution filtered, and treated with ammonia, lime, or magnesia. The alkaloid separates, and is purified by alcohol or ether. Most of these bodies are of importance in medicine.

Morphia ($C_{35}H_{20}O_6N+2HO$) is the active principle of opium, and was the first discovered of the alkaloids. It was isolated by Sertuerner in 1803. It may be prepared by mixing an infusion of opium with acetate of lead in excess; the meconate of lead is separated by a filter, and through the solution containing acetate of morphia a stream of sulphureted hydrogen is passed. The solution is warmed to expel the excess of gas, filtered, and mixed with ammonia, which throws down the morphia and narcotine: these are separated by boil-

What relation have they to the animal system? What is the general method for their separation?. What is morphia? Give the process for making it.

ing ether, which dissolves the latter. Turkey opium yields about an ounce to the pound of morphia.

Morphia is, when obtained from its alcoholic solution, in small crystals, six-sided prisms with dihedral terminations. It is almost insoluble in water, bitter to the taste, neutralizes acids, and forms crystallizable salts. It dissolves readily in dilute acids; the most common of its salts are the hydrochlorate, sulphate, and acetate. With nitric acid it gives a bright red color, with neutral perchloride of iron a blue color, and with iodic acid a reddish-brown color with the odor of iodine.

Narcotina ($C_{48}H_{24}O_{15}N$) is associated with morphia in opium. It may be obtained by digesting powdered opium in warm ether, which takes up little else than the narcotina, and yields it in crystals—rhombic prisms insoluble in cold water. It is soluble in volatile and fat oils, but insoluble in alkaline solutions: dilute acids form with it bitter solutions. By the action of peroxide of manganese and sulphuric acid, and by bichloride of platinum, an extensive class of bodies is produced, some acids, others bases.

Codeia ($C_{35}H_{20}O_5N+2HO$) is found in the hydrochlorate of morphia: it remains in solution when the morphia is precipitated by ammonia. It crystallizes in acicular or flat prisms, colorless and transparent. *Narceia*, *Thebaia*, *Papaverine*, and *Meconine* are other crystalline principles found in opium.

Meconic Acid ($C_{14}HO_{11}+3HO$), a tribasic acid associated with morphia in opium. It may be obtained from the meconate of lime, which precipitates in the preparation of morphia by chloride of calcium from infusion of opium. The precipitate is washed in water and hot alcohol, and warm dilute hydrochloric acid is added until all the lime is removed. It crystallizes in transparent micaceous scales, soluble in water and alcohol. When heated it loses 21½ per cent. of water, but if a strong solution be boiled it becomes dark-colored, carbonic acid is evolved, and oxalic acid and *Comenic* (Metameconic) acid are formed. The formula of comenic acid is ($C_{12}H_2O_8+2HO$). Meconic acid forms

What are its properties? How may narcotine be obtained? Give the properties of codeia. What is the composition of meconic acid? What bodies are derived from it?

with the persalts of iron an intensely red color. It forms several series of salts, like all tribasic acids.

Comenic acid, when heated, yields carbonic acid and *Pyromeconic Acid* ($C_{10}H_3O_5 + HO$), with a small quantity of *Parameconic Acid.*

Quinia—Quinine ($C_{20}H_{12}NO_2$).—This, which is one of the most valuable of the vegetable alkaloids, is obtained from *Cinchona Bark*, particularly predominating in yellow bark. The decoction of the ground bark in dilute hydrochloric acid is to be boiled in an excess of milk of lime, and the precipitate acted upon by boiling alcohol; on evaporation, *Cinchonia* is deposited in crystals, but the quinia remains in solution. It may be precipitated by the addition of water, and obtained in crystals from the spontaneous evaporation of its solution in absolute alcohol. Quinia neutralizes acids perfectly, giving rise to salts, of which the hydrochlorate, phosphate, sulphate, etc., are employed in medicine for the treatment of miasmatic disorders, and as tonics. It is sparingly soluble in water, but very soluble in alcohol or acids. The basic sulphate of quinine, a common preparation, is slightly soluble in water, the neutral sulphate much more so. For this reason, sulphate of quinia is dissolved in dilute sulphuric acid; the solution has a peculiar bluish opaline tint, the result of *fluorescence.*

Cinchonia ($C_{20}H_{12}ON$) is obtained, as just stated, in the preparation of quinia, with which it is associated in bark, and is found in the principal varieties of red and gray Peruvian bark. It crystallizes in prisms, requiring 2500 parts of water at 212° for their solution, and is sparingly soluble in alcohol, ether, and fixed oils. It has but little taste, but when mixed with an acid becomes intensely bitter.

Other analogous bodies exist in the different species of bark—*Quinoidine*, a mixture of quinine and *Quinidine* ($C_{40}H_{24}N_2O_4+4HO$), *Quinicine* ($C_{40}H_{24}N_2O_4$), and *Aricine* ($C_{48}H_{26}N_2O_8$).

Kinic Acid—Cinchonic Acid ($C_7H_5O_5+HO$)—is combined with the foregoing bodies in bark. It is ob-

---

Why is quinine valuable? What is its source? What salts are there of it? What peculiarity has the solution of sulphate of quinine? Describe cinchonia. What other bodies exist in bark? How is kinic acid obtained?

tained by decomposing the kinate of lime, obtained in the manufacture of sulphate of quinine by oxalic acid, filtering the solution from oxalate of lime; the kinic acid crystallizes on evaporation. It is very soluble in water.

Strychnia ($C_{44}H_{24}O_4N_2$) occurs in *Nux Vomica*, *St. Ignatius's Bean*, in the poison *Upas Tieute*, and other vegetable products. It may be extracted from nux vomica seeds by boiling them in dilute sulphuric acid, and then acting with lime and alcohol, as described in the case of quinia.

Strychnia requires 7000 parts of cold and 2500 parts of boiling water to dissolve it; water containing one forty-thousandth of its weight of it is rendered sensibly bitter. It forms a series of soluble, bitter, and poisonous salts, and may be precipitated from their solutions by the caustic alkalies as a white substance, soluble in ether and chloroform. It is a violent poison, causing intense muscular contractions; the antidote is tea. The best test is to immerse a frog partially in the suspected solution; he will become tetanized if strychnia be present.

Brucia ($C_{44}H_{25}O_7N_2$) is associated with strychnia, and, being very soluble in cold alcohol, is readily separated from it. It is also more soluble in hot water, requiring only 500 parts. It has one sixth the poisoning power of strychnia. These substances are found in union with a peculiar acid, *Strychnic* or *Igasuric Acid*.

The following table gives the names of other vegetable alkaloids, and bodies analogous to them:

Aconitine.	Daturine.	Picrotoxine.
Antearine.	Delphinine.	Piperine.
Asparagine.	Elaterine.	Phloridzine.
Atropine.	Emetine.	Populine.
Caffeine—Theine.	Gentianine.	Salicine.
Chelidonine.	Hesperidine.	Solanine.
Colchicine.	Hyosciamine.	Stramonine.
Conine.	Meconine.	Thebaine.
Curarine.	Narceine.	Theobromine.
Daphnine.	Nicotine.	Veratrine.

Nicotine is an oily liquid procured from tobacco; in-

What is the source of strychnia? What are its properties? What effect has strychnia on animals? Describe brucia. Name the vegetable alkaloids. What are the properties of nicotine?

flammable and poisonous, a single drop killing a dog. *Conine*, procured from hemlock, is a deadly poison, producing death by paralyzing the muscles of respiration. Two grains of it, neutralized with hydrochloric acid, and injected into the femoral vein of a dog, killed him in three seconds. *Theine* and *Caffeine* ($C_8H_5O_2N_9$) are found in coffee and tea. Their effect on the nervous system is well known.

The following bases are formed artificially:

Aniline—Kyanol ($C_{12}H_7N$)—is one of the ingredients of coal-tar. It may be obtained in large quantities by distilling nitrobenzole with a mixture of acetic acid and zinc or iron, in which case it is decomposed by nascent hydrogen. The decomposition is as follows:

$$C_{12}H_5NO_4 + 6H = C_{12}H_7N + 4HO;$$

that is, one atom of nitrobenzole and six of hydrogen produce one atom of aniline and four of water.

It is an oily liquid, boiling at 360°, specific gravity 1.020; soluble in cold water, alcohol, ether, wood-spirit, aldehyde, acetone, sulphide of carbon, and in fixed and volatile oils. It forms a series of salts with acids, and is now largely employed in the manufacture of the coal-tar colors.

Aniline Purple (*Mauve*).—One equivalent of a neutral salt of aniline is dissolved in water, and boiled for several hours with six equivalents of chloride of copper. When the reaction is complete the mixture is filtered, the black precipitate well washed and dried, and afterward digested repeatedly in dilute alcohol in order to dissolve out the coloring matter, which is aniline purple. By heating anhydrous hydrochlorate of aniline with nitrate of lead to 360°, a bronze-like, brittle mass is obtained, which contains aniline red mixed with aniline purple. The red may be separated from the purple by boiling water; one grain of it will strongly color a gallon of water. These coloring matters are fixed on cotton by preparing the goods with a solution of tannin and the coloring matter, and then passing them through a bath containing tartar emetic. The tannate of antimony thus produced fixes the dye.

What are the properties of conine? What is the composition of theine and caffeine. How is aniline made? What are the properties of aniline? Why is it valued?

Mauve is prepared by adding bichromate of potassa to sulphate of aniline: one tenth of a grain will form a rich violet-colored solution with a gallon of alcohol.

Aniline Red (*Magenta*).—Corrosive sublimate and aniline form by mixture a colorless paste: when heated it acquires an intense crimson color. From this product, which seems to be a salt, *Rosaniline* ($C_{20}H_{21}N_3O$) is procured; it acts the part of a base, and with acids, at a moderate heat, produces magenta.

Roseine results from the mixture of sulphate of aniline and peroxide of lead. *Fuchsine* is produced when aniline is heated with bichloride of tin, corrosive sublimate, nitrate of mercury, arsenic acid, or indigo. *Bleu de Paris* arises when aniline is heated with bichloride of tin to a high temperature in a close vessel.

The colors obtained by these processes depend on the oxidation of aniline, and vary with the degree of oxidation. The processes of Perkins, in which the oxidation is effected by bichromate of potassa, produce very perfect colors. Variation of tint depends on the proportion of the ingredients used; for example, 10 parts of aniline added to a mixture of 12 of arsenic acid and 12 of water, and heated to about 248°, yield a rich red with a violet tint. The same quantity of aniline, with 24 of arsenic acid and 24 of water, give a purple or violet.

The selection of a proper mordant materially affects the results. Of all yet proposed, the stannate of soda appears to be the most efficient.

Leukol ($C_{18}H_8N$), formed with aniline in oil of coal-tar, from which it may be separated by distillation. It is also an oily liquid, and can yield crystallizable salts.

Quinoline ($C_{19}H_8N$), made by distilling quinine or strychnine with caustic potassa. An oily liquid, very bitter, strongly alkaline, yielding crystallizable salts.

Besides these bodies, there are other artificial bases of an analogous nature, but which differ in the remarkable particular of containing platinum and arsenic; such, for example, as the platinum bases of Reiset and Gros, or

---

How is mauve produced? What mordant does it require? How is magenta made? How are roseine, fuchsine, and bleu de Paris prepared? What is the chemical explanation of the aniline colors? What is the best mordant for them? How is quinoline made? What other organic bases are there?

the arsenico-platinum radical kakoplatyle. The formation of these organic bases leads us to hope that the vegetable alkaloids themselves will hereafter be artificially formed.

LECTURE LXXXV.

THE COLORING BODIES.—*Their General Properties.—Dyeing.—The Non-nitrogenized Coloring Matters.—The Nitrogenized Coloring Matters.—Indigo.—Colorless Indigo.—Bodies derived from Indigo.—Litmus.—Chlorophyll.—Carmine.*

THE coloring principles derived from the organic kingdom may be conveniently divided into two classes, the non-nitrogenized and the nitrogenized. They may also be readily classed into groups, as blue, red, yellow, green. For the most part they are derived from vegetable productions.

For some coloring matters, the fibres of those tissues commonly employed for clothing have a sufficient affinity to hold the color so that it can not be removed by mere washing, and is permanently dyed. Such colors are called *substantive.* But in other instances this is not the case; the artist has then to avail himself of the properties possessed by intermediate bodies, such as alumina and the oxide of tin, called *mordants*, which at once possess the double quality of an affinity for the coloring matter and an affinity for the cloth fibre. Colors requiring a mordant are termed *adjective* colors. The attraction of these bodies for coloring matter may be illustrated by precipitating alumina in a solution tinged by litmus; the solution becomes perfectly clear, its color going down with the precipitate, and forming with it a lake.

NON-NITROGENIZED COLORING MATTERS.

The *blue* non-nitrogenized coloring matters are chiefly found in flowers and fruits. They are reddened by acids, and turned green by alkalies.

The *red* non-nitrogenized coloring matters are of some

What division is made of coloring principles? What is a substantive color? What is a mordant? What is an adjective color?

importance; among them may be mentioned *Madder* or *Garancine*. The plant is largely cultivated in Holland, and has a long, spreading root, which develops a red color during dyeing. *Alizarine* ($C_{20}H_6O_6$) is obtained by digesting garancine in boiling alcohol. *Madder Lake* is made by adding carbonate of soda to a decoction of madder root in alum.

Hæmatoxyline ($C_{16}H_7O_6+HO$) is the coloring matter of logwood. It is soluble in water and alcohol, and furnishes, with iron and alum bases, a black dye. The same principle is yielded by Brazil-wood and camwood.

Carthamine is a very beautiful red, obtained from the safflower. It is used in making pink saucers, and in the preparation of *Rouge*.

The *yellow* coloring matters. Among these may be mentioned *Quercitrine* ($C_{32}H_{15}O_{14}$), derived from the Quercus Tinctoria; *Gamboge*, the dried juice of the Garcinia Gambogia; *Turmeric*, used as a test for alkalies, which turn it brown, from the Curcuma Longa; and *Anatto*, employed for coloring cheese, from the seeds of Bixa Orellana.

Nitrogenized Coloring Matters.

The nitrogenized coloring matters, among which are some of the most valuable dyes that we possess, may be divided according to their tint.

Indigo is derived from the juice of several species of Indigofera, and is formed from a colorless or yellow compound, which is dissolved out from the leaves of these plants when they are allowed to ferment with water. A deep blue precipitate forms on the addition of lime-water and exposure to the air. It appears, therefore, to be a product of oxidation. It comes, in commerce, in small masses of a conchoidal fracture, which, when rubbed, exhibit a coppery aspect; is insoluble in water, alcohol, dilute acids, and alkalies; and may be sublimed, yielding a purple vapor, which condenses into crystals of pure indigo. It dissolves in about 15 parts of strong sulphuric acid, but still better in Nordhausen oil of vitriol. The mass produced is soluble in water;

What is madder? What are the reactions of hæmatoxyline? What is rouge? Name some of the yellow coloring matters. What is the source of indigo? What are its properties?

it is a solution of *Sulphindigotic Acid* ($C_{16}H_4O_2N+S_2O_5+HO$).

By contact with deoxidizing agents blue indigo becomes colorless, as may be shown by digesting powdered indigo, protosulphate of iron, hydrate of lime, and water together. In this state, as in its natural condition, it is soluble in water, and may be precipitated by hydrochloric acid. On exposure to the air, *Indigogene*, as this white indigo is called, absorbs oxygen rapidly, and becomes blue and insoluble.

When indigo is submitted to destructive distillation, it yields aniline, described in the last Lecture.

The relation existing between blue and white indigo is seen from their formulas:

Blue Indigo—Indigotine........................ $C_{16}H_5NO_2$
White Indigo—Indigogene..................... $C_{16}H_6NO_2$.

Under the action of heat and of reagents indigo yields an extensive class of bodies, to which much attention has been given. With dilute nitric acid it yields *Anilic* or *Indigotic Acid;* with strong nitric acid, *Picric* or *Carbazotic Acid* ($C_{12}H_2(NO_4)_3O, HO$), a substance of a yellow color, bitter taste, and forming explosive salts. Heated with bichromate of potassa, sulphuric acid, and water, it yields *Isatine* ($C_{16}H_5O_4N$), which crystallizes in reddish-brown prismatic crystals, inodorous, sparingly soluble in cold, but more so in hot water, readily soluble in alcohol, but less abundantly in ether. This body, under the influence of an alkaline solution, unites with one atom of water and changes into *Isatinic Acid.* Under the influence of chlorine, isatine yields *Chlorisatine*, an atom of chlorine substituting one of its hydrogen atoms; and *Bichlorisatine* by the substitution of two chlorine atoms for two hydrogen ones; and these again, as in the case of isatine itself, acted upon by alkaline solutions, yield each an acid. With bromine it produces *Bromisatine* and *Bibromisatine.* Caustic alkalies acting on indigo yield *Chrysanilic* and *Anthranilic Acids.*

Litmus is derived from the Rocella Tinctoria, Lecano-

How is indigogene prepared? What is the relation between blue and white indigo? How are indigotic and picric acids made? Describe isatine. What bodies arise from isatine? What is the source of litmus?

ra Tartaria, etc. These lichens give up to ether a crystalline substance, to which the name of *Lecañorine* is given. Its composition is ($C_{18}H_8O_8+HO$); it does not contain nitrogen. It is in white, inodorous, and tasteless stellated groups of acicular crystals, soluble in alcohol and ether. This substance, heated with baryta or alkalies, produces *Orcine* ($C_{16}H_8O_4+HO$) by losing two atoms of carbonic acid. Orcine crystallizes in flat, four-sided prisms, with dihedral summits: it has a sweet, repulsive taste, is vaporizable at 550°. Mixed with ammonia and exposed to the air, oxygen is absorbed, and the liquid assumes a deep purple tint. From this, acetic acid precipitates a deep red powder, *Orceine* or *Orceic Acid* ($C_{16}H_9O_7N$), which contains nitrogen, and is supposed to be the basis of the dye-stuff of litmus: with alkalies it gives a blue color. Litmus is extensively used in chemistry as a test for acids and alkalies. Litmus paper is white unsized paper, stained with an infusion of an ounce of litmus in half a pint of boiling water.

Chlorophyll ($C_{18}H_9O_8N$) is the green coloring matter of leaves. It is insoluble in water, but soluble in water and ether, and is a fatty substance. It is also found, under very interesting circumstances, in the animal system, as the coloring matter of bile. When an ethereal solution of it is long exposed to light, it acquires a yellow color, and leaves, on evaporation, a residue having all the characters of xanthophylline.

Xanthophylline is a term applied to the coloring matter extracted from the yellow leaves of autumn. *Erythrophylline* is obtained by digesting the leaves which redden, in alcohol.

Carmine is the coloring matter of the cochineal insect, Coccus Cacti. The coloring matter may be obtained from the insect by water or ammonia. The carmine of commerce is a lake prepared by adding alum to the freshly-filtered solution.

Aloes is the inspissated juice of certain species of Aloe, used as a purgative medicine. When heated with nitric acid, and water added, a yellow precipitate is thrown down, which, when purified, is *Chrysammic*

How are orcine and orceine made? What is chlorophyll? What substances are made from autumn leaves? What is carmine? Of what use is aloes? What is chrysammic acid?

Acid. It yields yellow crystals of a bitter taste, and furnishes a solution of a purple color. Its salts are crystallizable, by transmitted light of a red color, with a green metallic reflection, like murexide. The liquid from which this acid was precipitated contains picric acid.

LECTURE LXXXVI.

THE FATTY BODIES.—*Characteristics of the Fatty Bodies.—Fats.—Fixed and Volatile Oils.—Soaps.—Stearine and Stearic Acid.—Margarine and Margaric Acid. — Oleine. — Glycerine. — The Natural Oils, Palm Oil, etc.—The Volatile Oils.—The Camphors.*

THIS class of bodies is characterized by several well-marked peculiarities, and may be conveniently divided into two natural groups, oils and fats. They belong both to the vegetable and animal systems. In the former they usually abound in the seeds or fruits; in the latter, are deposited in the cellular structure of the adipose tissue. The natural fats are usually mixtures of two or more ingredients, differing from one another in consistency. In most instances they are stearine and margarine, along with a liquid, oleine. These oils can not be distilled without undergoing decomposition; exposed to the air they gradually absorb oxygen and evolve carbonic acid. Many of them, in which this change takes place with rapidity, turn into resinous bodies, and hence their application in painting as drying oils. Linseed oil, which is the most used for such purposes, has its drying qualities increased by boiling with litharge, and is also an important component of *Printers' Ink*, for which it is first heated and then set on fire, and allowed to burn for some time. When extinguished it is miscible with fresh oil or turpentine, and about a sixth of lamp-black is added.

Oily bodies may be divided into fixed and volatile; the former decompose when heated, the latter distill.

What are the general properties of the fatty bodies? What ingredients are there in natural fats? What change do the oils suffer in the air? Why is linseed oil valuable? How is printers' ink made? What division is made of oils?

A simple test suffices to distinguish them. When a few drops of an oily substance are put on paper, if it be a volatile oil it soon evaporates, and leaves the paper without a stain; if fixed, the paper remains greasy. The fixed oils have but little odor, the volatile oils commonly a characteristic one. They are all insoluble in water; many are soluble in alcohol; all are dissolved by ether.

By exposure to a low temperature the constituent principles of a mixed oil may often be separated from each other, the more solid substances separating first. When olive oil is thus treated, an exposure to 40° causes a deposit of *Margarine;* the fluid portion which is left is *Oleine.* Animal fats exposed to pressure between folds of blotting-paper communicate to it oleine, and the solid residue left behind is a mixture of margarine and *Stearine.* When the fixed fats are boiled with alkaline solutions, *Soaps* are formed. These substances, of extensive use in domestic economy and the arts from their detergent properties, are freely soluble in water. In the process of making them the fats undergo a change, true acids being liberated, which unite with the alkaline base. Stearine yields stearic acid; margarine, margaric acid; and oleine, oleic acid. They may be set free by decomposing the soap with an acid. At the same time, a sweet substance, *Glycerine*, appears; it is the base with which the acids were associated, oleine being, for instance, an oleate of glycerine. Of the varieties of soap met with in commerce, *Soft Soap* is made from potassa, combined with whale or seal oil; *Hard White Soap* from tallow and caustic soda; *Hard Yellow Soap* from soda, tallow, palm-oil, and resin. In the preparation of white soap the alkaline solution is made to boil, and tallow added in small portions until no more can be saponified. The solution now contains soap and free glycerine; the former is separated by the addition of common salt in a solution of which it is insoluble. It floats on the top of the liquid, and is then run into moulds, and cut into bars for commerce. In

---

How are oils distinguished? What is the process for analyzing mixed oils? What is soap? What is the reason of the appearance of glycerine? What is the difference between soft and hard soap? Describe the manufacture of soap.

this process the manufacturer does not add so much salt as to separate all the water. Commercial soap still contains from forty to fifty per cent.

Stearine ($C_{114}H_{110}O_{12}$) may be obtained from purified mutton fat by suffering a warm ethereal solution to cool. The stearine crystallizes, and margarine and oleine are left in solution. A repetition of the process purifies it. It is a white body, insoluble in water and cold alcohol, fusing at 140°; when saponified it yields glycerine and stearic acid.

Stearic Acid ($C_{36}H_{35}O_3+HO$) may be obtained by decomposing a soluble stearine soap by tartaric acid, and purifying the product by solution in boiling alcohol, from which it separates in crystalline flakes. It is white, inodorous, and tasteless; fuses at 160°, reddens litmus, may be distilled in vacuo, but is decomposed by a high heat in the open air. It forms monobasic and bibasic salts.

Margarine ($C_{108}H_{104}O_{12}$) is best obtained from olive oil by cooling it to 32°, pressing out the oleine, and dissolving the residue in boiling alcohol, from which the margarine separates in pearly crystals.

Margaric Acid ($C_{34}H_{33}O_3+HO$) is obtained by decomposing the soap of olive-oil and potassa by acetate of lead or chloride of calcium. The oleate and margarate of lead or lime is formed, the oleate is extracted by cold ether, and the remaining margarate decomposed by dilute hydrochloric or nitric acid. It crystallizes in white needles, the melting point being 140°.

Oleine ($C_{114}H_{104}O_{12}$).—When almond or rape oil is dissolved in ether, and the solution exposed to a low temperature, the margarine crystallizes, and oleine may be obtained by evaporating the ether. It remains liquid at a temperature of 0°.

Oleic Acid ($C_{36}H_{33}O_3$) is obtained from oleine by saponification and decomposition with hydrochloric acid, as in the foregoing instances. It solidifies at about 50°, and gives rise to a series of salts.

Margarone ($C_{66}H_{66}O_2$).—When a mixture of margaric acid and lime is distilled this substance is formed,

---

How may stearine and stearic acid be obtained? How may margarine and margaric acid be obtained? Describe the preparation of oleine and oleic acid. What is margarone?

and carbonic acid separates. It is a white solid, like spermaceti, and melts at 170°.

Glycerine ($C_6H_8O_6$) arises when any fatty matter is saponified with potassa, the soap being decomposed by tartaric acid and the glycerine dissolved out by alcohol. It is a sweet, colorless liquid, specific gravity 1.28, soluble in water and alcohol, but not in ether. It is not susceptible of vinous fermentation, but when left for some months in a warm place it produces *Propionic Acid.* Mixed with sulphuric acid, the two bodies unite, and *Sulphoglyceric Acid* ($C_8H_5O_7+2SO_3$) is the result, an acid having many analogies with sulphovinic. A corresponding *Phosphoglyceric Acid* is said to exist in the brain and yolk of egg. When glycerine is dropped into equal parts of strong nitric and sulphuric acids, *Nitroglycerine* ($C_6H_6(NO_4)_2O_6$), a very explosive and poisonous body, is formed.

Palm Oil is brought from Africa, and is used in the manufacture of yellow soap. It is of a reddish-yellow color, and contains, besides oleine, a solid fat, *Palmitine.* It is insoluble in water, slightly soluble in hot alcohol, but very soluble in ether. Its melting point is 118°. By saponification and decomposition with an acid it yields *Palmitic Acid*, the melting point of which is 140°. It is a bibasic acid.

Cocoa Tallow.—A solid fat obtained from the cocoa-nut, and used in the manufacture of candles. Its oleine and stearine may be separated by pressure or by boiling alcohol, from which the stearine crystallizes on cooling.

Among other fatty substances and allied bodies may be mentioned *Nutmeg Butter*, which yields, among other products, *Myristicine*, and, by saponification, *Myristic Acid. Elaidine* arises from the action of nitrous acid on oleine; it furnishes, by the common process, *Elaidic Acid. Suberic Acid* arises from the action of nitric acid on cork; *Succinic Acid*, by the destructive distillation of amber, or by the continued action of nitric on stearic acid; *Sebacic Acid*, by the destructive distillation of oleic acid. *Butyrine*, *Caproine*, and *Ca-*

When does glycerine arise? What bodies arise from glycerine? Describe palm oil. What is cocoa tallow? Name some other fatty substances and allied bodies.

prine, which are contained in butter, yield, by saponification and decomposition, *Butyric*, *Caproic*, and *Capric Acids*. Butyric acid can be made, as we have seen, artificially by fermentation. *Bees' Wax* is a mixture of three bodies—*Myricine*, insoluble in alcohol; *Cerine*, deposited in crystals as the solution cools; and *Ceroleine*, which is retained in solution. Vegetable wax is yielded by the *Myrica Cerifera* and some other trees. *Spermaceti* is obtained from certain species of whales; it yields, under the process for glycerine, a substance, *Ethal;* and this, under the action of hot potassa, gives *Ethalic Acid*, with evolution of hydrogen gas. *Cholesterine* is obtained from biliary calculi; it also occurs in the substance of the brain.

THE VOLATILE OILS.—These, for the most part, are found in plants, or are derived from them by simple processes. Many of them are extensively used in the arts in the manufacture of varnishes, and others in the preparation of perfumery. Their solutions in alcohol form *Essences*, and in water *Medicated Waters*. They are commonly obtained by the distillation of those parts of the plants in which they occur with water, and consist of two substances—a solid portion, or *Stearoptene*, or camphor, and an *Elaioptene*, or true oil. The former is an oxyhydrocarbon, the latter a hydrocarbon. The volatile oils may be divided into groups according to their constitution.

Volatile Oils containing Carbon and Hydrogen.

Turpentine.	Bergamotte.
Citron.	Cubebs.
Copaiva.	Etc.
Storax.	

Volatile Oils containing Carbon, Hydrogen, Oxygen.

Cajeput.	Pennyroyal.
Lavender.	Valerian.
Rosemary.	Spearmint.
Peppermint.	Etc.

Volatile Oils containing Sulphur.

Black Mustard.	Onions.
Horseradish.	Asafœtida.

What is bees' wax? What is spermaceti? Of what use are the volatile oils? What are essences and medicated waters? What is meant by stearoptene and elaioptene? What are the groups of volatile oils?

Of the first group, *Oil of Turpentine* may be taken as the type; the elementary composition may be regarded as (C_5H_4). It is principally procured from North Carolina, and is resolved by distillation into the volatile oil and *Yellow Rosin*. *Artificial Camphor* is made by passing dry hydrochloric acid gas into oil of turpentine: its constitution is ($C_{20}H_{16}+HCl$).

The stearoptens or camphors originate in several different ways; sometimes by the oxidation of the oils from which they are derived, sometimes they are hydrates of these oils, and sometimes they are isomeric with them. The stearoptens are best represented by common camphor, which is extracted from the *Laurus Camphora* of Japan, China, and Java by distillation with water. It is a white, tough, and semi-transparent mass; specific gravity .990; fuses at 370°, boils at 400°, and may be distilled without decomposition. It vaporizes at common temperatures, and its motion toward the light has been made the subject of interesting researches by Professor J. W. Draper.

Of the oils containing sulphur, the oil of black mustard seed ($C_8H_5NS_2$) is a good example: it arises from the action of *myrosine* upon *sinapine* in the presence of water. It has been regarded as the sulphocyanide of *Allyl* (C_6H_5). *Oil of Garlic* is an oxide and sulphide of allyl.

LECTURE LXXXVII.

THE RESINS, BALSAMS, AND BODIES ARISING IN DESTRUCTIVE DISTILLATION. — *Rosin.* — *Shellac.* — *Amber.* — *Caoutchouc.* — *Vulcanization.* — *Gutta Percha.* — *Products of the Destructive Distillation of Wood.* — *Tar.* — *Pitch.* — *Paraffine.* — *Creasote.* — *Destructive Distillation of Coal.* — *Coal Oil.* — *Carbolic Acid.* — *Products of Slow Decay.* — *The Varieties of Coal.* — *Petroleum.*

THE resins are bodies in many respects analogous to the camphors, but are distinguished from them by the

---

What is the composition of oil of turpentine? How do the camphors originate? Describe common camphor. What singular relation has it to light? What is the composition of oil of mustard? What are the resins?

circumstance that they are not volatile without decomposition. In many instances they act as acids; they all contain oxygen.

Colophony (*Common Rosin*) is a mixed resin, obtained by the distillation of turpentine with water, the oil of turpentine passing over. It is a mixture of two resins, *Pinic* and *Silvic Acids*, which may be separated by cold alcohol, in which the latter is insoluble.

Gum Lac, which is one of the resins, occurs under three forms—shell lac, stick lac, and seed lac. It is used in the preparation of lacquers, and is the chief ingredient in sealing-wax. Among other resins may be mentioned *Copal*, *Mastic*, *Dragons' Blood*, *Gamboge*, *Sandarac*, and *Dammar*.

Amber is a substance belonging to this class. It is found in beds of bituminous wood, and often incloses insects in a state of beautiful preservation. Its specific gravity is about 1.07. By distillation it yields succinic acid.

Caoutchouc—India Rubber—Gum Elastic—is the produce of the Jatropha Elastica, the Urceola Elastica, and several other tropical trees. It is found in small proportion in the poppy, lettuce, euphorbium, and other plants having a viscid, milky sap. The fresh juice is a yellow, milky fluid, which, when exposed to warm air, forms an elastic deposit of a dark color. Caoutchouc is a hydrocarbon, having the composition C_8H_7.

In its ordinary state it hardens at low temperatures, but does not become brittle; melts at 250°, and does not regain its former state on cooling. It is softened and dissolved by ether, chloroform, bisulphide of carbon, oil of turpentine, and coal naphtha. By the process of *vulcanization*, in which it is subjected to the action of sulphur at a temperature of about 300°, it is so modified that it resists the action of its ordinary solvents, and retains its pliability at both low and high temperatures. A hard compound of sulphur and rubber is called *Ebonite*. A large quantity of silicate of magnesia in fine powder is sometimes incorporated with the rubber before

---

How is common rosin obtained? What are the forms of gum lac? What is amber? What is the source of caoutchouc? What are its properties? Describe the process of vulcanization. Of what use is the silicate of magnesia in vulcanization?

vulcanization, to give a smooth and non-adherent surface. *Marine Glue* is made by dissolving a mixture of caoutchouc and shell lac in coal naphtha.

Gutta Percha is closely allied to caoutchouc, and is produced by the Isonandra Percha, a tree abounding in the islands of the Eastern Archipelago. It is a tough, unyielding, fibrous substance, of a black or brown color. When softened in hot water it admits of being moulded, and hardens again on cooling. It is an excellent insulator for submarine cables, and is applied to many purposes in the arts.

Gum Resins are natural mixtures of gum and resin, and often include volatile oils.

Balsams are compounds of resins with volatile oils: some of them also contain benzoic or cinnamic acid. Some, as benzoin, are solid; others, as *Balsam of Tolu* and *Peru*, and *Canada Balsam*, are viscid fluids.

The Products of the Destructive Distillation of Wood, etc.

When wood is submitted to distillation in close vessels, a thick, black, inflammable liquid, *Tar*, is formed. It contains a great many remarkable bodies, among which the following may be mentioned. The solid black residue which is left after the distillation or inspissation of tar is *Pitch*.

Paraffine (CH) was originally discovered among the products of distillation of wood-tar, but is more abundantly obtained from the distillation of bituminous schists and petroleums. It is a crystalline solid, without taste or odor. Its specific gravity is .87, melts at 112°, and distills unchanged. Few chemical agents act upon it; it remains unchanged by the acids, alkalies, etc., but is soluble in turpentine and naphtha. From its chemical indifference it has obtained its name (*Parum Affinis*). It is used in the manufacture of candles, and as a substitute for wax.

Eupion (C_5H_6) occurs abundantly in animal tar, from which it may be prepared by distillation, and sub-

Whence does gutta percha come? How may it be moulded? What are balsams? What products arise from the destructive distillation of wood? What is paraffine? What are its properties? Why is it so called? How is eupion prepared?

sequently purified by rectification from sulphuric acid. From paraffine it may be separated by exposure to cold, or, being more volatile, by distillation. It is a colorless liquid, specific gravity .074, boils at 339°, insoluble in water, but very soluble in alcohol.

Creasote ($C_{16}H_{10}O_2$) is extracted from the heavy oil of tar by a complicated process. It is an oily, colorless liquid, of burning taste, exhaling an odor of wood-smoke; specific gravity 1.04; boils at 400; burns with a sooty flame; is sparingly soluble in water, but very soluble in alcohol, ether, benzole, and acetic acid. It has the remarkable property of coagulating albumen and preserving flesh from putrefactive changes, whence its name. It is useful in toothache.

Among the allied substances may be mentioned *Picamar*, an oily liquid, of a bitter taste, which boils at 518°, and combines with bases to form crystalline compounds. *Kapnomar*, a colorless liquid, having an odor of rum; boils at 360°, and forms with oil of vitriol a purple solution. *Cedriret*, which forms red crystals, giving with creasote a purple solution, and with sulphuric acid a blue. *Pittakal*, a dark blue solid, yields blue precipitates with metallic salts, and contains nitrogen.

When coal-tar is submitted to distillation, like wood-tar it yields a volatile oil, which, by being submitted to rectification, becomes *Coal Oil*, or *Artificial Naphtha.* From it a variety of substances may be extracted; they either pre-exist in the oil, or are formed by the operation.

NAPHTHALINE ($C_{20}H_8$), obtained by distillation of coal-tar, is a white, crystalline substance; melts at 176°, boils at 420°, specific gravity 1.05, exhales an odor like the narcissus, is combustible, insoluble in water, soluble in ether and alcohol. It dissolves in sulphuric acid, and the solution, on being diluted with water and saturated with carbonate of baryta, yields a salt containing *Sulphonaphthalic Acid* ($C_{20}H_7S_2O_5+HO$).

Paranaphthaline ($C_{30}H_{12}$) is associated with naphthaline, but differs from it in being insoluble in alcohol, by which liquid they may therefore be separated.

What are the properties of creasote? Describe some of the allied substances. How is coal oil produced? Describe naphthaline. How does paranaphthaline differ from it?

Carbolic Acid—Phenylic Acid ($C_{12}H_5O+HO$)—is found in that portion of oil of tar which boils between 300° and 400°. This, being agitated with potassa, and the result decomposed by an acid, yields carbolic acid, which may be purified by rectification from caustic potassa. The pure acid forms a colorless, deliquescent, crystalline mass, fusing at 95°, and passing into vapor at 370°. It has a smoky odor, an acrid taste, and the antiseptic properties of creasote. It is much used as a disinfectant, and has a singular power in increasing friction; it is useful in boring glass.

When woody matter is gradually decomposed by contact with air and moisture, *Geine*, *Humus*, and *Ulmine* are produced. They arise from a partial oxidation, attended by the production of carbonic acid and water, the action being originally occasioned by nitrogenized matter in the wood. Corrosive sublimate, or any other body possessing the power of checking ferment action, may therefore be resorted to for preventing the dry rot of wood. These brown bodies, which are found in soils and moulds, combine with alkalies, and have been described as *Geic*, *Humic*, and *Ulmic Acids*. When the access of air is for the most part cut off, ulmine, etc., do not appear alone, but with them many other substances of the family of the hydrocarbons arise. Besides these, in the formation of vegetable soil and turf, azotized acids, such as *Crenic* and *Apocrenic*, appear. These originate in the decay of the nitrogenized constituents of the wood, an action which probably precedes its general disorganization. They are often found in mineral springs in combination with oxide of iron, forming ochery stains.

There is abundant proof that all the varieties of coal have originated from woody fibre. For the production of these it seems necessary that the wood should be immersed in water at a moderately high temperature, and without free contact of air. The ulmine bodies form from the decay of wood at the surface of the earth; the coal bodies under a heavy pressure. Of them we have

---

How is carbolic acid made? What are its properties? Why is it useful for boring glass? When do geine, humus, and ulmine arise? How may woody matter be preserved? What other bodies do soil and turf contain? What is the origin of coal?

many varieties, differing much in constitution. *Lignite* is of a brown color, and in it the structure of the wood is more or less perfectly preserved; the various forms of *Bituminous Coal*, as cannel coal, etc.; *Anthracite*, which contains but little hydrogen.

The essential elements of coal are carbon and hydrogen, but it also contains oxygen, nitrogen, sulphur, and various mineral matters, constituting the incombustible *ash*, chiefly silicious matter and unburnt carbon, with carbonate of lime and oxide of iron.

BITUMEN, ASPHALTUM, PETROLEUM, are substances closely allied to coal. Many kinds of coal may be regarded as carbonaceous matter impregnated with bitumen, and *Bituminous Schists* are earthy compounds similarly impregnated.

Asphaltum may be taken as the type of the bitumens. It occurs on the shores of the Dead Sea, in Trinidad, and many other places. It has a dark brown or black color, resinous fracture, burns with a smoky flame, is soluble in alcohol, ether, and benzole. Mixed with lime, chalk, sand, etc., it is used for pavements and water-proof cements.

Petroleum, a fluid substance found in America and the Burmese Empire to an enormous extent, and used as a fuel and for illumination, arises probably from the distillation of bituminous coal and shales by the internal heat of the earth. The annual production of the Rangoon wells in Burmah is 400,000 hogsheads; 80,000,000 of gallons were thrown out by the wells in the United States in 1863. It is also found in many other localities.

The wells in the United States are sunk from 100 to 450 feet through the sandstones of the Devonian series, or the coal measures which overlie these strata. In Canada the oil is found in shales and limestones. The quantity thrown out by some of these wells in a day has exceeded 2000 barrels of forty gallons. In many the aid of steam-pumps is required.

Petroleum may be regarded as a compound of various

What are the varieties of coal? What is the composition of coal? What substances are allied to coal? Where is asphaltum found? Where does petroleum occur most extensively? What is the annual production? In what formation is the oil found? What is the composition of petroleum?

hydrocarbons boiling at different temperatures. When subjected to fractional distillation it yields *Benzine*, used by painters as a substitute for turpentine; *Kerosene*, employed for illuminating purposes; *Lubricating Oils* of greater specific gravity, and *Paraffine.*

LECTURE LXXXVIII.

ANIMAL CHEMISTRY.—*The Animal System changes incessantly.—Chemical Processes interior to it.—Objects of Digestion.—Description of the Process.—Varieties of Food.—Absorption into the System by Lacteals and Veins.*

IN the preceding Lectures I have given the descriptive history of many of the more important organic compounds, and chiefly those belonging to or derived from the vegetable kingdom. It remains now to mention another class, which seems to bear a closer relation to animal beings. The appearance and destruction of these compounds lead by ready steps to the consideration of the physiological functions of the animal mechanism.

There are certain causes which tend constantly to change the weight of an adult, healthy individual—causes of increase and causes of diminution. Among the former may be mentioned food, drink, and atmospheric air; among the latter, urine, fæces, transpired and expired matters; and these, in the course of a year, amount to many hundred pounds, yet the resulting action of the mechanism is such that at the end of that time the weight remains unchanged.

This fact, the constancy of adult weight, can therefore only be explained by an examination of the action of the matter introduced into the interior of the system on each other, and an examination of the matter returned. Whatever is fit for food, when burned in the open air, with free access of oxygen, must yield car-

What substances may be obtained from petroleum? Why should the weight of an individual change? What is the amount consumed and secreted in a year? What is the explanation of constancy of weight?

bonic acid, water, and ammonia; and there are also the results of the action of the animal mechanism, as is shown by analyzing the excretions. Oxygen, introduced by the respiratory process through the lungs, effects eventually the destruction of the hydrocarbons and nitrogenized bodies which have been introduced by the digestive apparatus, and carbonic acid, ammonia, and the vapor of water, or substances in a transition state, which tend eventually to assume those forms, are the results. An elevated temperature must also, as a consequence, be obtained.

Before the introduction of chemical principles into the science of physiology, it was a favorite idea that the animal system possessed the peculiarity of resisting the influence of external agents. This is an error. There is no essential difference between the physical effects taking place in the body during life and after death, nor is there any principle of resistance to external agents possessed by living structures. The only distinction is, that during life the effete materials pass off by appointed routes—the kidneys, the lungs, or the skin; while after death, these passages being closed, they accumulate in the interior of the body.

The matters returned by an animal to the external world are all found to be oxidized bodies, or such as arise from processes of oxidation. The result, therefore, is forced upon us that the primitive action of the mechanism is the oxidation of the food in the system by air which has been introduced by the lungs.

The process of digestion appears to be exclusively for the object of effecting the minute subdivision of the food. By the action of the teeth, or other organs of mastication, it is first roughly divided, and simultaneously mixed with saliva. It is then passed into the stomach, and in that organ mixes with the gastric juice, a viscid and acid body. This mixture is perfected by certain movements which the food now undergoes, and, under the conjoint action of the saliva and the gastric juice, it is broken down into a gray, semi-fluid, homo-

What is the cause of animal heat? Is an animal emancipated from the influence of external agents? What becomes of food introduced into the body? What is the object of digestion? What fluids act on the food?

geneous mass, of the consistency of cream or gruel. A part of this, called the *Chyme*, passes out through the pyloric orifice of the stomach and enters the intestine, and a part is absorbed by the veins of the stomach.

It has been a question whether artificial digestion could be performed, but it now appears to be universally admitted that an acidulated water, containing animal matter in a state of change, has the power of impressing analogous changes on organized substances submitted to its action, just as the gastric juice, containing hydrochloric or lactic acid, with *pepsin*, an animal material undergoing metamorphosis, possesses the power of dissolving fibrin or coagulated albumen.

Soon after its entrance into the intestine, the chyme is mixed with bile, pancreatic juice, and the intestinal juices, and those parts which have escaped solution previously are digested. The valuable portions are then absorbed by certain organs in the intestine called *Lacteals*, and thrown into the current of the circulating blood.

Before we can trace the changes which then occur, it is proper to remark that, as respects the food itself, it may be distinguished into two varieties: 1st. Nitrogenized food, or that which repairs the tissues; and, 2d. Calorifacient food, or that used solely to produce heat.

The nutritive processes of carnivorous animals are very simple; they live on the herbivora, and find in the carcasses they consume the fats, the fibrin, and other such bodies, which are necessary for their own economy; these, therefore, simply require to be brought into a state of solution, or of extreme subdivision, and then are carried into the blood. In these cases the fats are the calorifacient, the muscles, etc., the nitrogenized element.

But the herbivora find in the vegetable matters they use the same essential principles; their fibrin, albumen, and fats are obtained directly from plants in which they naturally occur. In the digestive process of the two great classes of animals there is not, therefore, in reali-

---

What is the chyme? How may artificial digestion be performed? What is accomplished in the intestine? What varieties of food are there? Why is the digestion of the carnivora simpler than that of the herbivora?

ty, any difference; both find in their food the elements they require.

The two classes of food are introduced into the system by different routes—the fatty matters passing through the lacteals, and the nitrogenized and completely dissolved bodies being taken up by the veins of the stomach and intestines.

LECTURE LXXXIX.

ORIGIN AND DESTINY OF THE FATS AND NEUTRAL NITROGENIZED BODIES.—*Fat may be made in the Animal System.—Is generally derived directly from the Food.—Object of Calorifacient Food.—The Nitrogenized Bodies.—Fibrin.—Albumen.—Casein.—The Protein Group.—Food is Oxidized in the Body.*

FATS.—Two opinions have been entertained respecting the origin of the fat which occurs in the adipose tissues of animals: 1st. It has been supposed to be produced by processes taking effect in the system; or, 2d. Simply collected from the food.

In many various processes fatty bodies arise. Thus, when muscular tissue is left in a stream of water, a mass of *adipocire* is eventually found. During the action of nitric acid on fibrin, and in the preparation of oxalic acid from starch, oily bodies are produced. If a solution of sugar be mixed with powdered chalk, and a portion of casein added, on keeping the mixture at 80° for some weeks, *butyrate of lime* is formed, with the evolution of carbonic acid and hydrogen.

But, though the power of forming oily from amylaceous bodies be possessed by the animal mechanism, there can be no doubt that it is not resorted to in many instances, and that the fats found in the system are directly absorbed from the food. Often this absorption takes place with so slight a change impressed on the oil, that, without difficulty, we can detect its presence by the odor or taste. Thus the milk of cows fed on

---

How does the food get into the body? What is the origin of the fat of animals? How is adipocire produced? When does butyric acid arise? What is the most probable source of fat?

linseed cake tastes strongly of that substance; and at those seasons of the year when such animals feed on young shoots or leaves containing odoriferous oils, as the onion, the taste is at once detected in the milk.

The deposition of fat upon an animal, and the production of butter in its milk, bear a certain relation to the amount of oleaginous matter in its food. For this reason, Indian corn, which contains from eight to twelve per cent. of oil, furnishes one of the most valuable articles for feeding and fattening cattle. It must be admitted, however, that where foods without fat are used, the system possesses the power of effecting their production; thus bees will produce wax though fed upon pure sugar, and animals will grow fat though eating potatoes alone.

The fats which occur in plants pass into the systems of graminivorous animals, and there undergo changes, a series of partial oxidations occurring. It is only a part which is completely destroyed at once, so as to produce carbonic acid and water. The residue accumulates in the cells of the adipose tissues, to be used at a future time; or, if devoured by the carnivorous tribes, is destined to undergo in them those successive changes which bring it back to the condition of carbonic acid and water, and restore it to the atmosphere, from which it was originally derived by plants.

The amylaceous bodies and fats, or the non-nitrogenized bodies, are the calorifacient food. Their sole office is to unite with the oxygen introduced by the lungs, and, by the production of carbonic acid and water, keep up the temperature of the animal system.

The principal fatty bodies have already been described, and their general properties indicated.

NITROGENIZED BODIES.—When the expressed juices of plants, such as beets, turnips, etc., are allowed to stand, there is deposited, after a short time, a coagulum or clot, which does not appear to differ in any respect from animal *Fibrin.* If this be removed, and the temperature of the juice raised to 212°, it becomes turbid again, from the deposit of a second body, *Albumen.*

Give an instance in the case of the cow. Why is Indian corn valuable for feeding? What is the destiny of fats? What do they produce? Describe what occurs on warming a vegetable juice.

On separating this and slowly evaporating, a film forms on the surface identical with *Casein*. These three bodies contain nitrogen, and may therefore be looked upon as the representatives of the neutral nitrogenized class.

Fibrin may be obtained by beating fresh-drawn blood with twigs, and washing with water and ether the fibrous filaments which adhere thereto. As thus prepared, fibrin is whitish, elastic, insoluble in water, alcohol, and ether, but soluble in hydrochloric acid, with which it yields a blue solution. It possesses the power of decomposing deutoxide of nitrogen, and can coagulate spontaneously. When dried it shrinks very much in volume, but recovers its bulk when again moistened.

Albumen occurs abundantly in the serum of blood and the white of eggs, from which it may be obtained by neutralizing the associated soda with acetic acid. On dilution with cold water it falls as a white precipitate, soluble in water containing a minute quantity of alkali. Exposed to a sufficient heat, albumen coagulates, and becomes a white body wholly insoluble in water. The strong acids also unite directly with it and form insoluble compounds; acetic and tribasic phosphoric acids are exceptions. With metallic salts, as corrosive sublimate, it gives insoluble precipitates; hence its use as an antidote for that poison.

Casein is found abundantly in milk as cheese. It is insoluble in water, but, like albumen, is readily dissolved if free alkali be present. It may be obtained by coagulating milk with sulphuric acid, and dissolving the curd, after it has been well washed in water, in a solution of carbonate of soda. By standing, it separates into two portions, oily and watery. From the latter the casein is reprecipitated by sulphuric acid and the process repeated. The casein is finally washed with ether to remove any trace of fat. It is a white substance, soluble in an alkaline water, the solution not being coagulated by boiling, but a skin forming on the surface as evaporation goes on. It can, however, be coagulated by certain animal substances, as the interior coat of a calf's stomach. Casein can dissolve bone-earth.

The foregoing bodies are sometimes spoken of as the

What are the properties of fibrin and albumen? Where is casein found? What are its properties?

Protein group, from the circumstance that they were all supposed to be modifications of protein ($C_{48}H_{36}O_{14}N_6$). It may be extracted from them by dissolving in an alkaline solution and precipitating with an acid. It is a tasteless, white body, soluble in acetic acid and alkalies, but insoluble in water. This theory is not, however, generally received by chemists, the best authorities regarding the constitution of albumen as ($C_{216}H_{169}O_{68}N_{27}S_2$).

Gelatin is obtained by the action of boiling water on hide, hoofs, horns, etc. The solution is freed from fat by straining and skimming. It forms, on cooling, a soft jelly, which contracts as it dries. Solution of gelatin is precipitated by corrosive sublimate, tannic acid, or infusion of galls; with the latter it yields a precipitate which is the basis of leather. Glue is an impure gelatin.

The nitrogenized bodies introduced into the system pass through the same changes as the non-nitrogenized, partial oxidations giving rise to various tissue forms, and ending in perfect oxidation, with a production of water, ammonia, and carbonic acid.

Whether we regard the heat-making or the nutritive food, we see that the result is the same. Introduced through the blood-vessels into the system, it is brought under the destructive influence of oxygen arriving through the lungs; and, as has been already explained, the amount of oxygen is so adjusted to the amount of these classes of food combined, that in an adult and healthy individual the weight does not change, even after the lapse of a considerable period cf time.

Describe the protein theory. How may gelatin be prepared? What becomes of the nitrogenized bodies in the system? How is it that the weight of the individual does not change?

LECTURE XC.

INTRODUCTION OF HEAT-MAKING AND NUTRITIOUS FOOD INTO THE BLOOD, AND ITS TRANSMISSION THROUGH THE SYSTEM.—*Professor Draper's Explanation of the Circulation of the Blood.—Constitution and Properties of the Blood.—Plasma and Disks.—Function of each.—Coagulation.—Composition of Blood.*

THE ordinary principles of capillary attraction are amply sufficient to account for the absorption of nutritious matter from the digestive cavities, both by the lacteals and the veins. By these it is eventually brought into the general current of the circulation, and distributed to every part of the system.

With respect to the forces involved in the circulation of the blood, most physiologists have regarded the hydraulic action of the heart as amply sufficient to account for all the phenomena. It is now on all hands conceded that this organ discharges a subsidiary duty. The whole vegetable creation, in which circulatory movements of liquids are actively carried on without any such central mechanism of impulsion—the numberless existing acardiac beings belonging to the animal world—the accomplishment of the systemic circulation of fishes without a heart—and the occurrence in the highest tribes, as in man, of special circulations which are isolated from the greater one, have all served to demonstrate that we must look to other principles for the cause of these remarkable movements.

The cause of the circulation of the blood is to be found in the chemical relations of that liquid to the tissues with which it is brought in contact. On the principles of capillary attraction, a liquid will readily flow through a porous body for which it has a chemical affinity, but it will refuse to flow through it if it has no affinity for it. On this principle we can easily explain why the arterial blood presses the venous before it in

What principle accounts for the absorption of nutritious matter? Does the heart alone carry on the circulation? Give instances to show its insufficiency.

the systemic circulation, and why the reverse occurs in the pulmonary. This explanation of the circulation of the blood, which Professor J. W. Draper offered many years ago, is now generally admitted to be true.

The systemic circulation takes place because arterial blood has a high affinity for the tissues, and venous blood little or none. The pulmonary circulation takes place because venous blood has a high affinity for atmospheric oxygen, which it finds on the air-cells of the lungs, and arterial blood little or none. On the same principle we may explain the rise of sap in trees, the circulatory movements in the different animal tribes, and the minor circulations of the human system.

The most striking peculiarity of the blood is the incessant change which it undergoes. It is constantly being destroyed and as constantly being reproduced. It consists of two portions, the Plasma, a clear fluid of a yellowish tinge, containing albumen, fibrin, and fat, and in it floating disk-like bodies, of different shapes and magnitudes in different animals. In man they are circular, and about $\frac{1}{4000}$th of an inch in diameter. In the frog they are elliptical, as seen in *Fig.* 308, about $\frac{1}{1200}$th

Fig. 308.

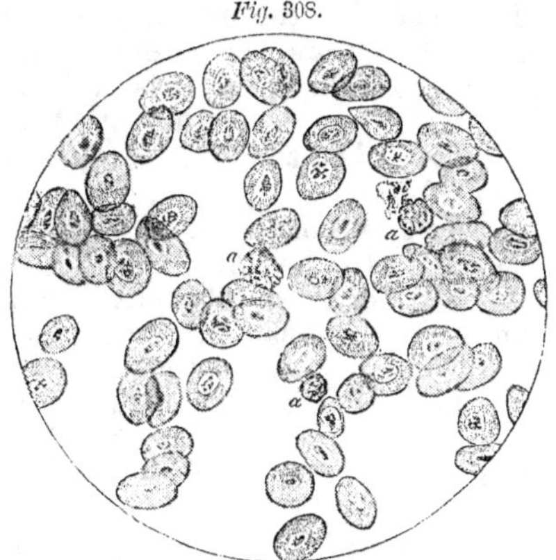

of an inch in diameter, and nucleated. They consist of

What is Dr. Draper's explanation of the circulation? Why does the systemic circulation take place? Of what portions does the blood consist? Describe a blood disk.

a cell-wall or sac of a substance like fibrin, containing in its interior *Hæmatin.* When the disks are old and about to die, the interior contents change into *Hæmapheine*, a yellow substance, the coloring matter of the urine and bile.

The disks serve for the carrying of oxygen. They absorb it in the air-cells of the lungs, and transmit it to all parts of the system. As they grow old and disappear, new ones are formed from the corpuscles that arise from the chyle, *a a a*, *Fig.* 308. The plasma serves for the purposes of nutrition and for the removal of waste bodies.

Although fibrin exists in plants, it is not absorbed directly, but passes through an allied form known as *albuminose.* Albumen and casein suffer the same modification. Besides the direct proof we have from the analysis of these bodies, we know that fibrin and albumen so closely resemble each other in constitution that they are mutually convertible. During the hatching of an egg, from its albumen the flesh (fibrin) of the young chicken is formed—a phenomenon accompanying absorption of oxygen from the air. In the human system abundant observation has proved that there is a direct connection between the quantity of oxygen introduced through the lungs, and the amount of fibrin in the blood. When the respiratory process is unduly active, the disks introduce oxygen with rapidity, and the amount of fibrin increases; but when the reverse takes place, the amount of fibrin declines.

The coagulation of the blood is a phenomenon which has excited much attention, physiologists generally looking upon it either as wholly inexplicable, or, what in reality amounts to the same thing, as due to the death of the blood; but it is plain that coagulation would also occur during life in the blood were it not for the circumstance that the muscles pick out the fibrin and appropriate it for their repair as fast as it is formed, and before it can solidify. The following analysis of blood is from Draper's Physiology; it will serve to give an idea of the constitution of that liquid. It must not be

What is their function? What body do fibrin, etc., change into? What is the relation of fibrin to albumen? What is the old and what the new hypothesis of coagulation?

forgotten, however, that such analyses, beyond mere general results, are of little value; the composition of the blood varies continually in the same individual: for instance, the mere accident of his having been thirsty, or having recently drank abundantly of water, will make an entire change in the analysis of the blood.

Water	784.00
Albumen	70.00
Fibrin	2.20
Disks	131.00
Fats	1.30
Salts	6.03
Extractive, etc	5.47
	1000.00

LECTURE XCI.

NATURE OF RESPIRATION AND SECRETION.—*Phenomena of Respiration.—Action in the Lungs.—Changes in the Blood.—Production of Animal Heat.—Removal of Effete Matters.—Composition of Milk.—Its Uses. —Chyle.—Lymph.—Mucus.—Pus.—Bile.—Saliva. —Gastric Juice.—Urine.—Diabetic and Albuminous Urine.—Calculi.—Bone.—Nervous Matter.*

DURING the starvation of an animal all its various secretions are still formed, a consideration which proves that the production of urine, bile, and other such bodies is in reality connected with the destructive processes going on in the animal system. These processes of decay originate in the action of oxygen admitted by the process of respiration.

The lungs, which constitute the organ by which air is introduced, are originally developed as diverticula from the œsophagus, and finally become an immense congeries of cells emptying into the trachea. In respiration they are generally passive, the air being introduced and expelled alternately by muscular contraction and the principle of the diffusion of gases. It is commonly estimated that on an average about seventeen inspirations are

What is the composition of the blood? What is the source of the secretions? How is this proved? How is air introduced into the body? What is the rate of inspiration and the amount introduced?

made each minute, and at each inspiration about seventeen cubic inches of air are introduced.

The blood presents itself in the air-cells of a deep blue color, and is then known as venous blood. Through the thin wall of the cell it obtains oxygen from the air and gives out carbonic acid. During the act of change its tint alters to a bright crimson, and it is now said to be arterialized, or to constitute arterial blood. The magnitude of the scale on which this operation is carried forward may be appreciated from the circumstance that, in a man of average size, in a single day about seven tons of blood have been exposed to 226 cubic feet of air.

The oxygen thus introduced acts directly either on the tissues themselves as it is distributed by the systemic circulation, or on the calorifacient material contained in the blood. In the latter case, carbonic acid and water are the result; in the former, carbonic acid, water, and ammonia. But these changes can not take place without an elevation of temperature. Carbon and hydrogen can neither burn in the air nor in the animal system without evolving heat. The high temperature which an animal can maintain is therefore directly proportional to the quantity of oxygen it consumes.

The tissues, being thus acted upon, give rise during their metamorphoses to new products, which require to be removed from the system. These, passing under the name of secretions, are discharged by glands or special organs. Thus the carbonic acid, for the most part, escapes from the lungs; the ammonia through the kidneys; the water from both these organs and the skin. The sulphates and phosphates found in the urine arise directly from the sulphur and phosphorus previously existing in the muscular fibre and nervous matter.

As an illustration of the principles here given in relation to the functions of nutrition and secretion, the constitution and properties of milk may be cited. The following is an analysis of it:

What change does the blood suffer in the lungs? How much blood is aerated in a day? What becomes of the air thus introduced? What is the cause of animal heat? What is the object of glands? What is the origin of the sulphates and phosphates found in the urine?

Water	873.00
Casein	48.00
Sugar of Milk	44.00
Butter	30.00
Phosphate of Lime	2.30
Other salts	2.70
	1000.00

Of the substances here mentioned, all are undoubtedly obtained directly from the food. In the herbage on which an herbivorous, milk-giving animal feeds, every one of these constituents occur. This has already been shown in the case of the fat, sugar, and casein; and the evidence is equally complete that all the salts of phosphoric acid and chlorine arise from the same source.

A young animal, which, in the first periods of its life, is nourished exclusively on milk, finds in that milk all the various compounds it requires for its own existence and growth. The respiratory food is there—it is the butter and milk sugar; the nitrogenized food is there—it is the casein; and we have already seen that albumen and casein are both convertible into fibrin. The casein thus in the mother's milk becomes converted into flesh in the young animal. To insure the growth of its bones, phosphate of lime, bone-earth, is present; there is also common salt to form the hydrochloric acid of its gastric juice, and to furnish the soda needed in the bile and pancreatic juice.

It remains now to furnish a brief description of the properties of the remaining leading animal substances, among which may be mentioned the following:

Chyle is a milky fluid found in the thoracic duct. It resembles blood in containing floating cells, and in the power of coagulation. It originates from fatty and albuminous matter absorbed by the lacteals.

Lymph is a fluid found circulating throughout the body in the lymphatics. It contains fibrin, which arises from waste albuminous matters picked up among the tissues.

Mucus exudes from the surface of mucous membranes. It is of a white or yellow color, of a viscid

---

Give the composition of milk. What is the origin of these constituents? Why is milk a perfect food? What are the phosphate of lime and salt for? Describe chyle, lymph, and mucus.

constitution, and insoluble in water. It dissolves in a solution of potassa, and is precipitated by acetic acid and alcohol.

PUS, a secretion from injured surfaces, resembling mucus in many respects, but distinguished by not being soluble in potassa solution, but converted by it into a gelatinous body, which can be pulled out into threads. Examined under the microscope, it contains white, colorless globules. When absorbed into the blood, pus appears to act as a powerful poison; it is a vehicle for the most virulent animal poisons, as those of glanders and small-pox.

BILE, a yellow liquid secreted by the liver from the portal blood; it turns green in the air, has a bitter taste, and an alkaline reaction, due to the presence of soda. It contains glycocholate and taurocholate of soda, cholesterine, fat, mucus, and coloring matter. The composition of glycocholic acid is ($C_{52}H_{43}O_{12}N$); of taurocholic acid ($C_{52}H_{45}O_{14}S_2N$).

SALIVA is a transparent, viscid liquid, secreted in glands in the neighborhood of the mouth. It contains an organic principle, *ptyaline*, which acts as a ferment to starch and sugar. The *tartar* which is deposited on the teeth consists of salts of the saliva, chiefly phosphate of lime, cemented with animal matter.

GASTRIC JUICE is a secretion from the mucous membrane of the stomach. Its essential ingredients are hydrochloric acid and *pepsin*, and its function is to dissolve nitrogenized food.

URINE, a yellow-colored fluid secreted from the kidneys, has an acid reaction; specific gravity, 1.005 to 1.030; putrefies at a moderate temperature, its urea passing into the condition of carbonate of ammonia.

Composition of Urine.

Water	933.00
Urea	30.10
Uric Acid	1.00
Lactic Acid and Extractive	17.14
Mucus	00.32
Salts (mostly phosphates and sulphates)	18.44
	1000.00

Where does pus form? What is bile? Give its composition. What is the object of saliva? What is tartar? What is the function of gastric juice? Give the composition of urine.

The constitution of the urine changes in disease. In *Diabetes* it contains grape sugar, as may be shown by adding to a small quantity of it, in a test-tube, a drop or two of a solution of sulphate of copper, and its own bulk of solution of caustic potassa. A blue liquid results, which, on being heated, deposits a green precipitate, turning red on prolonging the heating, because of the formation of suboxide of copper. Diabetic urine may even be fermented with yeast, carbonic acid and alcohol being produced.

The specific gravity of diabetic urine is high, as may be ascertained by the aid of a urinometer, *Fig*. 309.

Fig. 309.

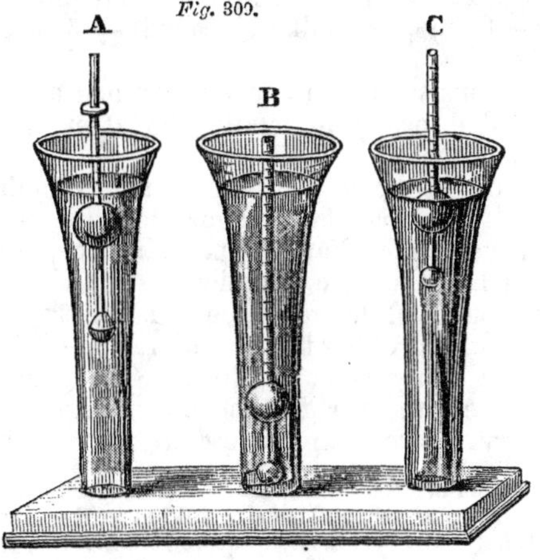

This consists of a glass bulb, prolonged above and below into a tube. The lower tube carries a smaller bulb, partly filled with mercury, which is intended to act as ballast. The upper tube contains a scale divided from 1000 to 1060, the latter figure being nearest the bulb that is lowest. On putting such an instrument into distilled water, it should float, as at B, so that the level of the liquid is exactly at 1000. If this is not the case, it indicates that the instrument-maker has put too much or too little quicksilver, as the case may be, in the low-

---

How does it change in diabetes? What are the tests for diabetic urine? Describe the urinometer.

er bulb. On immersing it in a liquid of high specific gravity, it floats, as at C; while in diabetic urine the liquid stands at some intermediate point, as seen at A.

Albumen is found in urine in Bright's disease, and may be detected by heating. A white precipitate forms, insoluble in nitric acid. The precipitate of phosphates formed by heating is soluble in that acid. Urine may also contain blood, pus, casts of the uriniferous tubuli, etc., and must in such cases be subjected to microscopic examination.

Urinary Calculi are stony concretions formed in the bladder of man and many animals. They are of different kinds: 1st. Uric acid; 2d. Urate of ammonia; 3d. Phosphate of lime, magnesia, and ammonia; 4th. Oxalate of lime, or mulberry calculus; 5th. Cystic and xanthic oxides.

Bones consist of two parts, an animal and an earthy matter. The former is gelatin, the latter phosphate of lime (bone-earth).

Nervous Matter consists of an albuminous substance, with several fatty principles, distinguished by the remarkable fact that they contain phosphorus, which oxidizes when they are in action.

As it would not, however, be consistent with the plan of this book to prolong the consideration of this subject, and become involved in physiological questions, the reader is referred for farther information to Draper's Physiology, where such matters are fully discussed.

When is albumen found in urine? How is it detected? What is the composition of urinary calculi? What is the composition of bone? What is the peculiarity of nervous matter?

INDEX.

A.

Absolute alcohol, 407.
Acetal, 419.
Acetification, 419.
Acetone, 422.
Acetyle compounds, 417.
Acid, acetic, 419.
" aconitic or equisetic, 455.
" aldehydic, 418.
" alloxanic, 451.
" althionic, 417.
" amygdalic, 443.
" anilic or indigotic, 467.
" anthranilic, 467.
" antimonic, 361.
" antimonious, 361.
" apocrenic, 470.
" apoglucic, 398.
" arsenic, 359.
" arsenious, 356.
" " tests for, 356.
" benzoic, 432.
" boheic, 457.
" boracic, 302.
" butyric, 473.
" caffeotannic, 457.
" capric and caproic, 473.
" carbazotic, 467.
" carbolic, 478.
" carbonic, 295.
" " liquefaction of, 297.
" chloracetic, 422.
" chloric, 282.
" chlorous, 282.
" chlorovalerisic, 433.
" chromic, 352.
" chrysammic, 468.
" chrysanilic, 467.
" cinchonic, 461.
" cinnamic, 439.
" citric, 455.
" comenic, 460.
" crenic, 478.
" croconic, 400.
Acid, cyanic, 444.
" cyanuric, 444.
" elaidic, 472.
" ellagic, 458.
" ethalic, 473.
" ethionic, 417.
" ferric, 344.
" fluoboric, 303.
" fluosilicic, 304.
" formic, 428.
" fulminic, 444.
" fumaric, 456.
" gallic, 372.
" geic, 478.
" glucic, 398.
" hippuric, 436.
" humic, 478.
" hydriodic, 288.
" hydrobromic, 290.
" hydrochloric, 283.
" hydrocyanic, 441.
" hydroferrocyanic, 446.
" hydrofluoric, 291.
" hydrofluosilicic, 305.
" hydromellonic, 449.
" hydrosulphocyanic, 448.
" hydrosulphuric, 271.
" hypochloric, 282.
" hypochlorous, 281.
" hyponitrous, 261.
" hypophosphorous, 275.
" hyposulphurous, 271.
" igasuric, 462.
" isatinic, 467.
" isethionic, 417.
" japonic, 457.
" kakodylic, 425.
" kinic, 461.
" lactic, 409.
" lithic, 450.
" maleic, 456.
" malic, 456.
" manganic, 338.
" margaric, 470.
" meconic, 460.

Acid, melanic, 438.
" melasinic, 398.
" mesoxalic, 451.
" metagallic, 458.
" metaphosphoric, 276.
" methionic, 417.
" mucic, 401.
" muriatic, 283.
" mykomelinic, 451.
" myristic, 472.
" nitric, 263.
" nitrohydrochloric, 286.
" nitrous, 261.
" œnanthic, 413.
" oleic, 471.
" oxalic, 398.
" oxalhydric, 400.
" oxaluric, 451.
" palmitic, 472.
" parabanic, 451.
" paramaleic, 456.
" paratartaric, 454.
" pectic, 396.
" perchloric, 283.
" permanganic, 339.
" phenylic, 478.
" phosphoglyceric, 472.
" phosphoric, 276.
" phosphorous, 275.
" phosphovinic, 414.
" picric or carbazotic, 467.
" pinic, sylvic, and pimaric, 475.
" propionic, 472.
" prussic, 441.
" purpuric, 452.
" pyrogallic, 458.
" pyroligneous, 419.
" pyromeconic, 461.
" pyrophosphoric, 276.
" pyrotartaric, 455.
" racemic, 454.
" rhodizonic, 400.
" rubinic, 457.
" saccharic, 400.
" sacchulmic, 398.
" salicylic, 437.
" sebacic, 472.
" silicic, 303.
" stearic, 470.
" suberic, 472.
" succinic, 472.
" sulphacetic, 422.
Acid, sulphamylic, 432.
" sulphindigotic, 467.
" sulphobenzoic, 434.
" sulphocyanic, 448.
" sulphoglyceric, 472.
" sulphomethylic, 428.
" sulphomephalic, 477.
" sulphosaccharic, 397.
" sulphovinic, 413.
" sulphuric, 269.
" sulphurous, 267.
" tannic, 456.
" tartaric, 453.
" thionuric, 452.
" ulmic, 397, 478.
" uramilic, 452.
" uric, 450.
" valerianic, 432.
" valeric, 431.
" xanthic, 424.
Acids, coupled, 455.
Aconitine, 462.
Adipocire, 433.
Affinity, chemical, 212.
Air-pump, 246.
Albumen, 485.
" vegetable, 484.
Albuminose, 489.
Alcargen, 425.
Alchemy, 1.
Alcohol, 407.
Aldehyde, 418.
Alkaloids, 458.
Alkarsine, 425.
Allantoin, 450.
Allotropism, 211.
Alloxan, 451.
Alloxantine, 451.
Alumina, 333.
" sulphates, 335.
Aluminum, 332.
Alums, 335.
Amalgamation process, 373.
Amalgams, 373.
Amber, 475.
Amidine, 392.
Amidogen, 305.
Ammelin and ammelid, 448.
Ammonia, carbonate, 440.
" nitrate, 440.
" preparation and properties of, 306.
" sulphate, 440.

Ammoniacal amalgam, 307, 440.
Ammonium, 307, 440.
" chloride, 440.
" sulphide, 441.
Amygdaline, 442.
Amyle compounds, 430.
Amylene, 432.
Amylum, 391.
Anæsthetic, 410, 429.
Anatto, 466.
Aneroid barometer, 255.
Aniline, 463.
Animal chemistry, 480.
Antearine, 462.
Anthracite, 479.
Antimony, 360.
" chloride, 361.
" oxide, 361.
" sulphides, 361.
Antozone, 229.
Aqua regia, 286.
Arabine, 396.
Arc, Voltaic, 156.
Argol, 408.
Aricine, 461.
Arrow-root, 392.
Arsenic, 355.
" sulphides, 360.
Arterialization, 491.
Asphaltum, 479.
Atmosphere, composition of, 242.
" physical constitution of, 244.
Atmospheric pressure, 245.
Atomic weights, 193.
" theory, 4.
Atoms, 4.
Atropine, 462.
Attraction, 126.
Aurum musivum, 351.
Azote, 241.

B.

Balloon, 16.
Balsam, 476.
Barium, 324.
" chloride, 325.
" oxides, 325.
" sulphide, 325.
Barley sugar, 393.
Barometer, 253.
Baryta, 325.
" carbonate, 326.
Baryta, sulphate, 326.
Basil Valentine, 2,
Bassorine, 396.
Battery, electrical, 137.
" Voltaic, 151.
Bell metal, 364.
Benzamide, 434.
Benzine, 435, 480.
Benzoine, 435.
Benzone, 435.
Benzoyle compounds, 433.
Bezoars, Oriental, 458.
Bile, 493.
Biscuit-ware, 334.
Binary hypothesis, 125.
Bismuth, 367.
" nitrate, 367.
" oxides, 367.
Bitumen, 479.
Bleaching powder, 330.
Blood, composition of, 490.
Boiling points of fluids, 45, 56.
Bone-earth, 330.
Bones, composition of, 495.
Boron, 301.
" nitride, 302.
Bouquet of wine, 406.
Brass, 364.
Breguet's thermometer, 30.
Bright's disease, 495.
British gum, 393.
Bromine, preparation and properties of, 289.
Brucia, 462.
Brush, electrical, 137.
Burning-glass, 92.
Bunsen's battery, 155.
Butyrine, 478.

C.

Cadmium, 349.
" compounds of, 349.
Cæsium, 96, 323.
Caffeine, 462.
Calcium, 327.
" chloride, 328.
" fluoride, 328.
" sulphide, 328.
Calculi, urinary, 495.
Calomel, 372.
Calorimeter, 32.
Calorimotor, 164.
Camphor, 474.

Camphor, artificial, 474.
Candle-bomb, 50.
Caoutchouc, 475.
Capacity for heat, 31.
Caproine, 472.
Caramel, 394.
Carbon, 291.
" chlorides, 416.
" compounds with oxygen, 293.
" sulphide, 300.
Carbonic oxide, preparation and properties of, 294.
Carbyle, sulphate of, 417.
Carmine, 468.
Carthamine, 466.
Casein, 485.
" vegetable, 485.
Cassava, 392.
Cast iron, 341.
Catalysis, 217.
Catechin and Catechu, 457.
Cedriret, 477.
Cellulose, 396.
Cerine, 473.
Cerium, 336.
Chameleon mineral, 339.
Charcoal, properties of, 293.
Chemistry, history of, 1.
Chloral, 423.
Chloric acid, 282.
Chlorine, 278.
" compounds with oxygen, 281.
" preparation and properties of, 278.
Chlorisatine, 467.
Chlorobenzide, 436.
Chlorocinnose, 355.
Chloroform, 429.
Chlorophyll, 468.
Chlorosamide, 438.
Chlorous acid, 282.
Chlorureted acetic ether, 423.
" formic ether, 423.
Chrome yellow, 353.
Chromic acid, salts of, 353.
Chromium, 351.
" oxide, 351.
Chyle, 492.
Chyme, 482.
Cinchonia, 461.
Cinnabar, 372.
Cinnamyle compounds, 438.
Circle, Voltaic, 148.
Circulation of blood, 487.
Clay, 338.
Clay iron-stone, 340.
Coagulation, 489.
Coal, 479.
" oil, 477.
Cobalt, 347.
" character of salts of, 347.
" chloride, 347.
" oxalate, 347.
" oxides, 347.
Cobaltocyanogen, 448.
Cocoa tallow, 472.
Codeia, 460.
Cohesion, 6.
Colchicine, 462.
Cold rays, 79.
Collodion, 121, 401.
Colloid, 390.
Colophony, 375.
Coloring principles, 465.
Colors, 103.
Columbium, 354.
Combination by volumes, 202.
" laws of, 199.
Combining numbers, 193.
Combustion, 383.
" furnace, 326.
" tube, 385.
Compound bar, 28.
" radicals, 379.
Condensation of gases, 52.
" of vapors, 54.
Conduction, 68, 125.
" in crystals, 72.
Congress Spring, 240.
Conine or Conia, 463.
Conservation of force, 9.
Copper, 363.
" alloys of, 364.
" arsenite, 364.
" carbonates, 364.
" nitrate, 364.
" oxides, 363.
" sulphate, 364.
Correlation of force, 9.
Corrosive sublimate, 372.
Creasote, 477.
Crown of cups, 151.
Cruickshank's battery, 151.
Cryophorus, 60.

Crystallization, Crystallography, 204.
Crystalloid, 390.
Culinary paradox, 56.
Cupellation, 368.
Cyanogen, 300, 441.
" chlorides of, 446.
Cystic oxide, 453.

D.

Daguerreotype, 120.
Dalton's theory, 4.
Dammara resin, 475.
Daniell's battery, 152.
Daphnine, 462.
Daturine, 462.
Davy's theory, 3.
Decomposition of water, 156.
Delphinine, 462.
Deutoxide of nitrogen, 260.
Developers, 123, 420.
Dew, 85.
Dew point, 63.
Dextrine, 392.
Diabetes, 494.
Dialysis, 390.
Diamagnetism, 176.
Diamond, 291.
" phosphorescence of, 115.
Dianium, 337.
Diastase, 392.
Diathermacy, 81.
Dicyanomelaniline, 383.
Didymium, 337.
Dielectrics, 142.
Differential thermometer, 18.
Diffusion of gases, 255.
" of liquids, 390.
Digestion, 481.
Dimorphism, 209.
Disks, 487.
Dispersion, 93.
Dragon's blood, 475.
Draper's photometer, 91.
" researches on light, 111.
" " on thermo-electricity, 186.
Drummond light, 224.
Dufay's theory, 136.
Dutch liquid, 415.

E.

Earthenware, manufacture of, 334.
Ebonite, 475.
Ebullition, 53.
Elaidine, 472.
Elaldehyde, 418.
Elaterine, 462.
Electricity, action of on magnet, 172.
" animal, 188.
" conduction of, 125.
" of steam, 146.
" statical, 124.
" Voltaic, 147.
Electro-chemistry, 160.
Electrolysis, 162.
Electrometers, 140.
Electrotype, 162.
Electrophorus, 146.
Emetine, 462.
Emulsine, 443.
Equivalent numbers, 198.
" " table of, 193.
Eremacausis, 383.
Essences, 473.
Ethal, 478.
Ether, 409.
" continuous process for, 415.
" hydrochloric, 411, 423.
Ethers, compound, 411.
Etherine and Etherole, 416.
Ethyle group, 411
Endiometer, 243.
Eupion, 476.
Evaporation, 66.
" at low temperature, 59.
" conditions of, 67.
Expansion of solids, 27.
" of fluids, 19.
" of gases, 15.

F.

Faraday's theory of polarization, 142.
Fatty bodies, 469, 483.
Fermentation, acetous, 405.
" alcoholic, 383, 403.
" lactic, 404, 409.
Ferricyanogen compounds, 447.
Ferrocyanogen compounds, 446.
Fibrin, 485.

Fibrin, vegetable, 484.
Fire syringe, 37.
Fixed air, 219.
Flame, structure of, 70, 113, 227.
Fluoride of boron, 303.
Fluorine, 290.
Foot-pound, 11.
Formomethylal, 492.
Franklin's theory, 135.
Freezing mixtures, 43.
" of water by evaporation, 61.
Fusel oil, 431.
Fusible metal, 29, 367.

G.

Galvanism, 147.
Galvanometer, 187.
Gamboge, 466, 475.
Gasometer, 220.
Gastric juice, 493.
Geber, 2.
Geissler's tubes, 183.
Gelatin, 486.
Gelose, 396.
Gentianine, 462.
Geoffroy's tables, 3, 215.
Glass, manufacture of, 334.
" soluble, 335.
Glucinum, 335.
Glucose, 395.
Glycerine, 472.
Gold, 373.
" compounds of, 373.
Goniometers, 208.
Goulard's water, 421.
Graphite, 292.
Gravity, specific, of gases, 249.
Green, Scheele's, 356.
Grouping, 196.
Grove's battery, 154.
Gum, Arabic, 396.
" British, 393.
" tragacanth, 396.
Gun-cotton, 401.
Gunpowder, 318.
Gutta percha, 476.
Gypsum, 329.

H.

Hæmaphein, 489.
Hæmatin, 489.
Hæmatite, 340.
Hæmatoxyline, 466.
Haloid salts, 195.
Hare's blowpipe, 233.
Heat affects measures, 13.
" animal, 481.
" capacity for, 31.
" conduction of, 68.
" dynamic theory of, 11.
" exchanges of, 83.
" expansion by, 12.
" latent, 39.
" produced by electricity, 145.
" " friction, 9.
" radiation, reflection, absorption, and transmission of, 76.
" varieties, of, 80.
Hesperidine, 462.
Homologous series, 381.
Hydrobenzamide, 434.
Hydrogen, antimoniureted, 362.
" arseniureted, 360.
" light carbureted, 298.
" peroxide of, 240.
" persulphide of, 273.
" phosphureted, 277.
" preparation and properties of, 229.
" sulphureted, 271.
Hygrometer, Daniell's, 63.
" Saussure's, 62.
" wet bulb, 64.
Hygrometry, 52.
Hyoscyamine, 462.
Hyponitrous acid, 261.
Hyposulphurous acid, 271.

I.

Ice, 241.
Ilmenium, 354.
Inca Indians, 17.
Indestructibility of matter, 9.
India-rubber, 475.
Indigo, 466.
Indium, 337.
Induction, 129.
Interference, 103.
Interstices, 4.
Inuline, 392.
Iodine, preparation and properties of, 286.
Iridium, 375.
Iron, 340.

Iron, carbonate, 345.
" cast, varieties of, 341.
" characters of salts of, 344.
" chlorides, 345.
" manufacture, 340.
" oxides of, 343.
" passive, 342.
" sulphates, 345.
" sulphides, 345.
Isatine, 467.
Isomerism, 211.
Isomorphism, 210.

K.

Kakodyle and its compounds, 424.
Kakoplatyle, 465.
Kapnomar, 477.
Kermes mineral, 362.
Kerosene, 480.
Koumiss, 395.
Kyanol, 463.

L.

Lac, 475.
Lacteals, 482.
Lactine, 395.
Lampblack, 292.
Lamp, safety, 71.
Lanthanum, 336.
Latent heat, 39.
Laughing-gas, 259.
Laws of combination, 199.
Lead, 365.
" action of water on, 365.
" alloys of, 367.
" carbonate, 366.
" characters of salts of, 366.
" chloride, 366.
" iodide, 366.
" nitrate, 367.
" oxides, 366.
Leaven, 403.
Lecanorine, 462.
Leiocome, 393.
Leukol, 464.
Leyden jar, 136.
Light, cause of, 87.
" chemical action of, 116.
" reflection, refraction, and polarization of, 108, 111.
" wave theory of, 100.
Lignine, 396.
Lignite, 479.
Lime, 327.
" carbonate, 329.
" chloride, 330.
" oxalate, 400.
" phosphate, 330.
" salts, characters of, 328.
" sulphate, 329.
Liquor of Libavius, 351.
Lithium, 323.
Litmus, 467.
Liquids, expansion of, 19.
" conduction of, 73.
Lunar photography, 116.
Lymph, 492.

M.

Machines, electrical, 127.
Madder, 455.
Magdeburg hemispheres, 248.
Magenta, 464.
Magnesia, 331.
" carbonate, 331.
" characters of salts of, 331.
" phosphate, 332.
" sulphate, 332.
Magnesium, preparation and properties of, 330.
Magnetic field, 177.
Magnetism, 168.
Magnets, artificial, 168.
Magneto-electricity, 180.
Malachite, 364.
Manganese, characters of salts of, 339.
" chloride, 339.
" oxides, 338.
" preparation and properties of, 338.
" sulphate, 340.
Marcet's boiler, 52.
Margarine, 470, 471.
Margarone, 471.
Marine glue, 476.
Marriotte's law, 50, 257.
Marsh's test for arsenic, 358.
Mauve, 463.
Maximum density, 24.
Meconine, 462.
Medicated tubes, 132.
" waters, 473.
Melam and melamine, 448.
Mellone, 449.

Melting points, 39.
Mercaptan, 412, 430.
Mercury, 371.
" characters of salts of, 372.
" chlorides, 372.
" iodides, 372.
" nitrates, 372.
" oxides, 371.
" sulphates, 372.
" sulphides, 372.
Mesityle, 423.
Metaldehyde, 418.
Metal, fusible, 29, 367.
Metals, general properties of, 309.
" classification of, 310.
Methyle compounds, 426.
Methylethylamylophenylium, 383.
Microcosmic salt, 322.
Milk, composition of, 492.
Mindererus spirit, 421.
Mineral chameleon, 339.
" waters, 239.
Moirè metallique, 351.
Molybdenum, 354.
Moon photographed, 116.
Moore's test, 398.
Mordants, 465.
Morphia, 459.
Morse's telegraph, 175.
Mosaic gold, 251.
Mucilage, 396.
Mucus, 492.
Multipliers, 173.
Merexan, 452.
Murexide, 452.
Muscovado sugar, 393.
Myricine, 473.

N.

Naphtha, 477.
Naphthaline, 477.
Narceia, 460.
Narcotina, 460.
Negative, 121.
Nervous substance, 495.
Nickel, 346.
" sulphate, 346.
Nicotine, 462.
Nihil album, 348.
Niobium, 354.
Nitric acid, 263.
" discovery of, 2.
Nitrobenzide, 436.
Nitrogen, chloride, 283.
" compounds with oxygen, 258.
" determination of, 388.
" preparation and properties of, 241.
Nitrous acid, 261.
" oxide, 258.
Nomenclature, 8, 191.
Norium, 354.
Nutmeg butter, 472.
Nutrition, function of, 481.

O.

Œnanthic ether, 413.
Ohm's theory, 165.
Oil of bitter almonds, 433.
" cajeput, 473.
" cinnamon, 473.
" copaiva, 473.
" horseradish, 473.
" lavender, 473.
" lemons, 473.
" mustard, 473.
" peppermint, 473.
" rosemary, 473.
" spiræa, 437.
" storax, 473.
" turpentine, 474.
" vitriol, preparation of, 269.
" wine, heavy, 415.
Oils and fats, 469.
" palm and cocoa, 472.
" volatile, 473.
Olefiant gas, 299.
Oleine, 471.
Orceine and orcine, 468.
Organic bodies, analysis of, 324.
" " classification of, 383.
" " decomposition of by heat, 382.
" " general characters of, 377.
" chemistry, 377.
Orpiment, 360.
Osmium, 376.
Oxalates, 399.
Oxamethane, 412.
Oxamide, 412.
Oxygen, preparation and properties of, 212.

Oxyhydrogen blowpipe, 233.
Ozone, 228.

P.

Palladium, 373.
Palmitine, 472.
Palm oil, 472.
Papin's digester, 54.
Paracyanogen, 441.
Paraffine, 476, 480.
Paranaphthaline, 477.
Parchment paper, 397.
Pascal's experiment, 254.
Pectine, 396.
Pelopium, 354.
Perchloric acid, 283.
Petroleum, 479.
Pewter, 351.
Phene, 435.
Phlogistic theory, 7.
Phloridzine, 462.
Phosphorescence, 114.
Phosphori, 114.
Phosphoric acid, 276.
Phosphorus compounds with oxygen, 275.
" preparation and properties of, 273.
Phosphureted hydrogen, 277.
Photography, 119.
Photometers, 89.
Picamar, 477.
Picrotoxine, 462.
Pile, Voltaic, 151.
Piperine, 462.
Pitch, 476.
Pit coal, 479.
Pittacal, 477.
Plasma, 487.
Platinum, 374.
" black, 375.
" chloride, 375.
" oxides, 375.
" power of determining union of gases, 374.
" spongy, 374.
Plumbago or graphite, 292.
Pneumatic trough, 219.
Polarization of light, 108, 394.
" of heat, 86.
Populine, 462.
Porcelain, manufacture of, 334.
Positive, 121.
Potassa, 315.
" bicarbonate, 317.
" bisulphate, 317.
" carbonate, 317.
" chlorate, 318.
" hydrate of, 315.
" nitrate, 318.
" oxalate, 399.
" salts, tests for, 316.
" sulphate, 317.
Potassium, chloride of, 317.
" iodide of, 317.
" peroxide of, 316.
" preparation and properties of, 313.
" sulphides of, 316.
Potato oil and its compounds, 431.
Printers' ink, 469.
Prism, 93.
Protein, 486.
Prussian blue, 447.
Pulse glass, 61.
Purple of Cassius, 373.
Pus, 493.
Putty powder, 350.
Pyroacetic spirit, 422.
Pyrometer, 26.
" Daniell's, 30.
Pyroxylic spirit, 427.
Pyroxyline, 401.

Q.

Quartation, 373.
Quercitron bark, 466.
Quicksilver, 371.
Quinia, 461.
Quinoidine, 461.
Quinoline, 464.

R.

Radiation, 76.
" interstitial, 12, 68.
Raphides, 400.
Rays of the sun, chemical, 118.
Realgar, 360.
Reflection, law of, 110.
Refraction, law of, 110.
Regelation, 44.
Reinsch's test, 357.
Repulsion, 126.
Resins, 476.
Respiration, 251, 490.
Rheostat, 167.

Rhodium, 375.
Rochelle salts, 454.
Rouge, 466.
Rubidium, 96, 324.
Rudberg's law, 15.
Ruhmkorff's coil, 181.
Rupert's drops, 29.
Ruthenium, 376.

S.

Sacchulmine, 398.
Safety-jet, Hemming's, 71.
Safety-lamp, 71.
Sago, 392.
Salicine, 437.
Salicyle compounds, 437.
Saliva, 493.
Sanctorio's thermometer, 16.
Scheele's green, 356.
Secretion, 490.
Seignette salts, 454.
Selenium, 273.
Silicon, 303.
Silver, 367.
" ammoniuret, 370.
" characters of salts of, 369.
" chloride, 369.
" German, 346.
" iodide, 369.
" nitrate, 369.
" oxides, 369.
" plating, 163.
" sulphide, 369.
Silvering on glass, 370.
Smalt, 347.
Smee's battery, 153.
Snow crystals, 236.
Soaps; saponification, 470.
Soda, biborate, 323.
" bicarbonate, 321.
" carbonate, 321.
" hydrate, 320.
" nitrate, 322.
" phosphates of, 322.
" sulphate, 42, 321.
" water, 296.
Sodium, chloride, 320.
" prepartion and properties of, 319.
Solanine, 462.
Solder, 351.
Spark, 145.
Specific gravity, 203.
Specific heat, 31.
" induction, 144.
Spectra, 98.
Spectres, 117.
Spectroscope, 97.
Spectrum, solar, 93.
Speculum, metal, 364.
Spermaceti, 473.
Spheroidal state of water, 57.
Spiræa ulmaria, oil of, 437.
Spirit of wine, 408.
Starch, 391.
Steam, elastic force of, 58.
" engine, 55, 66.
Stearine, 470.
Stearopten, 474.
Steel, 342.
Stone-ware, manufacture of, 334.
Strontia, 326.
" nitrate, 327.
" sulphate, 327.
Strontium, 326.
Strychnia, 462.
Sublimate, corrosive, 372.
Substitution, 380.
Sucrose, 393.
Sugar, cane, 393.
" eucalyptus, 396.
" from ergot of rye, 396.
" grape, 395.
" of lead, 421.
" of milk, 395.
Sulphobenzide, 436.
Sulphocyanogen compounds, 448.
Sulphur compounds with oxygen, 267.
" occurrence in nature, 265.
" properties of, 261.
Sulphureted hydrogen, 271.
Sulphuric acid, 269.
Sulphurous acid, 267.
Sun, atmosphere of the, 95.
Symbols, 195.
" table of, 193.
Sympathetic ink, 347.
Synaptase, 443.
System, crystallographical, 204.

T.

Tapioca, 392.
Tar, varieties of, 476.
Tartar, cream of, 454.

Tartar emetic, 454.
Telescope, silvered-glass, 116.
Tellurium, 362.
Thallium, 337.
Thebaia, 460.
Theine, 462.
Theobromine, 462.
Thermo-electricity, 183.
Thermo-electric pile, 82.
Thermometer, Breguet, 30.
" construction of, 20.
" differential, 18.
" Sanctorio's, 16.
" scales, 21.
Thorium, 336.
Tin, 349.
" chloride, 350.
" oxide, 350.
" sulphide, 351.
Tin plate, 351.
Titanium, 355.
Toning, 122.
Torpedo, 188.
Torula, 403.
Trade-winds, 75.
Transverse vibrations, 102.
Tungsten, 354.
Turmeric, 466.
Turpeth mineral, 372.
Type metal, 362.
Types, chemical, 379.

U.

Ulmine, 397, 478.
Undulatory theory, 100.
Uramile, 452.
Uranium, 354.
Urea, 445, 449.
Urinary calculi, 495.
Urine, composition of, 493.
Urinometer, 494.

V.

Vanadium, 353.
Vapor, elastic force of, 50, 58.
Vapors, density of, 66.
Vapors, nature of, 46.
Vaporization at low temperature, laws of, 47.
Varrentrapp and Will's bulbs, 389.
Veratrine, 462.
Verdigris, 421.
Vermilion, 372.
Vinegar, 419.
Vitriol, blue, 364.
" green, 345.
" oil of, 269.
" white, 345.
Voltameter, 164.
Volumes, combination by, 202.
Vulcanization, 475.

W.

Water, composition of, 156, 234.
" of crystallization, 239.
Waves, length of, 107.
Wax, 473.
Wines, 406.
Wire gauze, 70.
Wood ether, 426.
" spirit and its compounds, 426.
Woody fibre, 396.

X.

Xanthic acid, 424.
" oxide, 452.
Xanthophylline, 468.
Xyloidine, 401.

Y.

Yeast, 403.
Yttrium, 336.

Z.

Zaffre, 347.
Zamboni's piles, 142.
Zinc, 347.
" oxide, 348.
" silicate, 348.
" sulphate, 348.
Zirconium, 335.

THE END.

www.ingramcontent.com/pod-product-compliance
Lightning Source LLC
LaVergne TN
LVHW021318110826
845150LV00003B/606

* 9 7 8 1 4 2 5 5 5 8 1 9 2 *